P9-EEP-479

Applied Atomic Spectroscopy

Volume 1

MODERN ANALYTICAL CHEMISTRY

Series Editor: **David Hercules**
University of Georgia

ANALYTICAL ATOMIC SPECTROSCOPY
By William G. Schrenk

PHOTOELECTRON AND AUGER SPECTROSCOPY
By Thomas A. Carlson

MODERN FLUORESCENCE SPECTROSCOPY, VOLUME 1
Edited by E. L. Wehry

MODERN FLUORESCENCE SPECTROSCOPY, VOLUME 2
Edited by E. L. Wehry

APPLIED ATOMIC SPECTROSCOPY, VOLUME 1
Edited by E. L. Grove

APPLIED ATOMIC SPECTROSCOPY, VOLUME 2
Edited by E. L. Grove

TRANSFORM TECHNIQUES IN CHEMISTRY
Edited by Peter R. Griffiths

A Continuation Order Plan is available for this series. A continuation order will bring delivery of each new volume immediately upon publication. Volumes are billed only upon actual shipment. For further information please contact the publisher.

Applied Atomic Spectroscopy

Volume 1

Edited by

E. L. Grove

IIT Research Institute
Chicago, Illinois

Plenum Press · New York and London

Library of Congress Cataloging in Publication Data

Main entry under title:

Applied atomic spectroscopy.

(Modern analytical chemistry)
Includes bibliographical references and index.
1. Atomic spectra. I. Grove, E. L., 1913-
QD96.A8A66 543'.085 77-17444
ISBN 0-306-33905-6 (v. 1)
ISBN 0-306-33906-4 (v. 2)

A Division of Plenum Publishing Corporation
227 West 17th Street, New York, N.Y. 10011

Printed in the United States of America

Contributors

James W. Anderson	Consultant, Pleasantville, New York
Reuven Avni	Nuclear Research Center, Negev, Beer Sheva, Israel
A. H. Gillieson	Retired from the Department of Energy, Mines and Resources, Mineral Sciences Division, Ottawa, Ontario, Canada
J. W. Mellichamp	U.S. Army Electronics Command (DRSEL-TL-EC), Fort Monmouth, New Jersey
R. H. Scott	National Physical Research Laboratory, Council for Scientific and Industrial Research, Pretoria, South Africa
A. Strasheim	National Physical Research Laboratory, Council for Scientific and Industrial Research, Pretoria, South Africa
Geoffrey Thompson	Woods Hole Oceanographic Institution, Woods Hole, Massachusetts

Preface

From the first appearance of the classic *The Spectrum Analysis* in 1885 to the present the field of emission spectroscopy has been evolving and changing. Over the last 20 to 30 years in particular there has been an explosion of new ideas and developments. Of late, the aura of glamour has supposedly been transferred to other techniques, but, nevertheless, it is estimated that 75% or more of the analyses done by the metal industry are accomplished by emission spectroscopy. Further, the excellent sensitivity of plasma sources has created a demand for this technique in such divergent areas as direct trace element analyses in polluted waters.

Developments in the replication process and advances in the art of producing ruled and holographic gratings as well as improvements in the materials from which these gratings are made have made excellent gratings available at reasonable prices. This availability and the development of plane grating mounts have contributed to the increasing popularity of grating spectrometers as compared with the large prism spectrograph and concave grating mounts. Other areas of progress include new and improved methods for excitation, the use of controlled atmospheres and the extension of spectrometry into the vacuum region, the widespread application of the techniques for analysis of nonmetals in metals, the increasing use of polychrometers with concave or echelle gratings and improved readout systems for better reading of spectrographic plates and more efficient data handling.

Many of the far-reaching and on-going changes in industry and environment control would not have been possible without developments in spectroscopy, and committees of ASTM are continuing their work on evaluation and consolidation of procedures.

The available literature dealing with emission spectroscopy has until now been scattered among myriad sources and we in the field have long recognized an urgent need to gather the new ideas and developments together, in a convenient format. However, the enormous amount of work involved in preparing a comprehensive treatise on the subject has been a deterrent. Finally, this major collaborative effort was undertaken: *Applied Atomic Spectroscopy, Volumes 1 and 2* have been written by a group of authors, each of whom has an intimate and expert working knowledge of a special area within the discipline. Individual chapters are treatments in depth of new developments, placed within an historical perspective, in many instances incorporating much of the author's own experience.

I wish to extend my special thanks to all the collaborators for their cooperation and patience. The courtesy of the book and journal publishers who gave permission to reproduce figures and tables is gratefully acknowledged, with special thanks to the U.S. Geological Survey.

We also wish to thank the many practicing spectroscopists for their suggestions and help during the editing process, and last, though not least, Mrs. E. L. Grove and Nancy Robinson for editing, typing, and helping to keep detail in order.

E. L. Grove

Contents

Chapter 2

Laser Emission Excitation and Spectroscopy

R. H. Scott and A. Strasheim

Chapter 5
Preparation and Evaluation of Spectrochemical Standards
A. H. Gillieson

Chapter 6

Applications of Emission and X-Ray Spectroscopy to Oceanography

Geoffrey Thompson

Contents of Volume 2

Applied Atomic Spectroscopy

Volume 1

Photographic Photometry 1

James W. Anderson

1.1 INTRODUCTION

Photographic photometry is the process of measuring the intensity of radiant energy of specific wavelengths in spectra recorded on a photographic emulsion. Since the formation of a spectrogram takes a finite amount of time, the measurement is more properly the integration of intensity, or exposure.

Photography has played a major role in the development of spectroscopy and spectrochemical analyses. In his studies of the darkening effect of silver chloride by the sun's spectrum, Ritter[(1)] in 1803 noted that the maximum darkening action was just outside the visible spectrum—hence the discovery of the ultraviolet region. Shortly after the development of the Daguerreotype process in 1839,[(2,3)] which used sodium thiosulfate as the fixing agent, both Becquerel[(4)] in 1842 and Draper[(5)] in 1842 and 1843 obtained photographs of the solar spectrum.

The next important advance in photography was the development by Maddox[(2)] in 1871 of the dry gelatine plate, which very quickly found widespread use in spectroscopy. Its availability made possible the much improved wavelength measurements and improved catalogs of spectra, typified by Rowland's work[(6,7)] published in 1887 and 1893. This subsequently led to the wide use of the spectrograph.

Today, photography is one of the four methods for detecting and measuring radiant energy, the other three being photoelectric, visual, and thermoelectric or radiometric. Some characteristics of these four methods are compared in Table 1.1. Wavelength range in the table refers to the spectral region for which the method is useful. Contrast is the general slope of the curve in which the response of the detector is plotted as a function of the quantity of radiant energy, while linearity refers to how closely this plot approaches a straight line.

James W. Anderson • Consultant, Pleasantville, New York

Table 1.1 Summary of Methods for the Measurement of Spectral Intensities (Radiant Energy)

Method	Wavelength range (Å)	Contrast	Linearity	Neutrality	Cumulative	Panoramic
Photographic	10–11,000	High	Poor	Poor	Good	Excellent
Photoelectric	10–40,000	High	Good	Poor	Fair	None
Visual	3,900–7,500	High	Very poor	Poor	None	Limited
Thermoelectric	9,000–10^7	Low	Excellent	Excellent	None	None

A detector with high contrast is more sensitive to small changes of signal level but is likely to have a smaller dynamic range or latitude than a detector with low contrast. A detector is said to be highly neutral if the differences in its response to radiant energy of different wavelengths are negligible; that is, it responds in the same manner to the energy of one wavelength as to that of another. Because photographic emulsions have poor neutrality and are also nonlinear in response, they often require different calibrations in different wavelength regions. This is illustrated in Fig. 1.1 as shown by Harrison *et al.*[8] The cumulative property refers to the ability of the receptor to sum up exceedingly low intensities of light by increasing the time of exposure, while the panoramic property means the ability of a photographic emulsion to simultaneously record different wavelengths of radiant energy on different parts of the plate or film.

Pictorial photography is concerned with the linear recording of visually perceived illumination levels of objects under a heterochromatic light, whereas photographic photometry of the spectrum requires precise quantitative comparisons of much fainter and essentially monochromatic beams of radiation. The high sensitivity to small changes of signal level and the cumulative and panoramic properties of the emulsion are important for photometry, but linear recording (which can be realized only over a limited exposure range) is not. Important advantages of photographic photometry include the integration of light from sources of time-varying brightness and production of a permanent record.

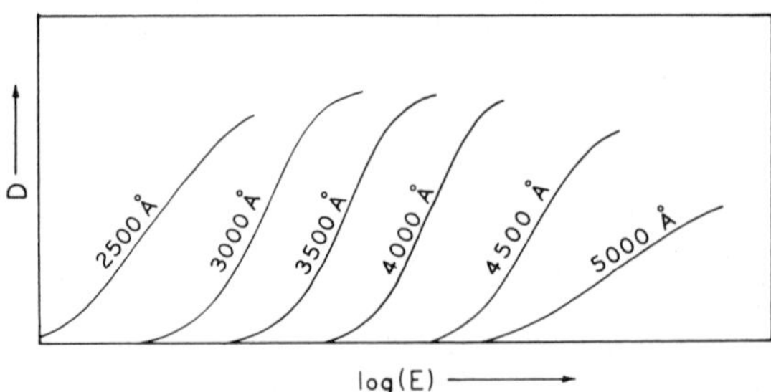

Fig. 1.1 Calibration curves for different wavelengths. The same scale, but different origins, were used to prevent overlap. (From Harrison *et al.*[8])

1.2 THE PHOTOGRAPHIC EMULSION

The photographic emulsion is a thin layer of gelatin containing a suspension of very fine, light-sensitive silver halide crystals or grains. While the gelatin is in the liquid state, it is coated on glass or on cellulose acetate or polyester base and allowed to dry.* Glass plates have an advantage with respect to dimensional stability but are restricted to spectrographs with flat or moderately curved focal planes. With standard plate widths of 2 and 4 in., they also provide more area for accepting a greater number of spectrograms, which permits more latitude in including exposures of standard samples for direct comparison to unknown samples. Although film is subject to expansion and contraction and presents some mechanical problems in processing and in being held flat in microphotometers, it can readily be bent to steeply curved focal planes. Film also avoids the obvious breakage damage to which glass is subject. In general, the emulsion layer on glass plates is slightly thicker than on film, which tends to make them more sensitive. On the other hand, the emulsions on film products have a thin clear gelatin overcoat of about 1 μm for protection against abrasion and handling. Kodak(9) specifically recommends that the emulsion surface of plates are not to be wiped, because they are very soft when wet.

The light-sensitive material is a mixture of silver bromide with some silver iodide and traces of nucleating compounds. The size of these crystals or grains is carefully controlled within narrow limits because many properties of an emulsion are grain-size-dependent. The average grain size may vary from about 5 μm in diameter for fast emulsions to submicroscopic for the slow Lippman emulsions. In general, the larger the average grain size, the more sensitive the film (partly because larger grains intercept more of the incident energy per grain) and the lower the contrast of the emulsion. The converse is also true, and thus one can expect that a fine-grained emulsion is generally slow with high contrast. This natural association of emulsion characteristics is unfortunate because the most desirable emulsion should have the finest grain possible to provide sharp resolution and yet be fast at the same time.

Another characteristic of an emulsion is the dynamic range over which it responds to radiation. The logarithm of the useful dynamic range or *latitude* varies inversely with the *contrast* or gamma of the emulsion. Both latitude and contrast also depend upon the minimum number of quanta a grain must absorb before it becomes developable and upon the dispersion of grain sizes about the average grain size of the emulsion. This is illustrated in Fig. 1.2, in which curve 2 represents a low-speed emulsion with high contrast, short latitude, and relatively poor sensitivity.

*In some special emulsions, more than one such coating may be applied. If two or more coatings are applied, they usually differ in grain size and sensitivity. The purpose of this procedure is to extend the dynamic range for visual photography.

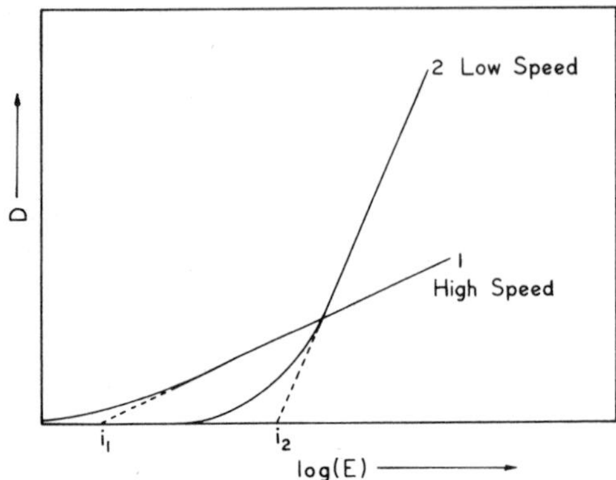

Fig. 1.2 Characteristic curves for typical emulsions with high and low speeds. (From Harrison *et al.*[10])

Speed, contrast, and latitude are all functions of wavelength. Both the absorbance of the photosensitive layer and the number of quanta that a grain must absorb before it can be developed depend upon wavelength. The absorbance of the photosensitive layer is by the gelatin substrate as well as by the silver halide grains embedded in the substrate. The absorption by gelatin, which begins below 2500 Å, affects the contrast of the emulsion, while the absorption by silver halide affects both sensitivity and contrast. These effects can be modified by various sensitizing dyes which are added to the emulsion to improve response at wavelengths above 5000 Å, where the silver halide itself is transparent. Gradient is a measurement of contrast in terms of the slope of the straight line between two specified densities on a characteristic curve. Eastman Kodak has described typical variations in gradients for different emulsions in which they

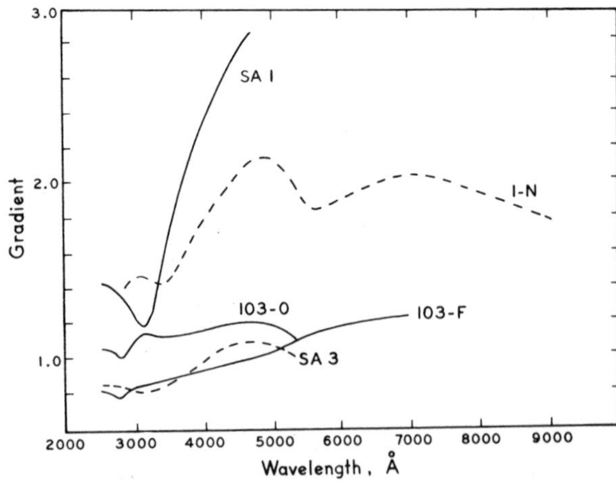

Fig. 1.3 Approximate gradient–wavelength curves for some typical spectrographic (plates) emulsions.

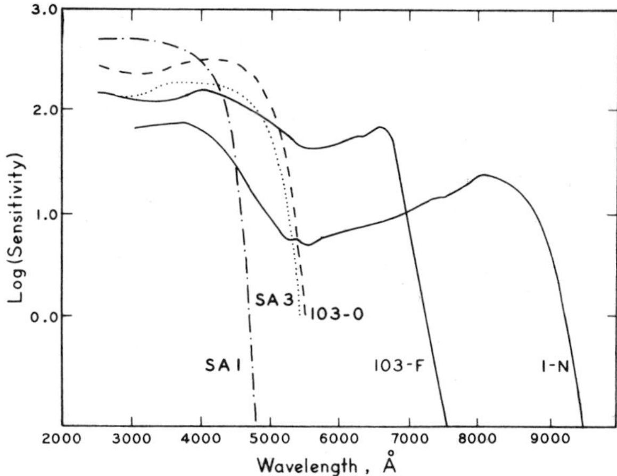

Fig. 1.4 Spectral sensitivities of some spectrographic (plates) emulsions.

use the specific densities of 0.3 and 1.0 above gross fog. Some examples of these appear in Fig. 1.3. These density points are equivalent to transmittances of 50% and 10%. In a similar fashion Fig. 1.4 shows the spectral sensitivity of these emulsions. Sensitivity here is defined as the reciprocal of exposure required to produce a density of 0.1 above gross fog, which is equivalent to 80% transmittance.

Conventionally, gamma has been defined as the slope of the straight-line portion of the characteristic curve, such as displayed in an H and D curve (Section 1.7.3.1). In a more recent description of contrast, Kodak[(11)] refers to contrast index as the slope, or average gradient, of a straight line drawn between two points on a density–log exposure curve that can usually represent the lowest and highest densities in a continuous-tone black-and-white negative, such as from 0.2 to 2.2 log exposure units. This index is of more value in pictorial photography, and gamma remains a more desirable expression for describing emulsions for spectrographic use.

1.3 THE LATENT IMAGE

The latent photographic image is formed when the emulsion is exposed to radiation of wavelengths to which it is sensitive. This latent image consists of the aggregate of all grains of silver halide that have been altered by absorption of incident radiation in such a way that selective reaction with development chemicals can take place. The latent image is completely invisible. On each latent-image grain there are one or more latent-image specks. The formation of each latent-image speck requires the absorption of one or more quanta of energy by the grain. The minimum number of quanta a grain must absorb to produce at least one developable latent-image speck is the grain activation number of the emulsion, denoted as q. This varies with the emulsion and wavelength and determines the shape of the emulsion response curve as well as the sensitivity. In the x-ray region q is always 1. In the optical region q is variable and is thought to be around 5 or 6 in the visible region.

Other photochemical processes are known where light action will produce a directly visible image by causing a color change in a chemical or chemicals such as in blueprint paper. All these reactions are of low sensitivity when compared to the photographic process because the energy from the absorption of one quantum of light affects only one molecule. By contrast, the development of a latent photographic image is, in fact, a chemical amplification of light in the sense that the absorption of only a few quanta of energy by a silver halide grain affects the very large number of molecules that constitute that grain. This latent image is stable: photographs have been developed several years after their exposure. The developing chemicals provide the energy to produce the structure and bring out the latent image.

Development of the latent image into a visible image consists of treating the exposed silver halide emulsion with a developing solution, usually containing an organic reducing agent. The developer differentially reacts only with grains possessing a latent image and reduces them to metallic silver. Once the development of a grain starts, it is an all-or-nothing process; that is, the grain either develops completely or not at all. Thus, growth of the image with increased development time is due to the increase in size of the individual silver grains and not to more exposed grains being developed as development time is increased.*

Development of the latent image into a visible image follows approximately a first-order reaction law. If the optical density of the developed image is measured as a function of development time, this relation may be expressed as

$$D_t = D_\infty(1 - e^{-kt}) \tag{1.1}$$

where D_t = value of density after development time t
D_∞ = value of density reached when development is indefinitely prolonged
k = development rate constant
t = development time

The value of the rate constant k depends upon the emulsion and upon the temperature and composition of the developer. The initial rate of development is rapid as shown in Fig. 1.5 in an illustration used by Brode.[12] Ignoring secondary effects, which may be important, little is gained by extending the development time. The temperature dependence of development rate follows

*A small amount of added development does take place with time, but this is mostly development fog due to attack of the unexposed grains. The kinetics for the development of fog are different from the kinetics for the development of the image, particularly with regard to the temperature dependence of reaction rates.

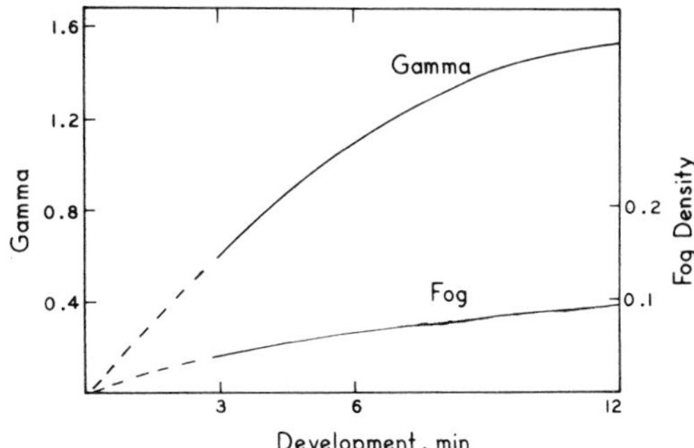

Fig. 1.5 Increase of gamma and fog with development time. (From Brode.[12])

the Arrhenius law and is logarithmic. A practical limit to extending the development time is usually set by the development of fog, which is also more severe at higher temperatures. These are dealt with in more detail in Section 1.5.

Since most developing agents are alkaline, the development process is stopped by the use of dilute acetic acid, approximately a 1-5% solution. The undeveloped grains as well as any undeveloped parts of developed grains are then removed by dissolving in sodium thiosulfate, or hypo. In the final negative photograph the latent image is observed in the form of the black silver grains which were affected by the exposure. The weight of silver per unit is proportional to optical density. The quality of the photographic image is influenced by many process variables, such as time, temperature, agitation, and activity of the developer. Useful information concerning these appear in the report by the ASTM Committee E2 on Emission Spectroscopy.[13] Photographic photometry depends upon the fact that a definite quantitative relationship exists between the number of developed grains in the negative image and the amount of radiation to which the emulsion was exposed. In order to obtain meaningful results, however, exposure, development, and image evaluation techniques must all be carefully controlled and proper calibration methods must be used.

1.4 EVALUATION OF PHOTOGRAPHIC IMAGE

The negative image is commonly evaluated by measuring the transmittance of the exposed area and relating this quantity to the exposure that caused the blackening. The transmittance of a spectrum line is measured by a microphotometer and is defined as the ratio of intensities

$$\tau = \frac{I}{I_0} \tag{1.2}$$

where I = transmitted light or light transmitted through the image

I_0 = incident light (i.e., light transmitted by an unexposed portion of the same negative)

More often, the term "percent transmittance" is used and may be designated as T, where

$$T = 100\,\tau \tag{1.3}$$

When developing conditions are constant, there exists for each emulsion a definite quantitative relationship between the transmittance of a line image as measured by a microphotometer and the intensity of radiation producing that image. This relation can be generally represented as an emulsion calibration curve relating the microphotometer reading to the relative exposure of the emulsion.

Photographic literature usually uses the term density, D, which is defined by

$$D = \log\left(\frac{1}{\tau}\right) = \log\left(\frac{100}{T}\right) = 2 - \log T \tag{1.4}$$

The characteristic curve of an emulsion, or the H and D curve, after Hurter and Driffield,[14] is a plot of density against the logarithm of exposure. A typical curve is shown in Fig. 1.6. Point A is the possible threshold exposure at which blackening first becomes perceptible. The curved toe, AB, is followed by the nearly linear or normal exposure region, BC, in which density is a linear function of log exposure. The knee, CD, is the region of decreasing response to exposure with saturation at D. The position i is known as the inertia. Emulsions with high inertia values are relatively insensitive to low exposures. The slope of the straight line portion is called gamma (γ). The degree of development influences the γ of a given emulsion. This is illustrated in Fig. 1.7. It will be noted that with increased development the γ tends to pivot about the inertia point. Development may be carried on until background fog prevents any further increase in contrast.

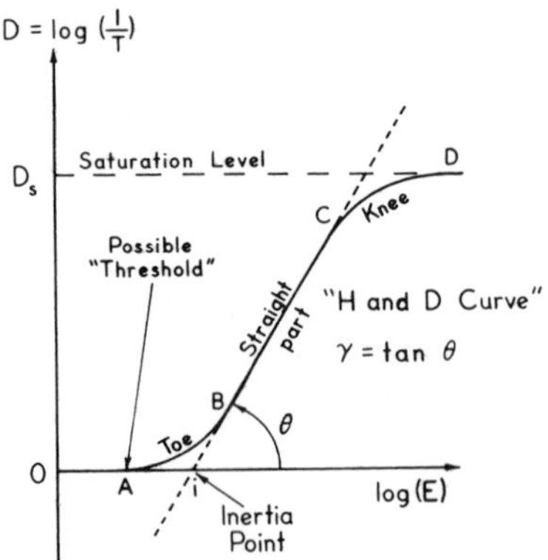

Fig. 1.6 Terminology for describing the emulsion calibration curve.

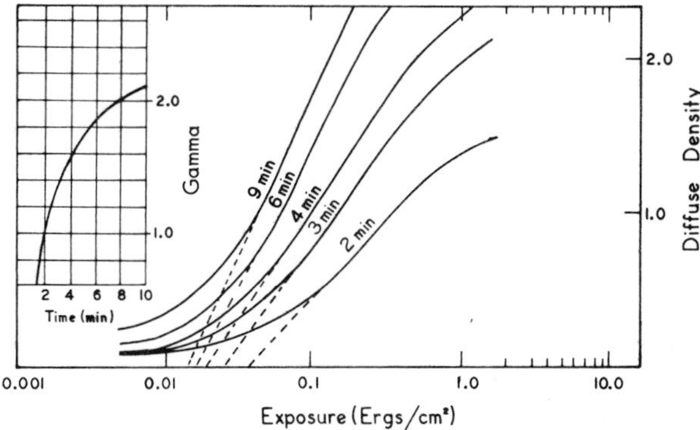

Fig. 1.7 Degree of development influence on gamma.

Density is directly proportional to the number of developed silver grains in the image since Beer's law applies to the developed negative image. It is also a direct measure of the psychovisual effect of a given amount of blackening as observed by a human observer since our sensory apparatus has a built-in logarithmic response. Both of these factors made D a logical variable for investigating photochemical or psychological aspects of the photographic process. For the purpose of photographic photometry, however, it is desirable to use a function that simplifies calibration procedures. The term D is not the best choice for this purpose because it is nonlinear except for a short segment between the toe and knee.

Density functions that are linear over wider exposure values are commonly used in spectrochemical calculations and will be described in Sections 1.7.3.2 and 1.7.3.3. Regardless of which function is used for the actual calculation, it is still necessary to understand some basic limitations on the accuracy with which a given transmittance or density value can be determined. These limitations are of two kinds, one depending upon the nature of the emulsion itself and the other being a function of the microphotometer.

1.4.1 Limitation: The Emulsion

The limitation imposed by the emulsion depends basically on the fact that a photographic image is composed of individual silver grains and the statistical fluctuation of the number of grains within the area being photometered determines the accuracy with which density can be measured. Since the same density value can be obtained from an image containing many small silver grains as from an image containing much fewer but larger grains, it is clear that the grain size or granularity of the emulsion determines a practical limit in respect to the accuracy with which densities of small areas can be measured.

To express the effect of granularity quantitatively, consider a section of emulsion which can be seen in a fixed area of view by a microphotometer. Let the maximum number of silver grains in this area that can be developed be designated G. For some density value, D, let the developed grains be g. From Beer's law relating absorption directly to concentration,

$$D = k\left(\frac{g}{G}\right) \quad \text{or} \quad g = \frac{DG}{k} \tag{1.5}$$

where k is the proportionality factor.

In reading many portions of an emulsion through the same aperture of view, the error of the density reading will be directly related to the error of the number of developed grains, g, which are seen by the microphotometer. Statistically this error is proportional to the square root of the number of grains viewed. Therefore, the relative error for a reading of D becomes

$$\sigma_D = \frac{\sqrt{g}}{g} = \frac{1}{\sqrt{g}} \tag{1.6}$$

Then substituting for g, Eq. (1.6) becomes

$$\sigma_D = \sqrt{\frac{k}{DG}} \tag{1.7}$$

Therefore, as $\sqrt{G}$ becomes larger, the relative error for D decreases. To have more grains, G, available in a finite space, the grains must be smaller. Therefore, if granularity increases (i.e., if the grains becomes coarser), the relative error in D increases.

For any given emulsion and developing condition, the grain size is fixed, so σ_D will then depend only on the microphotometer aperture and density D that is read. When density is very low, σ_D can become very large and adversely affect the precision of the measurement. To some extent this can be compensated for by increasing the measuring aperture provided that resolution is not lost in the process. If long uniformly exposed spectral lines are available and if a microphotometer slit several millimeters long can be used, the aperture area can be increased to compensate without reducing the spatial resolution below acceptable limits. The increase of σ_D at low densities, however, has an even more adverse effect on the detectable changes of exposure values because the slope of the characteristic curve in this low-density "toe" region becomes very small. In order to find the corresponding uncertainty in the logarithm of the exposure which produced the density D, it must be noted from the H and D plot that

$$\log E = \frac{1}{\gamma} D = \frac{d \log E}{dD} D \tag{1.8}$$

Multiplying in the error in the density measurement, σ_D, the relative error in the log exposure becomes

$$\sigma_E \log E = \frac{d \log E}{dD} (D) \sqrt{\frac{k}{DG}} = \frac{d \log E}{dD} \sqrt{\frac{kD}{G}} \tag{1.9}$$

Now, although the error in the determination of log exposure becomes less as the $\sqrt{D}$ becomes less, providing that the density range is still in the straight-line portion of the H and D curve, the large change in gamma as it flattens out toward zero soon overrides the whole error pattern so as to impose very large uncertainties in exposure.

1.4.2 Limitation: The Microphotometer

The second limitation is due to the practical problem involved in measuring the transmittance of the very narrow lines that constitute a line spectrum. Ideally, the microphotometer should be able to resolve area widths that are smaller than the central section of spectral lines. This, in itself, is hard to define, since the profile of a narrow line tends to have a Gaussian distribution of density; but, in general, resolutions in the order of 10 μm or better would be desirable. In practice this is usually not achieved. Even where it can be demonstrated that a microphotometer indicates peak readings for spectrographically resolved lines 10 μm apart, the amount of scattered light being accepted with these readings can seriously distort the indicated transmittance readings. This effect tends to be held to a minimum when the light that is used to read the line is focused down to a narrow beam for passage through the line. This holds provided that no other light is being used to illuminate the spectral pattern in the area of the line being read, unless some discrimination such as polarization or chopping of the reading light is done.

Emulsion characteristics described by Kodak are in terms of diffuse transmission densities such as are obtained in sensitometric measurements of relatively large exposed areas of emulsion. In spectrographic applications, the densities being read are specular densities, or at least semispecular. That is, the angular flux being measured in at least the direction across a spectral line image is a narrow angle. Kodak[(15)] defines specular densities as being those measured at solid angles of from 5 to 10 degrees in contrast to diffuse densities being measured at 180°. Specular densities are always somewhat higher than diffuse densities, but the disparity is reduced when the grain size is smaller.

As a result of the presence of scattered light, the limiting slope, γ, of the H and D curve may appear lower than its true value, and a false saturation of the characteristic curve, including a false knee and a saturation density considerably lower than the true saturation density, may be observed. This effect is dependent upon the line width because much of the scattered light originates in the immediate neighborhood of the line being measured. Like the turbidity effect, to be discussed later, the presence of scattered light will cause the observed characteristic curve to have a lower slope when narrow lines are being measured. When the source of scattered light is scattering by the emulsion close to the line being measured, this effect is known as the Schwarzschild–Villiger effect.[16] Usually a significant portion of this scattered light is due to multiple reflections between the optical surfaces in the path of the light beam illuminating the plate. Thus, if the collimator is close to the plate, the light reflected back from the surface of the negative is reflected back again through the plate from the front surface of the collimator lens, often through a slightly different path, and can be a significant source of scattered light. Likewise, any flat glass surface in close proximity to and parallel with the negative will provide opportunities for multiple reflections of light between the negative and the glass surface and create an additional source of scattered light.

From both the problems of scattered light in the microphotometer readings and from the problems about granularity of emulsions mentioned previously, it is apparent that the accuracy of transmittance readings will be improved if wider primary slit systems are used in both the spectrograph and microphotometer. Obviously, this cannot be done at the sacrifice of resolving the spectral lines of interest from other lines, but working in the order of 50 μm for primary slits in grating spectrographs or 25 μm in prism spectrographs would tend to minimize the problems in densitometry. In cases where such wide primary slits cannot be tolerated on the spectrograph, it must be recognized that separate emulsions calibrations may be needed for the different narrow slits and that background corrections, if required, may need to be referred to still another emulsion calibration.

1.5 DEVELOPMENT EFFECTS

Specific procedures and precautions to be taken in processing spectrum plates or films are described in the ASTM recommended practices.[13] Procedures are also available from manufacturers of photographic materials. When processing spectrographic plates or films, it should be borne in mind that the most critical step in the process is the actual development. The development rate constant, k, in Eq. (1.1) is sensitive to temperature and concentration of the developing agent, as well as to the oxidation products in the developer. These oxidation products build up as a result of continued use of the developer as well as air oxidation from simply standing around. Air oxidation is most rapid when new developer is first opened. For this reason many laboratories retain some of the old, oxidized developer, which is added to new developer in the developing tank. In order to minimize the effects of developer concentration, temperature, and time, development for as long a time as possible is desirable. Unfortunately,

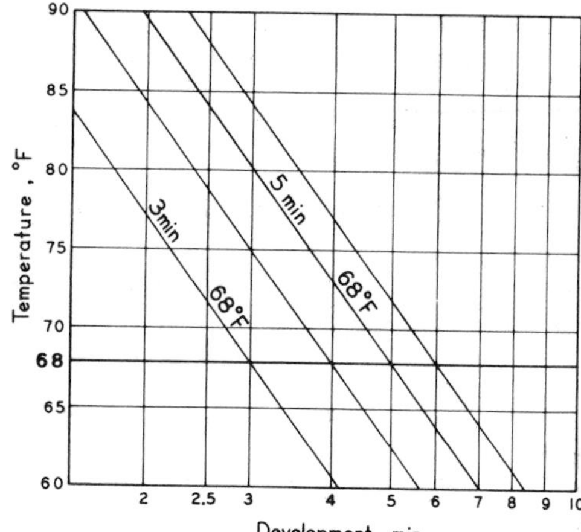

Fig. 1.8. Variation of development time with developer temperature for emulsions developed in D-19.

as previously pointed out, prolonged development as well as development at elevated temperatures produces fog. Graphs such as Fig. 1.8 can be used to correct the development time for small excursions of the developer temperature from its desired value or to estimate development time when an elevated temperature is used to speed up processing. The small amount of time saved by elevating development temperature, however, may be false economy since, in addition to the production of fog, the rapid reaction, which is exothermic, can cause localized melting and slight flow of the gelatin, which would distort the image. Subjecting the emulsion to appreciable changes in temperature in the subsequent processing stages, including the wash, can cause reticulation, a pattern superimposed on the spectral images.

1.5.1 Eberhard Effect

Rapid development can also result in localized depletion of the developing agent and buildup of its oxidation products. As a result, the development of the most dense portion of the image is restrained. This process is dependent upon the number of exposed grains in a given area, and is much more evident in areas of high density. At the boundary between a lightly exposed and a heavily exposed area, the lightly exposed area will have a higher concentration of developer near its surface, which will tend to diffuse toward the heavily exposed area while oxidation products from the heavily exposed area diffuse toward the lightly exposed area. The result is an edge effect known as the Eberhard effect. This causes the developed image of a heavily exposed area to be bounded by a dark fringe on the side immediately adjacent to a lightly exposed area, while at the same time the lightly exposed area has an even lighter fringe, Mackie line, on the side adjoining the heavily exposed area. The density in the center of the dense spectrum line is reduced, the edge is more well developed, and the density of the lower contour and adjacent background is reduced. This is clearly shown in the microphotometer tracing, Fig. 1.9. Eberhard[17] showed that this effect increases to a maximum in the course of development and then decreases again with increased development time. Eberhard also proved that the inorganic ferrous oxalate developer (which has fallen into disuse) is completely free of this effect. With the normally used organic developers this effect is more severe when the developer is dilute. The distance scale over which the Eberhard effect is most noticeable is of the order of 100 μm (0.1 mm) or more which puts the majority of spectral line images within its range. This is illustrated in Fig. 1.10.

When two closely spaced intense spectral lines are photographed they are subject to a compound Eberhard effect known as the Kostinsky effect.[18] In this case, the developer in the space between the two lines becomes exhausted to a greater extent than on the outside, and both images grow unsymmetrically and tend to exaggerate the actual separation between the lines.

Theoretically, both the Eberhard and Kostinsky effects could be eliminated by taking development to completion. Under normal conditions this is not practical because of the long development times required and because of the consequently high fog levels that would be produced. If the time can be permitted and precise dimensional and photometric results are all-important, development at a much lower temperature for even more extended periods should be considered. While this is not recommended for routine work, it is useful for processing those one-of-a-kind plates where maximum information must be derived from a spectrum taken under adverse conditions, with no possibility of obtaining a second shot. Nuclear track emulsions are generally processed at temperatures as low as 4°C for periods as long as 1 week.

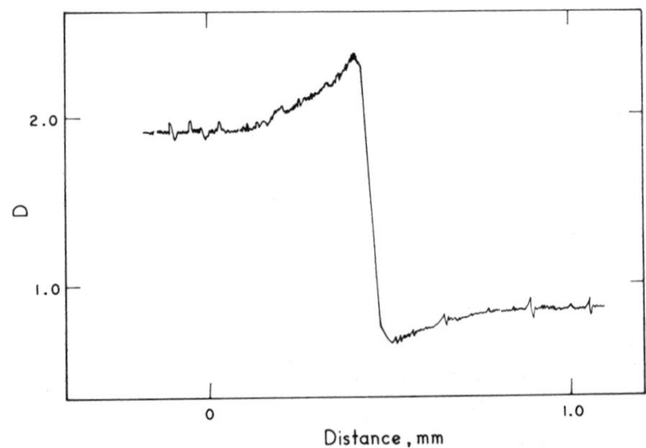

Fig. 1.9 Microphotometer tracing showing Eberhard effect at edge of dark image and Mackie line at the edge of the lighter image.

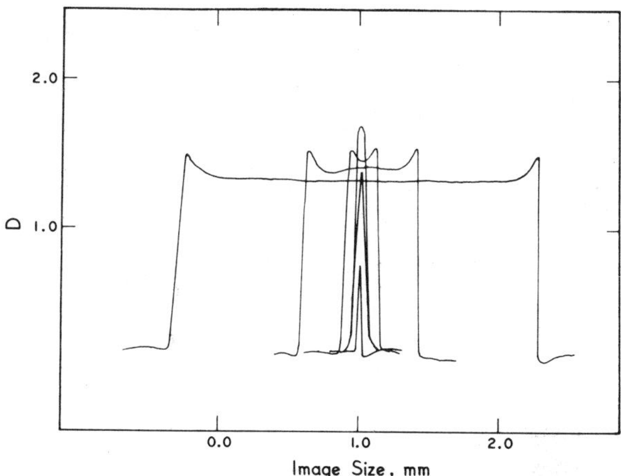

Fig. 1.10 Microphotometer tracings across images of uniformly exposed strips of different widths. The Eberhard effect as well as the turbidity effect (and/or Schwarzschild–Villiger effect) for the narrowest images obviously have a strong influence on the peak transmittance obtained when strong spectrum lines are to be measured.

1.5.2 Intensity-Retardation-of-Development Effect

Incomplete development has a greater effect on heavily exposed latent images than on lightly exposed images. Nachtrieb[19] discusses this, using work performed by Strock,[20] and calls it an intensity-retardation-of-development effect. It is based on the observation that the rate of development of a latent image is a function of the intensity of the exposure. The information collected by Strock compares the densities obtained in short development times with the maximum densities obtainable on longer development times. Data are given for a 2-min developing time in Table 1.2, in which the log intensity listing is translated into relative intensities. The term D_2/D_∞ is the factor of development obtained and is simply the ratio of the density reached in 2 min to the maximum density achievable. If there were not an intensity-retardation effect working, the ratios listed would be substantially the same throughout the entire list of intensities. These data are plotted in Fig. 1.11, which shows that there can be a range of intensities over which a very severe change in the retardation effect can occur. Depending upon developing conditions, if the range of intensities being photographed is such that they fall into this area of great change, a noticeable distortion will occur in the delineation of the characteristic curves. An example of how this would produce a noticeable effect will be shown in Section 1.7.2.4. As mentioned under the discussion of the Eberhard effect in the previous section, both the Eberhard effect and the intensity-retardation effect can be eliminated or at least reduced by increasing the time of development. The

Table 1.2 Degree of Development in 2 min[a]

Relative intensity	D_2/D_∞
1	0.965
2	0.965
4	0.96
8	0.92
16	0.82
32	0.76
64	0.68
128	0.62
256	0.59
512	0.575
2048	0.565
4096	0.56
8192	0.555

[a]From Strock.[20]

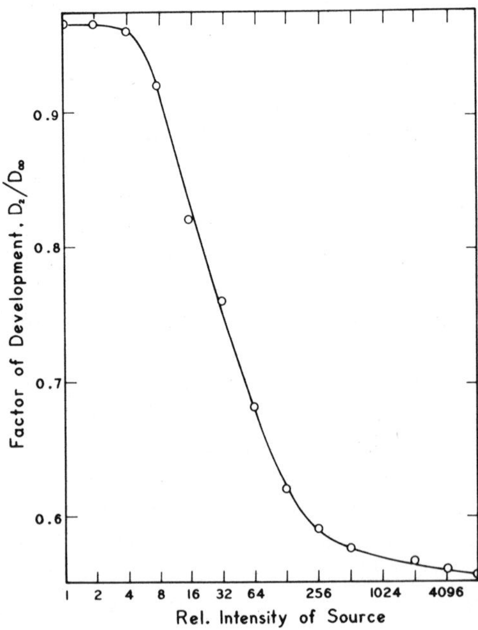

Fig. 1.11 Influence of relative intensity on intensity-retardation-of-development effect.

severe effect noted by Strock would not be expected to occur in most laboratories since developing times as short as 2 min are not commonly used.

1.5.3 Turbidity Effect

An effect known as the turbidity effect also influences the densities of narrow spectrum lines. This has been described by Arrak.[21] It is not a *development effect* but is due to the fact that the emulsion is a turbid or light-scattering medium. The effective exposure intensity at the center of a very narrow line is decreased by loss of light through lateral scatter into the surrounding emulsion, as shown in Fig. 1.12. This has the effect of both lowering the density of very narrow lines and lowering the effective slope of the emulsion calibration curve that must be used to evaluate these lines. It follows that when narrow lines to be measured are superimposed upon a continuous background, two different calibration curves must be used to determine the line and the background exposures; and an empirical correlation of line plus background exposure with line and background densities may be required as has been indicated by Slavin.[22]

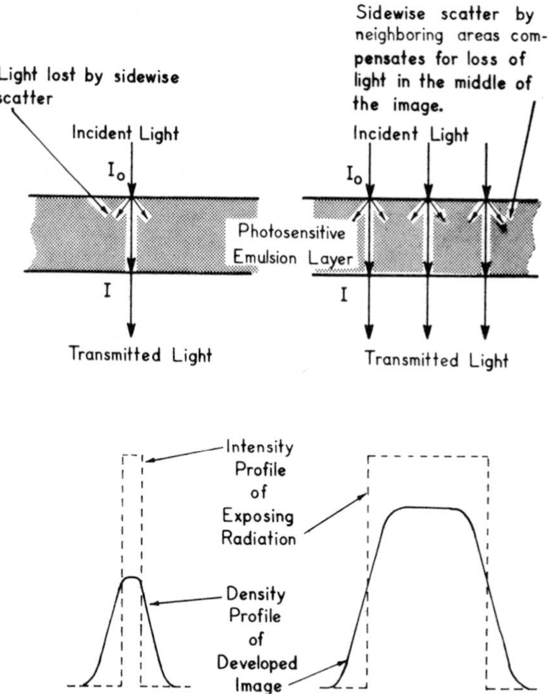

Fig. 1.12 Turbidity effect on density of spectrum lines.

Slavin also notes that the Schwarzschild-Villiger effect is a factor in the differences in these calibration curves.

The turbidity effect is also responsible for the increase in the width of strongly exposed lines and makes possible the correlation of line width with the logarithm of exposure. This relation has been known to astronomers for a long time and served for many years as the basis for measuring stellar magnitudes from astronomical photographs. The actual degree to which image broadening occurs depends upon the ratio of scattering and absorption coefficients of the photosensitive layer of the emulsion and is thus a function of wavelength. The law relating image width with intensity is likewise a function of wavelength, and caution must be exercised when it is used in the different wavelength regions.

1.5.4 Methods of Development

Problems such as the Eberhard and Kostinsky effects can be minimized, although not completely eliminated, by mechanical agitation of the developer or by movement of the plate or film. Although difficult to control, the simple holding of a film in a "U" shape and drawing it back and forth through a tray of developer provides good agitation. Developing tanks permit the reel holding the film to be twisted back and forth in the developer, or permit the entire tank assembly to be alternately turned gently upside down and right side up to keep fresh developer in contact with densely exposed areas. In the case of plates in a tray a very good method of agitation is the use of a camel's-hair brush wider than the plate and slowly brushing back and forth over the surface of the plate as it develops. Since these methods require some degree of skill on the part of the operator to keep them uniform, automatic developing machines are favored. The actions used here involve rocking trays in which the developer moves back and forth across the emulsion, or deep trays in which the emulsion is moved up and down in a harmonic motion or in which the developer is pumped past the emulsion. To handle film in these units, it must be either stretched into a flat section or wrapped around a frame. The deep trays offer some advantage in that a relatively small area is exposed to air and therefore the oxidation rate of the developer is reduced. The requirement of wrapping film around a frame for these units, however, is awkward and tends to put the sharply curved sections of film into a different developing pattern than those sections which are held flat. Rocking trays handle film or plate about equally well. There is the danger in trays, however, that the flow of developer back and forth over the emulsion may set up a pattern of static waves which will cause uneven development. This tendency is sometimes reduced by working with sufficient developer to keep the emulsion well under the surface. On the other hand, Harvey[(23)] noted that there will be nonuniform agitation if the developing solution is too deep in the tray. He specifies keeping the minimum amount of solution necessary to keep the film or plate covered at the end of each rocking stroke. Where rocking trays are too narrow, they may set up a different agitation at the edge of the film or plate. Since sprocket holes in film may set up differences in the agitation pattern, the area near them is usually avoided for quantitative microphotometer readings but may be used for photographing reference spectra. The area of $\frac{1}{4}$ to $\frac{3}{8}$ in. away from the edges of a plate should also be avoided both because of development variations and because there can be variations in emulsion characteristics there. A Kodak report[(24)] even suggests that an area up to 1 in. from an edge should be avoided in the most critical applications, which would make 2-in.-wide plates of only limited use.

With the aid of computers to evaluate photographic spectra, Török *et al.*[(25–29)] carefully evaluated deviation in the blackening of uniformly exposed plates in terms of methods of development. The methods used were mechanical and manual shaking, brushing, and laminar and turbulent flow of the developer. Minute errors (i.e., errors within a small area of the plate) were independent of the method used. Deviation was smallest with the use of laminar and uniform turbulent flow. The edge effect persisted.

Light-tight darkrooms and proper safe-lights are obvious requirements for the proper developing of emulsions. Proper care, however, is not always observed in practical setups, and Kodak[(30)] warns about the Sabattier effect, in which an image reversal can occur when development is carried on in the presence of some light leak. (This can be done deliberately for an artistic effect by interrupting the developing process with another exposure.) The completeness of reversal depends on the relative exposures before and during the development process. In discussing darkroom layouts, Slavin[(31)] suggests that the walls and ceilings should be white or of a very light color. Although it is more typical to see the inside of a darkroom painted dull black, the lighter coloring assists in finding out whether or not a darkroom is truly light-tight and that it remains so. With dull black paint, there may be light leaks that are not noticed by the operator.

1.6 PROBLEMS IN GENERATING LATENT IMAGES

The formation of a latent image is considered to be ideal when an intensity of radiation, I, reaching an emulsion over a period of time, t, acts to produce an exposure, E, that is the product of $I(t)$. If the illumination is with visible light, the exposure may be expressed as meter-candle-seconds. In more general radiation terms, the exposure may be expressed as ergs cm^{-2}. The reciprocity law of Bunsen and Roscoe[(32)] states that the photographic effect is a function of E only and not I or t separately. In practice this holds approximately only for a limited range of intensities and can diverge radically from reciprocity for either very weak or very strong signals.

1.6.1 Failure of Reciprocity

For very low intensities, as involved in astronomical photography where exposure times may be in hours, the reciprocity law fails and consistent photographic response is obtained when the exposure product, $I(t)$, is replaced by the Schwarzschild[33] expression in which the effective exposure becomes

$$E_{\text{eff}} = It^p \tag{1.10}$$

For astronomical exposures the exponent, p, has a value of approximately 0.8, but as the intensity is increased and the exposure time decreased, the value of the exponent gradually increases until its value is equal to unity, at which time the reciprocity becomes optimum.

Variation from the reciprocity law occurs in both long exposures of very low intensity radiation and in short exposures of very high intensity radiation. The pattern of the reciprocity variation can be seen in a three-coordinate isodensity plot such as Fig. 1.13 for an SA-1 emulsion as shown by Nachtrieb,[34] based on Eastman Kodak data. In this figure the scales are logarithmic so that wide ranges in intensity and time could be shown. The abscissa is expressed as intensity in meter-candles (m-C). The series of 45° lines designate the exposure time coordinate in seconds. The ordinate is the exposure scale in It (meter-candle-seconds). If there were no reciprocity failure, an intensity, I, exposed for some time, t, would yield exactly the exposure product, It. Thus for an intensity of 1 m-C and an exposure time of 1 s, the product would be 1.0; then, ideally, an equal density would be obtained if the intensity was only 0.1 m-C as long as the exposure time was raised to 10 s. This would show in Fig. 1.13 as a straight line horizontal to the abscissa and going through $It = 1.0$. In the example given for SA-1, the curve drawn approaches such a horizontal straight line in the region where the intensity is 1.0 m-C or more. The case shown is for about 13 m-C exposed for 0.1 s, yielding 1.3 m-C-s. With an intensity much weaker than 10 m-C, more than the ideal exposure time is required to produce an

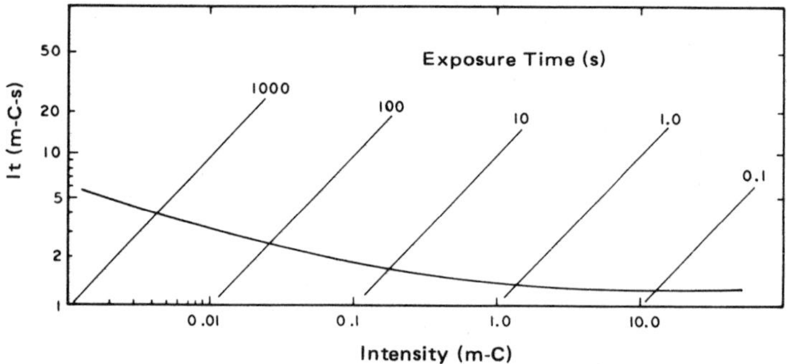

Fig. 1.13 Reciprocity failure of SA-1 emulsion.

Table 1.3 Reciprocity Characteristics for SA-1 Emulsion

Time (s)	I (m-C)	It
0.03	43	1.3
0.1	13	1.3
0.3	4.5	1.35
1.0	1.4	1.4
3.0	0.5	1.5
10	0.18	1.8
30	0.07	2.1
100	0.027	2.7
300	0.011	3.3
1000	0.0043	4.3
3000	0.0018	5.4

equally dark photographic image. Thus, when the intensity is reduced to only 0.026 m-C, we see that the curve intersects the 100-s line: theoretically, it should only have taken 50 s to get an equally dark image, but in this case divergence from the reciprocity law was so severe that the ideal exposure time had to be doubled. Table 1.3, based on values from the curve in Fig. 1.13, shows the much larger products of I and t that are needed at low-intensity signals. The table has been expanded to include exposure times intermediate to those shown in the plot.

Kodak(35) notes that although reciprocity failure effects have not been observed for very highly energized radiation such as x-rays and gamma rays, photographic emulsions usually show some loss in sensitivity at extremes in radiation intensity for the ultraviolet, visible, and infrared regions. These characteristics vary with emulsion type. Nachtrieb(36) shows some of these variations which are depicted in Fig. 1.14. Notable in this display is the very flat curve for

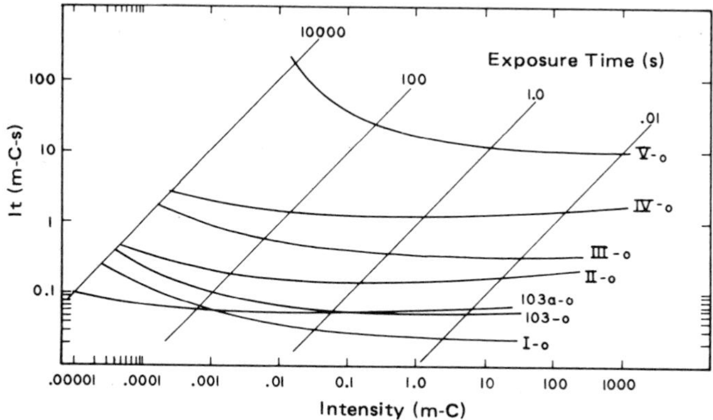

Fig. 1.14 Reciprocity failure of various spectrographic emulsions.

the 103a-0 emulsion, which approaches an ideal agreement to the reciprocity law. This emulsion was made especially for long exposures.[37]

In his discussion on forming preliminary calibration curves that are straight, or nearly so, Green[38] showed how discrepancies from the reciprocity law could be clearly shown. Green's equation is similar to that of Schwarzschild, Eq. (1.10), since it provides for an exponential difference between the intensity and time factors in the exposure product. This equation will be discussed in more detail in Section 1.7.4.2.

1.6.2 Intermittency Effect

A special case of the reciprocity law failure is the intermittency effect. If two equivalent exposures are photographed, one continuous and the second in installments, and if the image density is less in the second, it is due to the intermittency effect. That is, the $\Sigma_0^t I\Delta t$ does not produce the density equivalent to the product of It, even though $\Sigma_0^t \Delta t = t$.

The intermittency effect has been of considerable interest because of the wide use of rotating sectors for emulsion calibration. In their work, O'Brien and Parks[39] and Webb[40] observed that the intermittency effect disappeared if the frequency of the flashes was above a certain critical frequency and that this critical frequency varied with illumination, emulsion type, and wavelength. Sawyer and Vincent[41] however, found that for some very fast emulsions, even very high flash rates did not eliminate the effect. However, Pierce and Nachtrieb[42] found that the γ of time-scale characteristic curves did not vary for sector speeds from 100 to 1800 rpm and found it to be identical with that of intensity-scale characteristic curves. Malpica[43] also reported no difference in the γ for sector speeds that were varied from 1 to 3600 rpm. In practice, sector speeds of 1800–2000 rpm yield satisfactory results.

Radiation from dc arc excitation is quite steady or varies in brightness in a random fashion. However, for ac arcs, and particularly for sparks in which there is a definite frequency in the flashes of radiation, there is danger of the rotating sector developing some kind of synchronization with the frequency of the spark which will cause a variation in emulsion darkening. One ploy to avoid this is to use a series-wound motor and change the rheostat setting just before beginning the exposure so that the speed of the sector is changing constantly during the exposure.

The validity of exposure control with rotating sectors can be verified by cross checking with emulsion calibrations obtained from stepped filters or from known homologous line pairs which will be discussed later.

1.6.3 Other Image Effects

Kodak[30] describes other effects on the formation of the latent image. The Clayden effect is a desensitization of the emulsion following a very short exposure of very high intensity. In its extreme exhibition, when the first exposure has effected only a part of the emulsion and the second, lighter exposure is uniform over all the emulsion, the image of the first may appear reversed upon development. This has been observed in photographs of lightning flashes as "black lightning." A similar result occurs from the Villard effect, but this is considered to be a loss of developability and not merely a desensitization. Solarization is another effect of reversal in the darkening of the emulsion which can occur when an exposure is greatly increased beyond that which can cause the generation of a maximum density. In spectrographic applications this effect would more likely be masked by spectral line reversal–when an element line image is reversed through self-absorption of the signal by the atmosphere of atoms of the element surrounding the radiation source.

The Herschel effect is one in which a latent image formed by one type of radiation may be destroyed by another type. For example, if an emulsion has not been sensitized to red or infrared radiation and is exposed to blue or white light, the latent image formed may be subsequently erased by exposure to the longer wavelength. The energy needed to destroy the latent image, however, is several orders of magnitude greater than that required to form the image in the first place. It is unlikely, therefore, that the overlapping of a first-order red radiation on a second-order ultraviolet would show any evidence of this effect.

1.6.4 Halation Effect

A halation effect can occur in cases of overexposure in which radiation is not completely absorbed by the silver halide grains in the emulsion layer and is reflected back from the support to reenter the emulsion at some distance away from its original entry point. Thus a bright spot of radiation hitting the emulsion can appear as a circle or a halo rather than as a point. If not as pronounced as a halo, it might still appear as an enlargement of the spot. Kodak[(44)] points this out and notes that the thicker support base of spectrographic plates, typically 0.04 in., can cause a wider-spreading halation effect than films which can run from 0.004 to 0.005 in. thick. Many plates and most films are manufactured with some sort of an antihalation coating.

1.7 THE CALIBRATION CURVE

Final analytical curves relate how the intensity or exposure of a particular spectral line varies with changes in concentration. Measurement of the degree of darkening of an emulsion, however, does not directly show the variation in intensity. The translation of the darkening measurement, usually percent transmittance, into relative intensity is done by use of an emulsion calibration curve.

If only an approximate result is needed and if the general darkness of the spectrogram can be controlled closely, it is possible to run a series of standards and to make a useful approximate plot of percent transmittance, or optical density, versus concentration. The spectrochemical method, however, is capable of an analytical precision of 1-2% coefficient of variance for many elements under certain desirable excitation conditions. To achieve this precision, all aspects of the spectrochemical method must be controlled as closely as possible, including the use of an accurate emulsion calibration curve.

In some practical analytical systems the requirement for accuracy in the calibration curve may be relaxed as long as corrections are made by the extensive use of control standards. This would not be true for systems in which background corrections were being made. It might be valid only for the narrow range of 2-30% transmittance. Light line readings of 80% transmittance or more could be badly misinterpreted without an accurate calibration curve. If it were assumed that optical density had some arbitrary proportionality to log relative exposure, which is tantamount to drawing an arbitrary straight line as an H and D calibration curve, the relative intensities obtained would be raised to some fixed, but unknown, exponential value. This would result in final analytical curves that were either steeper or flatter than they should be. If the same arbitrary curve was used for unknown samples, the errors would be completely self-compensating for readings that were in the normally straight-line portion of the H and D curve. The errors would not be compensated, however, when the arbitrary display remained linear beyond the point where a true calibration was linear. Attempts at background corrections would be meaningless not only because background readings are likely to be relatively light but also because no computation can be made for the difference between two values when they are raised to some unknown exponent. Taking ratios of values raised to some exponent at least gives an answer which bears the same exponent. Thus $A^x/B^x = (A/B)^x$ but $A^x - B^x$ cannot be reduced further mathematically.

Corrections with control standards overcome errors in calibration curves, even when light lines are involved, provided that the control standards are close in composition to the unknowns or provided that the general shape of the assumed calibration curve is reasonably similar to the true calibration. In fact, the author has been surprised at the many laboratories he has visited in which satisfactory analytical results were being obtained even though the calibration curves in use were so far out of date that the newer technicians did not know how to construct them. In these cases, the errors of calibration were overridden by the control standards.

A true calibration curve is valid only for some specified developing conditions and may be restricted to certain spectral line widths if very narrow entrance slits are used on the spectrograph. Except for the differences in resolution of narrow lines, the calibration should be independent of the spectrograph or source unit. This independency on the generation of the spectral line, however, assumes that the calibration is made from lines which have been uniformly exposed along their lengths or from reading equivalent positions along homologous line pairs.

In the case of narrow spectral lines, differences in the amount of background being generated or in the amount of scattered light encountered would upset this independency. Although different batches of the same emulsion type should be very similar, the assumption is made that there can be enough difference between batches to require recalibration. Where there is a need for close control, the calibration is often rechecked, at least for individual boxes of plates or individual rolls of film. In cases demanding very precise control, calibration would have to be performed on each plate or piece of film developed.

Since the normal requirement in spectroscopy is to determine concentrations by observing the relative exposures of certain line pairs, the calibration curve is delineated by some expression of transmittance or density versus log relative intensity or log relative exposure. If it can be assumed that there was no reciprocity failure or intermittency effect in obtaining the calibration data, the terms "intensity" and "exposure" can be used interchangeably.

Actual calibration data may also include distortions caused by the microphotometer. Besides the obvious distortion of taking scattered or background light into readings, the amplifier or readout on the microphotometer may exhibit some nonlinearity. This same distortion would be present in reading the analytical lines and therefore would be self-compensating.

1.7.1 Methods for Obtaining Emulsion Calibration Data

Although Harrison[(45)] describes an imposing list of ways to determine the relationship of photographic darkening to relative intensities or exposure, only a few practical methods are currently used. In each method the $E = It$ relationship applies; either I or t is varied while the other is constant.

1.7.1.1 Rotating Step Sector

Rotating step sectors, which are restricted to stigmatic spectrographs or to astigmatic spectrographs in which the secondary or Sirk's focus is attainable, have a long history of application. These sectors have the distinct advantage of being completely neutral to wavelength and not subject to change unless damaged through mishandling. They are basically very accurate, limited only by the mechanical precision of cutting out angular sectors in the rotating vane. As mentioned in the section on intermittency effects, they may easily be rotated at variable, high speeds to avoid both the intermittency effect and any stroboscopic effect or synchronization with a pulsating spark source.

For calibration, the factor between steps has often been set from 1.5 to 2.0, with from four to eight steps. A useful factor is 1.585, which has the common logarithm of 0.200 for an easy display when working with logs of exposure. From the reading of all the steps in such a multistep sector, it is possible to directly draw the calibration curve from just one line. With some trial-and-error fitting, it is also possible to fit in the readings of other lines by trying for a direct match of points in the straight-line portion of the curve, such as would be found around 10% T in an H and D plot. This will be shown in Section 1.7.3.1 A primary requirement of this system is that the jaws of the entrance slit be exactly parallel and the illumination be made so uniform that no appreciable differences in readings are observed over the length of the lines used when they are photographed without being interrupted by the sector. If uniformity cannot be achieved for the entire length of the line or if averaging out, which is possible by first drawing a preliminary curve, is desired, neighboring steps can be used for the Churchill two-line method(46) as described in Section 1.7.2. Care must be exercised so that these motor-driven sectors do not set up vibrations which could reduce the sharpness of lines obtained on the spectrograph.

1.7.1.2 Neutral Step Filters

Multistep neutral density quartz filters can be used in place of rotating sectors. They differ from the sector in that the variations in exposure are by intensity change instead of by time change. These filters, of course, would not impose any intermittency effect. Two or three step filters are most commonly used and can readily be adapted to the two-line method of calibration. Two-step filters are commonly used on astigmatic spectrographs, where they are located near the focal plane of the camera as part of the aperture system. These filters are usually 1 inch wide and have a section of clear quartz and a section of approximately 50% filter. Although the term "neutral" is applied to these filters, they are not strictly neutral to all wavelengths but may be calibrated to obtain the effective filtering factor as a function of wavelength. Difficulty may be encountered in maintaining exact factors of filtering between steps, and the actual factors may have to be established for each step. When these filters are made by vacuum flashing aluminum on quartz, there is some danger that they will change with time either by the uneven collection of dirt or by the gradual decay of the aluminized coating. This attrition is more apt to occur to filters held on the optical bench, where they are subjected to the intense ultraviolet generated by the arc or spark. Focal plane filters used in astigmatic instruments are subject to less dirt and to much less intense ultraviolet radiation. In cases where the filtering is incorporated within a quartz lens, as is done with the field lenses on the Dual Grating Spectrograph,* no change in the filter factor should be anticipated.

1.7.1.3 Homologous Lines

A third very useful method is the use of multiplets or homologous lines. This approach was suggested by Dieke and Crosswhite in 1943(47) and the

*Initially manufactured by Bausch and Lomb, Inc.; presently manufactured by Baird-Atomic, Inc.

lines used are often referred to as Dieke lines. The implication of truly homologous lines is that they behave relatively the same for any kind of excitation. As pointed out by Ahrens and Taylor,[48] who did some further evaluation of specific sets of such iron lines, a major concern was self-reversal in low-level lines. They found they could work with a multiplet set having a low-level energy of 2.4 V but cautioned that calibration be restricted to either low concentrations of iron or low-amperage arcs. The author's uses of these lines has not indicated any particular distortion from self-reversal, but it does appear to be necessary to calibrate these lines for each spectrograph. Although this would seem to make these lines less useful, they have provided information that permitted evaluations of neutral filters or checks on emulsion development control. Examples of these uses were reported by Anderson and Lincoln.[49]

Once the relative intensities have been established for a set of homologous lines, a very complete picture of emulsion calibration is available from one or a few spectrograms. An example of this is given in Section 1.7.3.1. If just a pair of homologous lines is used, the Churchill[46] two-line method can be used. Futhermore, these spectrograms can be obtained by running as closely as possible to actual operating conditions, without the use of sectors or filters. Calibration by the use of homologous lines and with either the sector or split-field filters provides a means of checking the validity of the overall emulsion calibration and also provides a cross check on whether the developing conditions are in control, that is, if uniform development occurred along the length and width of the plate or film which was processed.

1.7.1.4 Line Uniformity

For calibration purposes, the common requirement is that there must be uniform darkening of the lines to be measured when the filtering or sectoring medium is not in position. If this were not so, the nonuniformity would upset the assumed intensity factors of the steps, making the effective intensity ratios either more or less than they were supposed to be. The degree of nonuniformity can also be different for different lines, which could mean that a series of lines would not even be consistent in their apparent intensity factors. This is not required for homologous lines as long as the microphotometer readings are taken consistently on the same area along the lines.

Although it is often suggested that calibration be done under normal operating conditions, it is only important that the entrance slit be set at its normal opening or checked at any different slit widths which might be used. Either spark or arc excitation can be used unless the normal excitation causes line broadening. Low-current dc iron arcs of 2 or 3 A are often used. To obtain line uniformity these arcs are run at wide gaps of 6 mm or even 9 mm. It is interesting to note that in 1922, under the auspices of the International Astronomical Union, the gap of the Pfund arc for iron secondary standard lines was increased from 6 to 12 mm. Both upper and lower electrodes can be iron rods; however, a positive lower iron rod opposed by a negative upper graphite electrode will provide a steady burning arc. Rods $\frac{1}{4}$ in. in diameter have been used successfully, but a more massive iron rod will tend to avoid the formation of a molten bead during the course of the arcing. Another approach, of course, is to permit the iron rod to form the bead before opening the shutter. For the massive electrode approach, Harvey[50] describes an electrode made by Sherman from 1-in.-diameter rods faced with a 120° cone. The other end was turned down to $\frac{1}{4}$-in. diameter to permit its being held in an electrode holder. The massive, wide-diameter portion is about 1 in. long.

Optical methods, several of which are described in the summary by Ahrens and Taylor[51], may be used to achieve uniformity. One precaution relates to focusing in the horizontal plane. Whatever optical system is used for calibration, it must be consistent with the operating conditions in respect to focusing in the horizontal plane. If only a cylindrical lens in a practical lens system is used to focus the electrode system in the vertical, then, for achieving uniformity, no other lens system, spherical or cylindrical, should be used for focusing in the horizontal. This precaution becomes necessary since, as Nachtrieb points out,[52] the optimum slit width for maximum intensity and resolution is half as much for systems which focus on the slit as they are for a remote point source which has coherent illumination entering the slit.

The entrance slit on a concave grating spectrograph may require a very critical adjustment to make it parallel to the centers of each of the ruled lines of the grating when the incident beam is at a fairly wide angle to the normal.

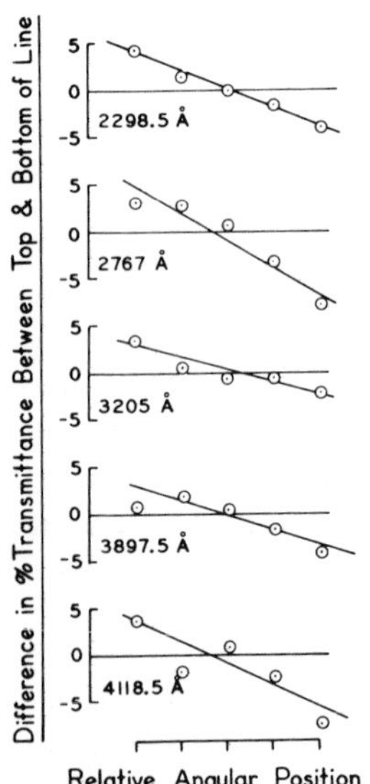

Fig. 1.15 Entrance slit rotation related to line uniformity for a concave grating spectrograph. (From Bond.[53])

If this is not done carefully, the lines may appear to be well focused but not uniform in darkness. This was observed by Bond[53] at International Nickel working with an ARL 2M spectrograph in which the angle of the incident beam to normal is 32°. This condition has not been observed as critical on all such spectrographs, but it can be a serious problem. In one instrument the series of lines being checked for uniformity required very small changes in rotation of the entrance slit. The angular changes made were in the order of magnitude of 3 min (0.05°). Figure 1.15 shows some typical patterns of how uniformity of line shifted as the entrance slit passed through its optimum rotational position.

1.7.2 The Preliminary Curve

A preliminary curve shows how values of emulsion darkening relate to each other when the intensity of the radiation causing the darkening changes by a specific ratio. For example, with a step filter in which one section is clear quartz and the other section has enough aluminizing to reduce radiation in half, a preliminary curve can be made by plotting, for each line read, the darkening measurement of the filtered portion of the line against this measurement for the unfiltered portion. In this case the curve would be for the intensity ratio of 0.5 or its reciprocal, 2.0, depending on which way the curve is used. Thus, if any one darkening measurement is given, this curve will directly show what line reading would be obtained from radiation having double the intensity, as well as another line reading for radiation of half the intensity. Each new reading, in turn, can generate another one in which the relationship of relative intensity is maintained. A whole series of readings can therefore be established for a series of related intensities, and this list can be used to construct a final calibration curve.

Preparing an emulsion calibration curve by observing the darkening measurements of line pairs is called a two-line method, or a two-step method if the line pairs are established from a filtering or sectoring method. Churchill[46] demonstrated how a two-line method could establish an emulsion calibration curve by working with a preliminary curve. Details on applying this method appear in the ASTM proceedings.[54]

By the use of some of the methods described in Section 1.7.1. to obtain emulsion calibration data, one may bypass the preliminary curve to generate the calibration curve. Although it represents an additional step, the two-line preliminary curve has the distinct advantage of being able to assimilate a large display of calibration data and smoothing it out so that any faulty microphotometer readings will be deemphasized or even ignored.

Once the preliminary curve has been drawn, an unlimited number of lists of transmittance readings can be extracted in which each neighboring transmittance would represent the darkening from radiation which had the specified factor of intensity. Since any arbitrary starting point can be used to set up the lists of transmittance values, an infinite number of lists could be evolved. In practice it would be valuable to set up only two or three such lists reasonably separated from each other, and work them into an overall display on the final emulsion calibration. Bringing an additional set of readings into an emulsion calibration display is relatively easy if one of the new points is made consistent with the first set of data in the area where the final calibration can be expected to be linear. A multistep filter having a factor of 1.5 between steps would be considered reasonably fine to establish the final calibration curve, whereas a factor of 2.0 would normally be considered to be too coarse for any fine delineation of the calibration curve. However, if two coarser displays were offset from each other so that a final plotting of points had one set falling about halfway between the other set, this would be close to working with the square root of the intensity factor. Thus two well-chosen sets of data from a step factor of 2 could be equivalent to one set having a factor of $\sqrt{2}$ or approximately 1.4. If three such well-chosen lists were extracted from the preliminary curve, the final calibration curve would have points with intensity spreads of only about $\sqrt[3]{2}$, or about 1.25.

1.7.2.1 Transmittance Preliminary Curve

A simple, direct method for drawing a preliminary curve is to use transmittance values on rectilinear coordinates and to designate one axis as the light line of a pair and the other axis the dark line. This is demonstrated in Fig. 1.16 from data in Table 1.4A, using the filtered and unfiltered readings of the lines in each spectrogram as the line pairs. Two points of this curve must be met regardless of the data: the curve must go through (0, 0) and (100, 100). There is usually no problem in bringing the curve into the (0, 0) point, but if the points do not line up well for going through (100, 100), this is evidence that the calibration data was faulty, at least in establishing the 100 reading on the microphotometer. If this were observed, the obvious advice would be to go back and reread the calibration data.

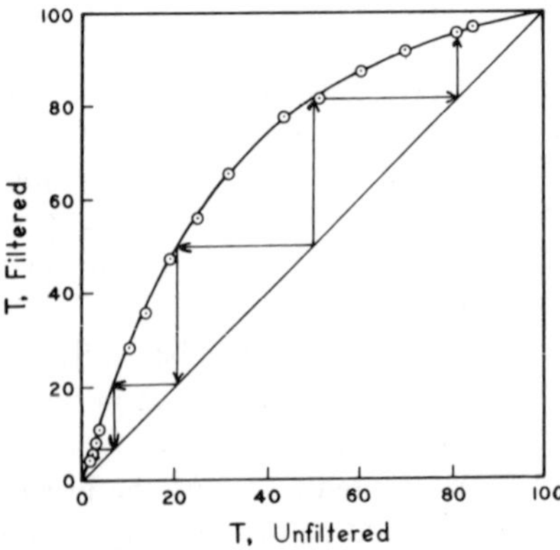

Fig. 1.16 Transmittance preliminary curve.

Table 1.4 Data for Four Homologous Iron Lines[a]

3161.9 Å		3175.4 Å		3205.3 Å		3222.0 Å	
Unfiltered	Filtered	Unfiltered	Filtered	Unfiltered	Filtered	Unfiltered	Filtered
A: % Transmittance readings from microphotometer							
81.0	95.5	43.5	77.8	19.0	47.6	2.9	8.5
84.4	97.0	51.2	81.6	24.9	56.0	3.5	11.0
70.0	92.0	31.5	65.5	13.5	36.0	2.0	5.5
60.6	87.6	25.0	56.0	10.0	28.5	1.6	4.5
B: Density values translated from % transmittance data from A							
0.0915	0.0200	0.362	0.109	0.721	0.322	1.538	1.071
0.0737	0.0132	0.291	0.0883	0.604	0.252	1.456	0.959
0.155	0.0362	0.502	0.184	0.870	0.444	1.699	1.260
0.218	0.0575	0.602	0.252	1.000	0.545	1.796	1.347
C: Seidel values translated from % transmittance data from A							
−0.630	−1.327	0.114	−0.545	0.630	0.042	1.525	1.032
−0.733	−1.510	−0.021	−0.647	0.479	−0.105	1.440	0.908
−0.368	−1.061	0.337	−0.278	0.807	0.250	1.690	1.235
−0.187	−0.849	0.477	−0.105	0.954	0.399	1.789	1.327

[a]Four different exposure conditions were used with a two-step filter having a nominal filtering factor of 50%.

Since this curve will be used by starting at some arbitrary point and going back and forth from the curve to the 45° line, horizontally and vertically, to evolve a list of transmittance values related to a geometric series of relative exposures, it is important that the curve be spread out on a large graphical display to give precise readings. Figure 1.16 shows how the list could be developed by starting at 50% transmittance and going in both directions, as shown by the arrows. A summary of these readings appears in Table 1.5 in the set

Table 1.5 List of Readings Obtained from Preliminary Curves for Exposures Differing by a Factor of 2[a]

Relative E	Seidel	% T	Density	Relative E	Seidel	% T	Density
1.414	−1.350	95.7	0.019	1	−1.727	98.2	0.008
2.828	−0.645	81.5	0.089	2	−0.990	90.7	0.042
5.657	0.000	50.0	0.301	4	−0.315	67.4	0.171
11.314	0.590	20.5	0.688	8	0.302	33.3	0.478
22.628	1.130	6.9	1.161	16	0.866	12.0	0.921
45.255	1.625	2.3	1.638	32	1.383	4.0	1.398
				64	1.856	1.4	1.854

[a]The second set was selected so that its readings would be displaced from the first set by a factor of $\sqrt{2}$; therefore, one set starts out with a relative exposure of 1.414.

headed by the relative exposure of 1.414. Actually, this list as well as the second list, headed by the relative exposure of 1, was calculated from the Seidel preliminary curve described in Section 1.7.2.3, but the readings should be consistent with what could be read directly from the curve.

1.7.2.2 Density Preliminary Curve

A direct plot may also be made of the line pairs of the previous section where the darkening measurement is in terms of density. This would be particularly valuable if the microphotometer being used gave its reading directly in terms of density values. To represent this, the transmittance values of Table 1.4A were translated into density terms, and these values appear in Table 1.4B. The resulting preliminary curve appears in Fig. 1.17. This figure also shows how a series of density values could be developed, starting from 0.301, which is the density value of 50% transmittance. These values are listed in Table 1.5. A related plot which involves using the natural logarithm of 10 times the transmittance value is discussed in Section 1.9.1.

Although this plot does well in spreading out the readings for dark lines where there is usually very little trouble in defining the emulsion calibration, it is badly cramped in the area of light lines. Like the transmittance preliminary curve, this curve should go through the point (0, 0). With the poor definition in this area, however, it might not show up faulty microphotometer readings. There appears to be little, if any advantage in bothering to translate transmittance values to these terms. An improved display could be made by plotting the

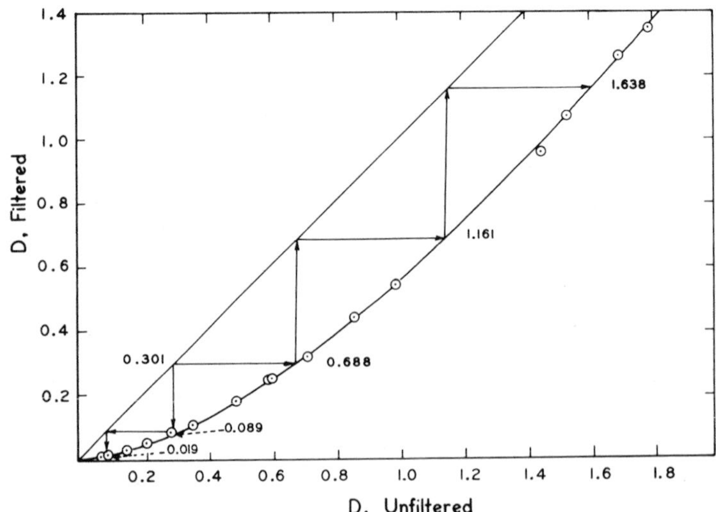

Fig. 1.17 Density preliminary curve.

density values on log scales. This was described by Green[38] and is discussed in Section 1.7.4.2. The preliminary plot on this basis appears in Fig. 1.25.

1.7.2.3 Seidel Preliminary Curve

An increasingly popular preliminary curve is one in which the microphotometer readings appear as Seidel values. This can be done by converting transmittance readings into the Seidel transform by using a table, such as Table 1.11 at the end of this chapter, and plotting the resulting values on a large sheet of rectilinear coordinate paper. If certain calculators such as the Jarrell-Ash Seidel Calculating Board or the Dennert and Pape Respectra are available, the plot may be made directly without translating transmittances into Seidel values. Details on the Seidel transform appear in Section 1.7.3.2. The manipulation of these boards for preliminary plots is discussed in Section 1.8.3.

The advantage of the Seidel preliminary plot was pointed out by Wheeler,[55] who claimed that the resulting plot should be linear. The validity of this claim has been questioned, but wide experience has demonstrated that is is valid for the range of microphotometer readings encountered with the commonly used SA-1 emulsion. Although there is some question as to how far the linearity of this relationship extends, when linearity is not achieved it may indicate that the microphotometer is giving biased answers or that scattered light is playing an excessive part in the densitometer readings. Consideration of these effects appears in the next section.

When this preliminary plot is linear, it is easy to describe the line mathematically and then calculate lists of readings for setting up the final emulsion calibration curve. This permits the lists to be determined more precisely than can be done graphically from the plot. From just one well-defined preliminary curve, additional plots or algebraic equations may be developed for obtaining transmittance lists in which the factor of intensity ratio is arbitrarily set as low as desired. This latter feature will be dealt with in Section 1.7.4.3.

The calibration data in Table 1.4A are translated to Seidel values in Table 1.4C. These are then plotted in Fig. 1.18, where a straight line was drawn through the points. As was shown on the other preliminary curve plots, calibration data could be obtained by starting from an arbitrary point and following the paths of the arrows to pick up each reading. The actual calibration data, which are tabulated in Table 1.5, were derived from an algebraic statement of the line. The data were subjected to a least-squares fitting to a straight line and the following equations were derived:

$$S_{\mathrm{Lt}} = 1.0928 S_{\mathrm{Dk}} - 0.6446 \quad \text{and} \quad S_{\mathrm{Dk}} = 0.9151 S_{\mathrm{Lt}} + 0.5899 \qquad (1.11)$$

where the subscripts refer to whether the line is lighter or darker or, in the case at hand, whether filtered or unfiltered, respectively.

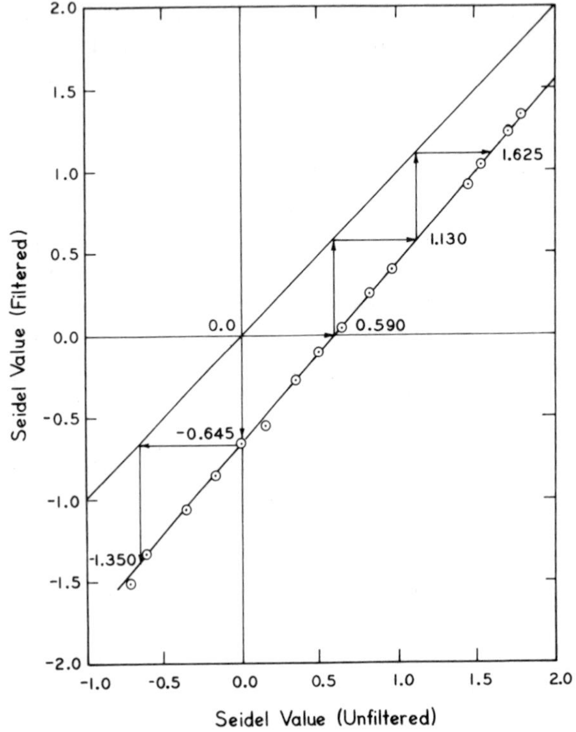

Fig. 1.18 Seidel preliminary curve.

Basically, the same equations would have evolved from making the best fit by eye and then describing the line algebraically. For example, inspection of the ends of the plot show that there is a point for a filtered line reading 1.54 Seidel and an unfiltered line reading 2.00. Similarly, at the low-density end we see a point for -1.52 and -0.80. Putting these into the general statement for a straight line, $y = mx + b$, we obtain the two equations

$$1.54 = 2.00m + b \quad \text{and} \quad -1.52 = -0.80m + b \tag{1.12}$$

Solving these simultaneous equations, we find that $m = 1.0929$ and $b = -0.6458$ which is the close agreement with Eq. (1.11).

The second set of calibration data in Table 1.5 was made to have the $\sqrt{2}$ factor of relative exposure by a calculation discussed in Section 1.7.4.3. Without resorting to this refinement, an approximate starting point could have been chosen for the second set so that is would have approximately the same factor. The exact factor need not be known for fitting the two sets of data into the final emulsion calibration, as described in Section 1.7.3.1.

1.7.2.4 Nonlinear Seidel Preliminary Curve

A nonlinear Seidel preliminary plot may be due to a characteristic of a particular emulsion in some wavelength region. In such a case the preliminary curve could be handled graphically for obtaining the calibration curve data. These curves would at least be expected to have only slight curvature.

The nonlinearity, however, may be due to errors in the microphotometer readings. When linearity is not observed, it would be advisable to check for some distortion, such as an incorrect setting of the 100 reading, or poor resolution in the microphotometer, or excessive pickup of scattered light. An intensity-retardation-of-development effect, as discussed in Section 1.5.2, can also cause a nonlinear plot. A treatment of a case that approximates this effect is discussed in Section 1.9.2.

If the 100 setting for a clear portion of the plate was made in an area that was not truly clear, all transmittance readings would be made to appear proportionately high. To see how this would affect the preliminary curve, assume that the calibration data in Table 1.4A had been read with a 100 setting made on a slightly fogged area amounting to 98% transmittance. The data would have appeared proportionately higher as shown in Table 1.6A. (In this case, the second pair of readings for the 3161.9-Å line would have been excluded, since there was no significant reading on the filtered line.) These data, plotted in Fig. 1.19 with the scales arranged according to the Seidel transformed density, show how the lighter lines curve away from the straight-line plot of Fig. 1.18, which is the dashed line in this plot.

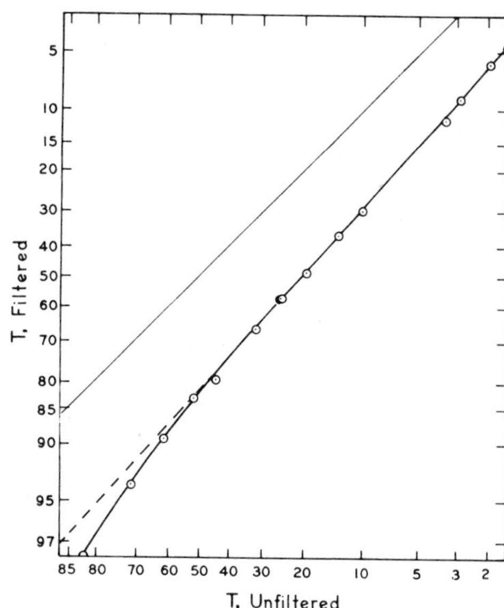

Fig. 1.19 Curvature in light line section of Seidel preliminary curve.

We can also approximate how the calibration data would be affected by manipulating the data of Table 1.4A to indicate what lighter readings would have resulted from poor resolution, from scattered light, or from the intensity-retardation-of-development effect. Imagine that each reading had some pickup of stray light or included some background reading from either side of the line. To illustrate, assume that only 95% of the reading was truly the transmittance of the line and that the other 5% of the reading was effectively from an area of 100% transmittance. This is, of course, a simplification, since the inclusion of additional light would not be as severe if the reading was from a spectrogram with a dark background or from one which had a cluster of dark lines that would cut down on the scatter. Futhermore, if the problem were one of poor resolution, the fact that dark lines also become broader would have mitigated the effect on these lines. To represent the effect, however, we can take the transmittance readings in Table 1.4A and modify them by taking 95% of their T value and adding 5. This was done to obtain the readings in Table 1.6B. Similarly, we can project how readings would have appeared lighter from too short a development time by applying some progressively greater reductions in line density as the line becomes darker using the scale of density reductions in Fig. 1.11. This was done to the data in Table 1.4A by first converting to density terms, applying the intensity reduction factor, and then converting back to transmittance values. These are shown in Table 1.6C.

The effect of the data in the last two tables is shown in the curves in Fig.

Table 1.6 Percent Transmittance Data from Table 1.4A Modified as a Result of Errors

3161.9 Å		3175.4 Å		3205.3 Å		3222.0 Å	
Unfiltered	Filtered	Unfiltered	Filtered	Unfiltered	Filtered	Unfiltered	Filtered
A: Data with 2% error in setting the 100 reading on microphotometer							
82.6	97.5	44.4	79.4	19.4	48.6	3.0	8.7
86.1	99.0	52.2	83.3	25.4	57.1	3.6	11.2
71.4	93.9	32.1	66.8	13.8	36.7	2.0	5.6
61.8	89.4	25.5	57.1	10.2	29.1	1.6	4.6
B: Data with inclusion of 5% scattered light							
82.0	95.7	46.3	78.9	23.0	50.2	7.8	13.1
85.2	97.2	53.6	82.5	28.7	58.2	8.3	15.4
71.5	92.4	34.9	67.2	17.8	39.2	6.9	10.2
62.6	88.2	28.8	58.2	14.5	32.1	6.5	9.3
C: Data modified as if affected by intensity retardation of development							
83.0	96.0	50.0	80.0	28.9	53.6	10.8	18.4
86.0	97.5	56.8	83.0	33.9	60.8	11.8	21.1
71.5	93.0	39.4	69.5	23.7	43.3	9.1	14.7
63.0	88.5	34.0	60.8	20.1	36.9	8.2	13.3

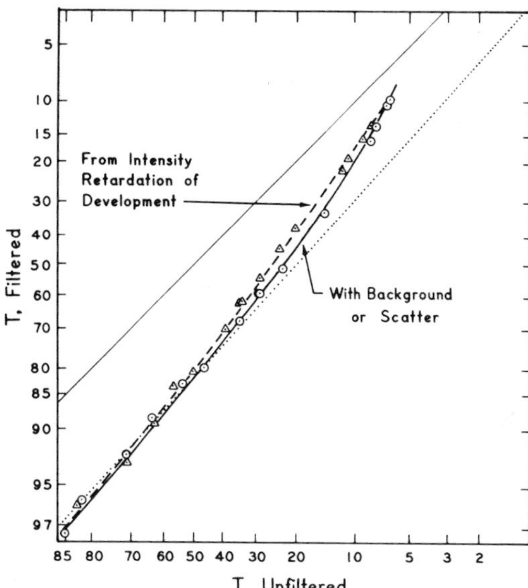

Fig. 1.20 Curvature in dark-line section of Seidel preliminary curve.

1.20. The straight-line plot from Fig. 1.18 is also repeated here by the dotted line to show that these errors displace the darker line readings toward the reference 45° line. The plot for the scattered light effect obviously is not linear. The plot for the example of the intensity retardation of development was more nearly linear, but the curve is actually two straight lines with a flex point at about the halfway point.

The ASTM recommended practices[54] describe the effect of the improper setting of the 100 reading, properly calling it a distortion of the Seidel preliminary curve by a variation on microphotometer sensitivity. The discussion in the ASTM recommended practices also includes the effect of having an error in the zero setting of the microphotometer.

1.7.3 Final Emulsion Calibration Curve

For delineating the final emulsion calibration curve, three expressions for transmittance are commonly used: density, Seidel, and Kaiser. In all the usual curves the relative intensity values are either translated into logarithms or plotted on log scales. In general the transmittance function is made the ordinate and the log intensity is made the abscissa.

1.7.3.1 The H and D Plot

Historically, the H and D plot should be considered first, and indeed still does have wide application. It could also be designated as a density plot since it is plotted in terms of the log $(100/T)$, which meets the definition of Eq. (1.4). Figure 1.21 shows the H and D curve based on the data from Table 1.5. The encircled points are from the first list and the triangles are from the second set. Since this plot notoriously yields poor definition of the weak lines, Fig. 1.21 also includes the Harvey reverse scale plotting[56] in which the transmittance function abruptly changes at 50% T to log $[100/(100 - T)]$. As the figure shows, this brings the curved toe portion of the H and D curve into a gentle curve with a negative slope comparable to the γ of the linear portion of the plot. In applying the reverse plot, it is convenient to mark another set of values on the ordinate scale for the $100 - T$ operation, to avoid having to continually do this simple arithmetic.

When a multistep sector or filter is used, the curve can be defined directly by plotting the set of transmittance readings obtained on any one line versus the log relative exposures which can be assigned to each step. If a set of homologous lines is used the same direct approach can be used except that there would not be equal distances along the abscissa for each pair of points. This could have been done from the data in Table 1.4 by assigning relative intensities of 0.249 for Fe 3161.9, 0.552 for Fe 3175.4, 1.000 for Fe 3205.3, and 3.38 for Fe 3222.0, which are factors obtained by Anderson and Lincoln.[49] In this plotting one set of points is laid down and other sets fitted in by trial and error until all the points fall into general agreement. This manipulation is made easier

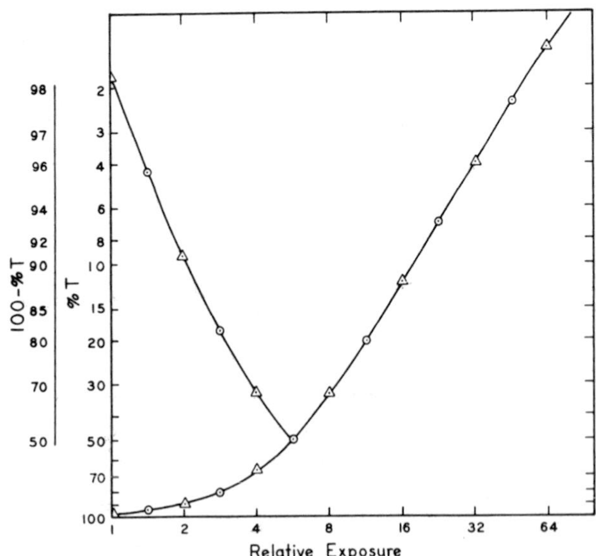

Fig. 1.21 H and D calibration plot.

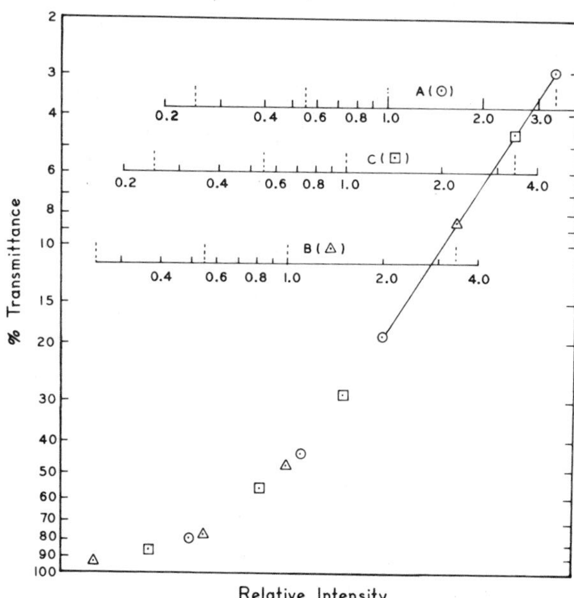

Fig. 1.22 Plotting a calibration curve directly from readings of four homologous iron lines.

if the abscissa is shifted horizontally, as is usually done on available calculating boards. Figure 1.22 demonstrates how this may be done. The intensity scale is set at some arbitrary position, *A*, and the unfiltered readings in the first spectrogram of Table 1.4A are plotted according to their assigned relative intensities as marked on the scale. These points are circled in the demonstration plot. A straight line is then drawn between the points of the two darkest readings as a first approximation on how the curve will go. The filtered readings in this spectrogram are now added by shifting the intensity scale to position *B*, where the darkest reading will fall on the straight line. Its points are plotted from this setting and are shown as triangle dots. Similarly, the scale can be shifted to position *C*, where filtered line readings from the last spectrogram can be plotted, using square dots on the example. Again the scale was set so that the darkest line point would fall on the straight line. This can be carried further until all the points of Table 1.4A are located. If any of these sets did not seem to have their lighter line points in a consistent display, the scale setting for these should be shifted to give the best compromise setting for all its points.

Calculating boards may use a different scaling for ordinate and abscissa to spread out the H and D plot. This helps in giving an improvement in reading of intensity ratios. It should be noted, however that the gamma of the emulsion would be greater than the observed slope of this kind of plot.

1.7.3.2 Seidel Transformed Density

A simple modification of the density function was reported by Baker[57] and Sampson[58] in the latter part of the 1920s. Using D_s to designate this function of transmittance, we get what is commonly called the Seidel function:

$$D_s = \log\left(\frac{100}{T} - 1\right) = \log\left(\frac{100 - T}{T}\right) = \log\left[\frac{100/T}{100/(100 - T)}\right] \qquad (1.13)$$

The final expression above has been contrived to show the density term, Eq. (1.4), in the numerator. We can then see that D_s is an expression of the ratio of the amount of darkening which has occurred to the amount of darkening still possible. Expressed this way, it can be seen as a simple probability term. McCrea[59] points out that the Seidel transform is merely a special case of the logistic function originated by Verhulst[60] around 1838 in his studies related to population growth.

The Seidel function has appeared in many applications, including preliminary curves, as previously mentioned. It has been the basis for special graph paper for directly setting up analytical curves such as done by Noar[61] and by Trandafir,[62] who worked with his parametric method. It also has been the basis for developing other modifications to yield straight-line calibration curves. Examples are those by Kaiser,[63] Anderson and Lincoln,[49] and Severin and Rossikhin.[64]

In an early use of the Seidel function in preliminary curves, Schmidt[65] showed that when the preliminary curves are 45° lines, the final plot of the emulsion calibration would be straight if the Seidel transformed density was used. In addition, he showed that displacements of these preliminary lines from the origin of the graph were proportional to the logs of the ratios of intensity of the line pairs. Schmidt also discussed how preliminary plots which were not 45° lines indicated how the curvature would be affected in the final calibration curve.

The Seidel transform may be used directly for plotting a calibration curve. In special cases, as mentioned, it will yield a straight line. In any case it will at least show no more than gentle curving. If the calculating board being used does not have a Seidel scale, plotting may be done by first translating transmittance readings into Seidel values and plotting on either rectilinear or semilog paper. A table of conversion values appears at the end of this chapter. For the convenience of having these appear as larger numbers to improve the precision of the values of the function without increasing the number of significant digits, these are shown for the Seidel expression in terms of natural logarithms. To modify these to the conventional expression on log base 10, the values shown should be divided by 2.303. The examples given in this report have been kept to the common log basis, but all the examples can be readily performed with natural logs. Electronic calculators often permit working with natural logarithms even more readily

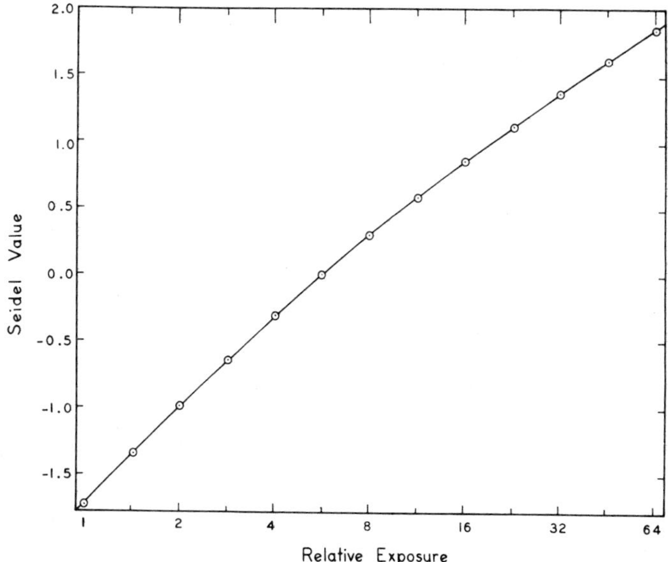

Fig. 1.23 Seidel calibration plot.

than with common base 10 logarithms. The data from Table 1.5 are shown in Fig. 1.23 plotted according to the Seidel transform.

1.7.3.3 The Kaiser Transformed Density

Kaiser[63] developed a modification of the Seidel transform to take out the curvature which is usually found in these calibration curves. The modification is simply to use a combination of the standard density function with the Seidel function so that the slopes in both the darker and lighter line regions are made equal. Using D_k to represent the Kaiser function, we may define it as

$$D_k = kD_s + (1 - k)D \tag{1.14}$$

where k is the transformation constant needed to achieve linearity. If k were taken as zero, the function would become just the standard density equation. If k were taken as 1, the function would be just the Seidel equation. The k value can be obtained by trial and error or it can be obtained by a calculation that involves making a determination of slopes in selected areas of light and dark readings that would be obtained on the H and D plot and on the Seidel plot. One precaution that should be observed is that if the light and dark areas are taken too far apart, the final calibration curve may take on an S shape. This can be avoided if one transmittance is used as the dark readings for the light pair of lines and also as the light reading for the dark pair of lines. In this case it is expedient that the ratio of intensities represented by the line pairs be about 2.0.

An example of how an optimum k value can be determined can be seen from the data in Table 1.5. From the lists of transmittance values, the readings of 67.4, 33.3, and 12.0% T can be handled as

$$\begin{array}{rl} \log 33.3 = 1.522 & \log 67.4 = 1.829 \\ \log 12.0 = \underline{1.079} & \log 33.3 = \underline{1.522} \\ \text{diff. } 0.443 & \text{diff. } 0.307\text{; then } 0.443 - 0.307 = 0.136 = \Delta H \end{array}$$

$$\begin{array}{rl} \text{Seid. } 12.0 = 0.865 & \text{Seid. } 33.3 = 0.302 \\ \text{Seid. } 33.3 = \underline{0.302} & \text{Seid. } 67.4 = \underline{-0.315} \\ \text{diff. } 0.563 & \text{diff. } 0.617\text{; then } 0.563 - 0.617 = -0.054 = \Delta S \end{array}$$

Now

$$k = \frac{\Delta H}{\Delta H - \Delta S} = \frac{0.136}{0.136 - (-0.054)} = \frac{0.136}{0.190} = 0.72 \qquad (1.15)$$

A derivation of this equation appears in Anderson's description of the Respectra calculator.[66] Applying this value for k in Eq. (1.14), the data in Table 1.5 can be translated to the Kaiser transform. These new data, including their makeup from the Seidel and density terms, are shown in Table 1.7. The final calibration plot appears in Fig. 1.24. A straight line was drawn, but it can be seen that there is a small displacement from this straight line for very light lines. In general, it is advisable to determine a transformation constant which favors the middle and low value transmittance lines and let some curvature occur for the less used high transmittance values. The Kaiser transform is the basis of the Respectra calculator, made by Dennert & Pape, Hamburg, West Germany.

Table 1.7 Kaiser Transform Values from the Transmittance Values of Table 1.6, Using the Transformation Constant of 0.72

% T	0.72 × Seidel	0.28 × Density	Kaiser
95.7	−0.970	0.005	−0.965
81.5	−0.464	0.025	−0.439
50.0	0.000	0.084	0.084
20.5	0.424	0.193	0.617
6.9	0.814	0.325	1.139
2.3	1.172	0.459	1.631
98.2	−1.251	0.002	−1.249
90.7	−0.712	0.012	−0.700
67.4	−0.227	0.048	−0.179
33.3	0.217	0.134	0.351
12.0	0.623	0.258	0.881
4.0	0.994	0.391	1.385
1.4	1.331	0.519	1.850

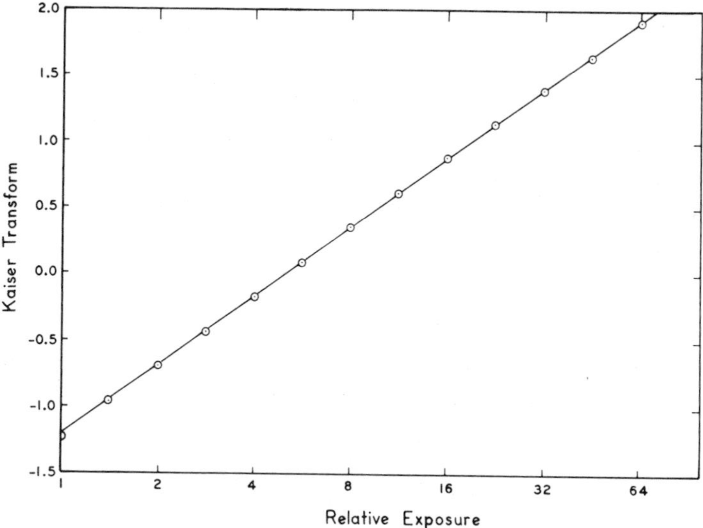

Fig. 1.24 Kaiser calibration plot.

1.7.4 Calibration Equations

Equations for describing an emulsion calibration are particularly desirable since they permit setting up computer programs in which transmittance readings can readily be translated into relative intensities or intensity ratios. Some of these applications are primarily empirical fitting to emulsion data obtained in conventional ways and may include fragmentation of curves to fit sections separately or may involve high-order equations in which a computer is used to determine the constants of the equation. Although these may be very practical applications, they shed little basic information on the nature of the emulsion response.

1.7.4.1 The Kaiser Equation

Equation (1.14) described in Section 1.7.3.3, is one which can be programmed in a computer to evolve calibration data.[67,68] From these data intensity ratio calculations can be made and fitted into final working curves. The relative exposure in the various calibration curves, which have been discussed, can be described by the expression

$$\log E = \frac{1}{\gamma} f(t) \tag{1.16}$$

as it was used in Eq. (1.8), where the function of transmittance was optical density. Then the linearity of the emulsion calibration for the Kaiser transform permits the general statement of relative exposure for a wide range of transmittance readings to be

$$\gamma \log E = kD_s + (1 - k)D \tag{1.17}$$

when the function of transmittance is Eq. (1.14). Ratios of exposure or intensity can readily be determined from Eq. (1.17) after the k and γ constants are known. This equation was used to derive the formula for determining k in Eq. (1.15).

1.7.4.2 Green's Equation

In Section 1.6 it was pointed out that Green[38] had evolved an expression relating intensity with photographic density which allowed for a difference between the effect of intensity and the effect of time of exposure. This is expressed in the equation

$$\log\left(\frac{D_{\text{sat}}}{D}\right) = \left(\frac{I_{10}}{I}\right)^n \left(\frac{t_{10}}{t}\right)^q \tag{1.18}$$

where D_{sat}, I_{10}, t_{10}, n, and q are constants depending upon the emulsion, wavelength, and developing conditions. The subscript 10 on the I and t merely indicates that these would be certain values when the base of the logarithm was 10. A similar equation could be set up in terms of natural logs, in which these constants would have different values. The term D_{sat} represents a saturation density obtained in the extrapolation of a straight-line preliminary curve to the point where the densities of a pair of light and dark lines would be equal. The equation is derived from the observation that a substantially straight preliminary curve is obtained when the coordinates are expressed in terms of log densities. In Section 1.7.2.2 on the density preliminary curve it was observed that points in the low-density area were cramped together and yielded a curve. In taking the logs of density, Green was able to spread out the display at the low density end and make it nearly linear.

In this study of the reciprocity effect, Green obtained two sets of data in which the ratio of exposure was kept constant as twofold; in one, the time was kept constant and the intensity was varied; in the other, intensity was kept constant and time was varied. Specifically, he used a hydrogen lamp as a constant light source in both cases. To vary intensity, the distance of the lamp to the emulsion was changed according to the inverse-square law. The log density preliminary curves had different slopes in the two sets of data, which established different values for the exponents n and q.

In a later discussion of his equation, Green[69] goes into more detail on the nonlinearity of the log density preliminary curve. Although there would be some error if the equation were used for very light lines, the preliminary curve is well defined. Logarithms of the density data in Table 1.4B were taken and plotted

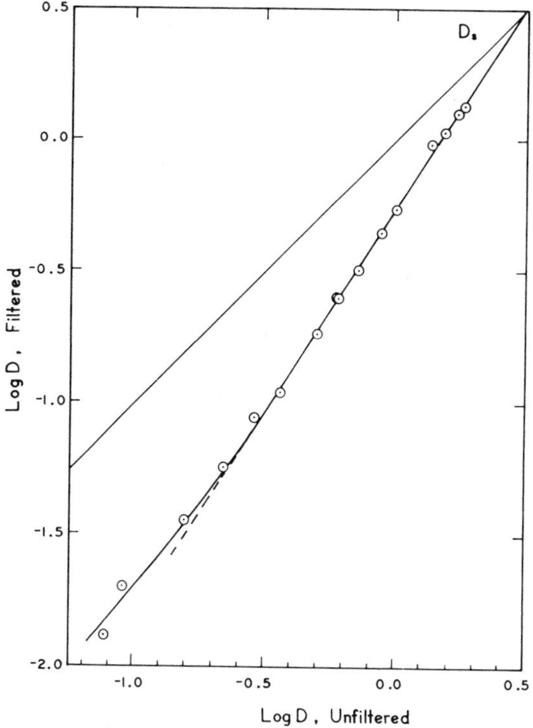

Fig. 1.25 Preliminary plot of Green.(38)

in Fig. 1.25. Curvature begins to show on this plot when the lines become lighter than 80% transmittance. The slope of this plot is steeper than the slope of the Seidel preliminary curve, Fig. 1.18, and is shown with its extension to an intersection with the 45° line to establish the point D_{sat}.

1.7.4.3 Seidel Equation of Anderson and Lincoln

When the emulsion response is such that a Seidel preliminary curve is a straight line, the plots for all ratios of intensity intersect at a common point on the 45° line. From this observation, Anderson and Lincoln(49) proposed the equation

$$\log\left(\frac{I_i}{I_0}\right) = k \log\left(\frac{S_s - S_0}{S_s - S_i}\right) \tag{1.19}$$

where S_s is called a saturation point and is the intersection of the preliminary curve with the 45° line. The equation relates relative intensity to Seidel readings with only two constants for characterizing the emulsion: the point S_s and the proportionality factor k. The key intersection point, S_s, was utilized by Noar[61] in his nomograph and is similar to the D_{sat} point of the Green equation in the preceding section. An example of the computation of the value of k appears in Section 1.9.2.

When one Seidel preliminary curve has been established, a whole family of curves is implied in which slopes vary exponentially with intensity ratios. This can be demonstrated in the calibration data of Table 1.4C to evolve a second preliminary curve having a different intensity ratio. The first part of Eq. (1.11) stated that

$$S_{Lt} = 1.0928 S_{Dk} - 0.6446$$

When the light and dark Seidel values become equal, they are defined as S_s; therefore,

$$S_s = 1.0928 S_s - 0.6446$$

from which S_s is calculated to be 0.6446/0.0928 = 6.9461. The slope of the line which was found for an intensity ratio of 2 was 1.0928. It follows that the slope of a line having a ratio of intensity of $\sqrt{2}$ would be such that its log would be one half of log 1.0928 or 0.01927. Therefore, the slope itself would be 1.0454. The new line would then become

$$S_{Lt} = 1.0454 S_{Dk} - b$$

The intercept, b, is calculated by substituting S_s for the Seidel values, which makes $b = 6.9451\ (0.0454) = 0.3154$. The new preliminary curve can therefore be stated in two forms as

$$S_{Lt} = 1.0454 S_{Dk} - 0.3154 \quad \text{and} \quad S_{Dk} = 0.9566 S_{Lt} + 0.3017 \qquad (1.20)$$

Starting with a Seidel value of 0.0, a series of Seidel values was calculated from the equations above for an intensity ratio of 1.414. Since alternative values in this list had the intensity ratio of 2.0, the list was separated to form the lists in Table 1.5.

1.7.4.4 Linearizing the H and D Curve

Since the H and D curve has a reasonably linear section for the darker lines, some consideration has been made on modifications that would straighten the curved portion sufficiently to yield a straight-line relationship that could be used for a calibration equation.

An early approach, referred to as the P transformation, was developed by Kaiser,[70] Seith,[71] and Haftka.[72] A correction was made in the density value by subtracting a function of the Gaussian subtraction logarithm of the density value. Thus, the P transformation can be written as

$$P = D - xD_G \qquad (1.21)$$

where

$$D_G = D - \log(\text{antilog}\, D - 1) \tag{1.22}$$

and x is the transformation constant.

When densities are large, the log of the subtraction of 1 from the antilog yields a value almost as large as the density itself, making D_G small and the P transform almost as large as the initial D value. This tends to retain the linear character of the darker lines. For the lighter lines, the D_G correction becomes increasingly large, and a transformation constant, x, can be determined which will effectively linearize the curved toe of the H and D curve.

Török and others(73–76) developed a modification of the P transformation to avoid the tendency of this plot to end up as two straight lines having similar slopes but with a flex point where the two lines are joined. The modification was basically the retention of the actual straight-line portion of the H and D curve with the application of the correction only to the curved portion. The modification developed through the steps of the P_L transform,

$$P_L = D - x'D_G \tag{1.23}$$

where the transform factor, x', was no longer constant but went through a pattern of changing values as the density changed. In this case the x' factor becomes zero at that point where the H and D curve becomes linear. This led to the expression

$$L = D - x'D_{G'} \tag{1.24}$$

where $D_{G'}$ is the value of the Gaussian subtraction logarithm of density when density has been divided by the gamma of the H and D curve. In this treatment, the variable transformation factor, x', takes on the specific definition of

$$x' = x_0 \left(1 - \frac{D}{D_L}\right) \tag{1.25}$$

where D_L is the value for the least density that fits on the linear section of the H and D curve and x_0 is an extrapolated value for the transformation factor when D is zero.

It was finally observed that x_0 equaled D_L and that the simple equation needed was

$$\ell = s - (k - s)d \tag{1.26}$$

where $\ell = L/\gamma$, $s = D/\gamma$, $k = x_0/\gamma = D_L/\gamma$, and $d = D_{G'}$.

In listing this series of transformation equations, only Eq. (1.26) is shown with the same symbols used in the reference articles. The symbols in the earlier equations have been modified to avoid conflicting with symbols used in other equations in this chapter.

The author has had some difficulty in justifying the claims made by Török on the nature of the x' transformation constant in Eq. (1.23), since no support is found for the linear character of the x' plot when the gamma of the H and D curve is 1.0. On the other hand, Ref. 74 includes an example of a table of ℓ values for a constant of $k = 0.40$ in Eq. (1.26). Lists of densities with fixed ratios of intensity obtained from this table showed good linear agreement when displayed on Seidel preliminary curves. Only a very close scrutiny of the slopes at different sections revealed a pattern of nonlinearity with a flex point where the ℓ transformation correction begins. The claim that one equation is more accurate than another in describing emulsion calibration in curves such as these is probably meaningless since they involve measurements in a practical system which are generally not capable of showing these fine differences. In spite of the fact that methods of linearizing the H and D curves are criticized for ignoring the reverse curvature which occurs in the dark-line knee portion, the various forms of linearization are practical computation tools. This is particularly true for the work of Török and Zimmer, which has been applied to a wide variety of emulsions. A summary of their equations with working tables has recently appeared in the United States.[77]

1.8 ANALYTICAL CURVES

The relative intensity values of selected analytical lines determined from microphotometer readings are plotted against element concentrations to form final analytical curves. This may be done by plotting log relative intensity or log intensity ratio against log concentration. Relative intensity values which are not referenced to another intensity value should be restricted to spectrographic radiation sources which are stable and reproducible, such as some plasma arcs or gaseous discharges. More often, in arc and spark sources, a reference is made to the intensity of a spectral line of a matrix element or of an element which has been purposely added to the sample and these intensity ratios are used for the analytical curve. Ideally, such a matrix line would be close to the analytical line. When a section of background of a spectrogram is dark enough for precise densitometry and when the background varies proportionately with the effective excitation of the sample, background can be used as the matrix reference. When a reference is made to a signal generated from within the sample in which the signal is independent of any expected change in matrix composition, the method is known as ratioing to an internal standard.

In well-controlled spectrographic techniques in which the element line is relatively strong in comparison to background or to interfering lines, and in which the element line does not show reversal, the plot of intensity or intensity ratio to concentration should be a straight line. Curvature due to background or line interference can be eliminated if the relative intensity of these interferences can be subtracted from the total element line signal. Furthermore, when the emulsion calibration curve is correct for the element line wavelength, as well as for background or for the reference line wavelengths when these are involved, the plot of intensity against concentration will be a 45° line when log scales are used. If the gamma of the emulsion calibration curve is different than it should be for the wavelengths of any of the lines being measured, the log intensity versus log concentration plot will vary from the 45° plot, although it may remain linear.

1.8.1 Background Corrections

Precautions on the determination of background corrections are mentioned in Sections 1.4.2 and 1.5.3, in which calibration curves are subject to change due to line widths. Slavin[22] indicated that a common calibration curve may not be valid for both background and line when the spectrogram is generated from a spectrograph with a primary slit width smaller than 100 μm. Nevertheless, it is common practice to use the same calibration curve for a line and its background.

If the intensity of a line plus background is A and the background intensity is B, then the correction of $A - B^x$ by the use of a single calibration curve, rather than the true $A - B$ correction, produces the error:

$$B - B^x = B\left[1 - B^{(x-1)}\right] \tag{1.27}$$

Since background, B, would usually be low and since the exponent x would be expected to be fairly close to 1 (that is, that the value of the slope of a calibration curve for background would be expected to be close to the slope of the curve for a discrete line) the common practice of using a single calibration curve for line and background appears to be acceptable.

A precise treatment of background correction using separate calibration curves for line and background is difficult since it requires a determination of how the curves would relate to each other against a common relative intensity scale. Considering that these curves would have to match at 100% transmittance, where the relative intensities of both curves could be properly defined as being zero, is of no help since the log 0.0 is not a finite value. One solution would be to make an arbitrary match of curves and then empirically determine a correction factor for the background intensity to put it into the same relative intensity scale as the element line curve. If there were a poor match between the calibration curves or if an incorrect factor was used, the resulting error could easily exceed that of assuming a single curve for both line and background.

The usual way of correcting for background using a single calibration curve on a calculating board is to obtain the ratio of the intensity of line plus background to the intensity of the background. When 1.0 is subtracted from this ratio, the resulting statement is the ratio of intensity to background of the line alone. This, in turn, can be ratioed to an internal standard. Thus, if we set the relative intensity scale of a calculating board at 1.0 for the background reading, we can read some relative exposure for the line plus background as being $E_L + E_B$, where the subscripts refer to line or background. Showing this in terms of relative exposure, as is done on the calculating board, we have

$$\frac{E_L + E_B}{E_B} = \frac{E_L}{E_B} + 1 \tag{1.28}$$

from which the ratio of the corrected line to background follows directly with the subtraction of 1.0. If a table of relative intensities is used, the correction is the simple subtraction of the background intensity from the reading of the intensity of the line and its background. Such a simple subtraction cannot be done with transmittance readings, since they do not have a direct relationship to spectral intensity.

The ASTM-recommended practice(54) notes that background may vary either directly or independently with the internal standard reference line. In the former, a background correction need not be made unless it is desired to straighten out the final analytical curve. In the latter, the correction is important for accurate results. This discussion also includes cases where background is not observed because the radiation intensity is too weak to overcome the inertia of the emulsion. Such background would still add to the line but would be important only if very light lines were being read. To correct for such background, the emulsion must be fogged enough to exceed the threshold of the inertia. Another solution, of course, would be to simply increase the exposure by either increasing the exposure time, if this did not upset other relationships, or by reducing any masking in the optical system.

Chamberlain(78) points out that a correction for background is justified only when the error attributable to the correction does not increase the standard deviation of the log of the resulting relative intensity proportionately greater than the increase in the slope of the analytical curve. The possibility of such an adverse effect is supported by the observation that the coefficient of variation of the relative intensity measurement increases for higher transmittance readings, as discussed in Section 1.8.2.

An additional problem is how background readings are taken. Line readings are made by slowly scanning over a line and noting the lowest transmittance (or highest density) which is read. Background can be variable and the lightest reading obtained in a line-free portion on either side of the line may not be the best measurement for the background under the line. This would be particularly true if an area close to a dark line was lighter as in the Mackie line effect (Section 1.5.1) or because light scattered more from dark lines into the lighter region in a Schwarzschild-Villiger effect (Section 1.4.2). A good estimate of background can be made if a recorder trace is made of a scan and if it includes sufficient travel before and after a line to show an averaged-out reading for background. If a recorder trace cannot be made, the observer can look for a median reading by remembering the high and low readings that he observed during the scan. With practice this may be done well with an analog readout such as is obtained with a galvanometer or the servodrive on a transmittance drum. Digital readout systems would be more difficult to judge in this manner.

Because of the difficulty in obtaining meaningful background readings and in handling them in calculations, there is a temptation to make an approximate background correction by resetting the 100 reading in the background region. This is not a valid correction. It falls far short of making the background correction which was intended, and should never be used. The temptation might be particularly strong for cases in which the background is light. As the following appraisal shows, however, this is where the faulty correction is very much in error.

Setting the 100 reading on a background area is tantamount to taking transmittance readings and dividing the line reading by the background and multiplying by 100. This would be valid only for a calibration curve which was linear between transmittance and exposure. Examples of the errors which occur in this oversimplification can be seen by using the data from Table 1.5 to set up a series of line and background readings as shown in Table 1.8. These are shown for special cases where the intensity of line plus background is exactly twice the intensity of background. Thus, when a true background correction is made, the corrected reading for the line minus its background would be exactly the same as the background reading itself. Another way of stating this would be to say that the relative intensity of the line alone would be the same as the relative intensity of the background. The table shows the false correction the ratioing method makes and translates the "corrected" line into a relative intensity value based on the emulsion calibration curve evolved from Table 1.5. This is compared to the true relative intensity of the line alone and the percent error is shown. In all the cases shown, the line readings by the faulty method remain darker than the truly corrected readings, and the error is greater for lighter backgrounds.

Table 1.8 Error in Making Background Correction by Setting Background to Read "100"

Line		Background		Correct	False correction		
% T	Relative I	% T	Relative I	line–background	T_L/T_B	Relative I	% Error
90.7	2.000	98.2	1.000	1.000	92.4	1.822	82
81.5	2.828	95.7	1.414	1.414	85.2	2.516	78
67.4	4.000	90.7	2.000	2.000	74.3	3.426	71
50.0	5.657	81.5	2.828	2.829	61.4	4.523	60
33.3	8.000	67.4	4.000	4.000	49.4	5.721	43
20.5	11.314	50.0	5.657	5.657	41.0	6.772	20

1.8.1.1 Analysis of Residuals

In the preparation of a series of standards for analytical curves, such as mixing powdered material in various combinations to give the required ranges of concentrations for the elements of interest, a material representing the matrix may be used, although no prior determination had been made of its residual elements. The problem is closely associated with background, not only because a low level of element signal may be involved, but also because the effect of ignoring the concentrations of the residuals can cause the same type of curvature in the analytical curve as background. Nachtrieb[79] illustrated the residual effect with an example by Duffendack and Wolfe[80] in which the data are plotted on log-log scales and the concentration of a residuum is determined by successively approximating the residuum until a straight line is obtained. The plot is shown in Fig. 1.26. The uncorrected curve is the solid line starting at 0.4 intensity ratio for 0.001%. Trial-and-error assumptions gave the two dashed curves when 0.001% and 0.005% residuals were assumed; the reverse curvature occurred when too much residuum was assumed. In this case, assuming 0.0025% residuum made the working curve linear. However, this is valid only if there is no background or if a proper background correction is made before the initial

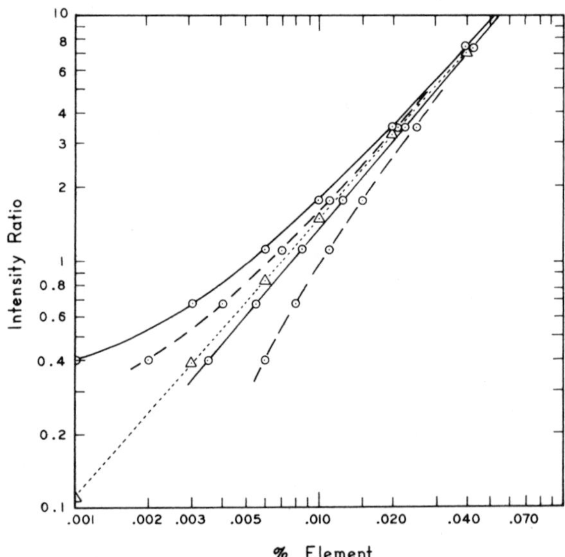

Fig. 1.26 Correcting analytical curves for residuals or background. The solid curve is modified to the solid line by assuming 0.0025% residuum in the matrix material. Trials with the assumptions of 0.001% and 0.005% residua are shown by the dashed curves. The same original curve can be corrected to the dashed straight line going through the triangle dots if it is assumed that the background contributed 0.29 relative intensity to the total signal.

plot. For example, the data in Fig. 1.26 would be displayed as a different straight line if it was assumed that there was no residuum but that there was background intensity of 0.29 relative to the internal standard. This value also was determined by trial and error and the reduction of intensities by 0.29 is shown as the dotted line through the triangles. An added confusion could be the presence of an unknown interfering line from the matrix that contributes to the intensity of the total signal. A background correction would not eliminate its effect.

If the problem is simply to determine a residual in the sample the analysis may be done by an addition such as described by Ahrens and Taylor.(81) Standards having the sample matrix are not necessary. Two or three portions of the sample are spiked with known, small amounts of any element of interest. These, as well as a portion of the sample itself, are excited. A simple way to handle the exposure data, after correcting for any observable background, is to plot the intensity ratios of the series against the added element concentrations on a rectilinear plot. If a straight line is defined, it can be extrapolated to an intensity ratio of zero. The negative concentration read at this point should be the residual concentration in the sample. This can only be used, however, for low-level concentrations with small additions of the element being sought. If extended too far, there will be perceptible curvature in the plot which would make the extrapolation uncertain. Even a small curvature in this plot might easily be misconstrued as being a straight line which would bias the answer to a lower concentration. Although not as simple, the addition method can be applied to higher concentrations as well as low to obtain a wider concentration range of the added elements if the plot is done on log-log paper and the Duffendack and Wolfe(80) method is used to convert the curve to a straight line.

1.8.2 Limitations on Line Readings

In the linear portion of the H and D calibration curve, small changes in the low transmittance readings of the darker lines represent relatively large changes in intensity. Slight errors in these readings cause relatively large errors in intensity. Because of the bending of the H and D curve, a second area is found where intensity changes markedly for small changes in readings when the readings become very light. This is also the area where transmittance readings are obtained with less precision. Mathematical statements for these relative errors were developed in Section 1.4.1 and expressed by Eqs. (1.8) and (1.9).

The pattern of error probability has been examined in determining the ideal ranges of transmittance readings for practical photometry. Slavin(82) discussed the effect of an error of 1% in transmittance reading on the various sections of a typical emulsion calibration curve. With an SA-1 emulsion he noted that a minimum error of 1% in spectral energy occurred at about 30% T and that the range of transmittance readings within a 2% error in spectral energy fell between 8.5%

and 63% *T*. Expanding to a 3% error in spectral energy, the transmittance range went from 6.3% to 75%.

Chamberlain[78] expanded on Slavin's work to include a consideration of other factors in addition to the measurement error. He showed that other factors of the determination would contribute more to the overall coefficient of variation in a practical spectrographic system than the measurement uncertainty. These include sampling, electrode preparation, excitation, selection of analytical line pairs, and the accuracy of emulsion calibrations and working curves. Chamberlain showed a somewhat wider practical range of acceptable transmittance readings than did Slavin,[82] even though he assumed that he had an identical emulsion. He also noted that the range of acceptable readings expanded as gamma increased since a given error in transmittance represents a smaller error in line intensity. Chamberlain noted that a maximum range was favored when the internal standard line was around 30% transmittance. He generally did not work above 80% transmittance and indicated that a practical limit for low transmittance readings was 5%.

Boswell *et al.*[83] took cognizance that extreme microphotometer readings are not as reliable as the more centrally located readings and used this as a means of weighing the data in a computerized determination of a preliminary curve. A large display of transmittances was studied on a fit to a Seidel preliminary curve to determine the repeatability of readings in various ranges. From this it was concluded that the readings from 10% to 70% *T* should be weighed more than other readings, which is another way of saying that it would be preferable to stay in this range in any final working system. They showed a curve in terms of the standard deviation in Seidel units that occurred at different transmittance values. Assuming that the values for the ordinate scale had been listed at 10 times their actual value because of a typographical error, and translating these Seidel deviations, it can be observed that standard deviations of low transmittances are about ±0.2% and that they gradually spread out to about ±1.0% at about 80% *T*. (The lightest readings in this plot show some reduction in deviation in transmittance, but this may not be a significant fact.) The general pattern would appear to be a standard deviation of about 1% of the transmittance reading plus a minimum deviation of 0.2% *T* absolute.

1.8.3 Calculating Boards

A basic calculating board provides two scales at right angles to each other in which one scale, usually a two-cycle log scale, is used for either relative intensities or for concentrations and the other is for transmittance readings. Sometimes the latter scale shows density readings. The transmittance scale is usually the ordinate and may be a two-cycle log scale, with the scale from 1 at the top to 100 at the bottom in order to have the low-reading, denser lines plot higher on the board. This scale can also be displayed as the Seidel transform. In the usual arrangement, the abscissa log scale may be moved laterally in some slots and the whole ordinate transmittance scale may also be moved laterally in a manner that keeps it at right angles to the abscissa. The vertical transmittance scale cannot be moved up or down. This arrangement permits points to be located on any part of the board surface or for any point to be read off a curve on the board's surface. Since the intensity scale can be easily shifted, ratios of intensity can be directly determined.

When the board has a Seidel scale, it can be used to plot a Seidel preliminary curve by first drawing a 45° line on the board. To spot a point for a pair of a light line reading against dark, first move the transmittance scale until the light line reading matches the 45° line. Then shift the log intensity scale until its central 1.0 is indexed to the transmittance scale either at a designated point on the scale or at one of its edges. Now move the transmittance scale until it is indexed to the value of the ratio of intensity of the dark line to the light line. The point is then located where the transmittance scale reads the dark line. All other pair points are located the same way, and the best preliminary curve drawn. To find a series of transmittance readings from this plot, start with some arbitrary transmittance reading matched to the curve. Record this and also the transmittance reading at the 45° line. Move the scale until this second reading matches the curve. Record the next reading at the 45° line and continue these shifts and readings as far as the preliminary curve permits. To go in the reverse direction, merely start at the 45° line and read out the transmittance value on the curve.

The main function of a calculating board is to hold a display of an emulsion calibration, or of several calibrations if all are needed. This permits a graphical determination of the ratio of intensities of two lines. The simple manipulation for doing this is: (1) move the transmittance scale to make the internal standard line reading match the calibration curve, (2) shift the intensity scale so that its 1.0 index matches a reference or edge of the transmittance scale, (3) move the transmittance scale to make the element transmittance reading match the calibration curve, and (4) read the intensity ratio on the intensity scale.

Some boards contain two identical log intensity scales that can be moved separately to make it easier to hold intermediate ratios for making background corrections. These corrections can be on either the element line or the internal standard line, or both. Analytical curves are often drawn on the working surface of the calculator. This is particularly convenient with boards that have both horizontal and vertical log scales. Some boards also have provision for transposing a working curve to a scale. With a series of these scales running parallel to the intensity scale, concentrations can be read directly without taking an intermediate intensity ratio reading.

Figure 1.27 is an overall view of the Jarrell-Ash Seidel calculating board. The whole horizontal scale assembly can be shifted laterally and includes a second set of log scales that shift in the main assembly in slide-rule fashion for performing background corrections. The photograph also shows a series of 15 blank hori-

Fig. 1.27 Overall view of calculating board; design by Jarrell-Ash.

zontal scales for holding working curve information. An enlarged view of the vertical scales in Fig. 1.28 shows the Seidel display on the right-hand side and a log display on the left. The vertical log scale is needed for drawing working curves on the board preparatory to translating them on to the horizontal working strips.

The Respectra calculator referred to in Section 1.7.3.3 is shown in Figs. 1.29 and 1.30. It has its ordinate and abscissa transposed from normal positions in order to use a horizontal drum to hold a family of transmittance scales for applying the Kaiser transformed density. This board has duplicate log scales, and equivalent log values on a linear scale, plus a special fan scale for making background corrections. The enlarged view shows how final working scales can be drawn in the sliding section marked "C." Additional scales can be kept on hand to expand this program.

The Dunn-Lowry calculator was developed to provide a way to display a large number of analytical scales in a readily accessible manner by having the scales mounted on a large drum. Each scale is brought into view by rotating the drum. Curves are not normally on display. Some units, in fact, do not include any drawing surface at all. A transmittance scale which displays the emulsion calibration runs parallel to the analytical scales and a short cursor is able to be

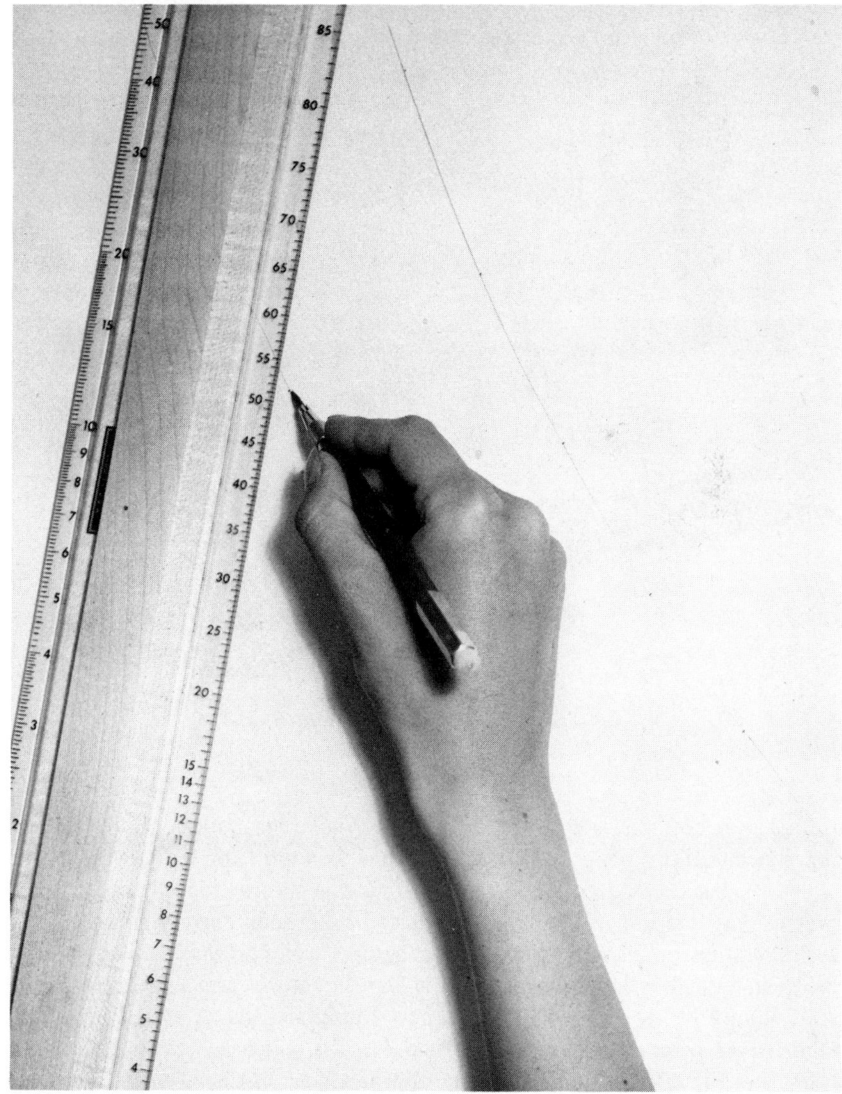

Fig. 1.28 Close-up view of vertical scale, Jaco calculating board.

Fig. 1.29 Respectra calculator, overall view.

shifted over both scales to permit a slide-rule type of manipulation for determining concentrations. The Dunn-Lowry is mainly useful for repetitive analyses where no background correction is required. The special transmittance scale is prepared by the user from his own emulsion calibration curve. Preparing these scales requires time and patience. The analytical scales normally do not have to be reestablished as long as a method is kept in control, but the transmittance scale should be redrawn each time a new emulsion batch is used. In practice, laboratories often retain old transmittance scales which are tried against new calibration data to see if the drawing of a new scale can be avoided. Although

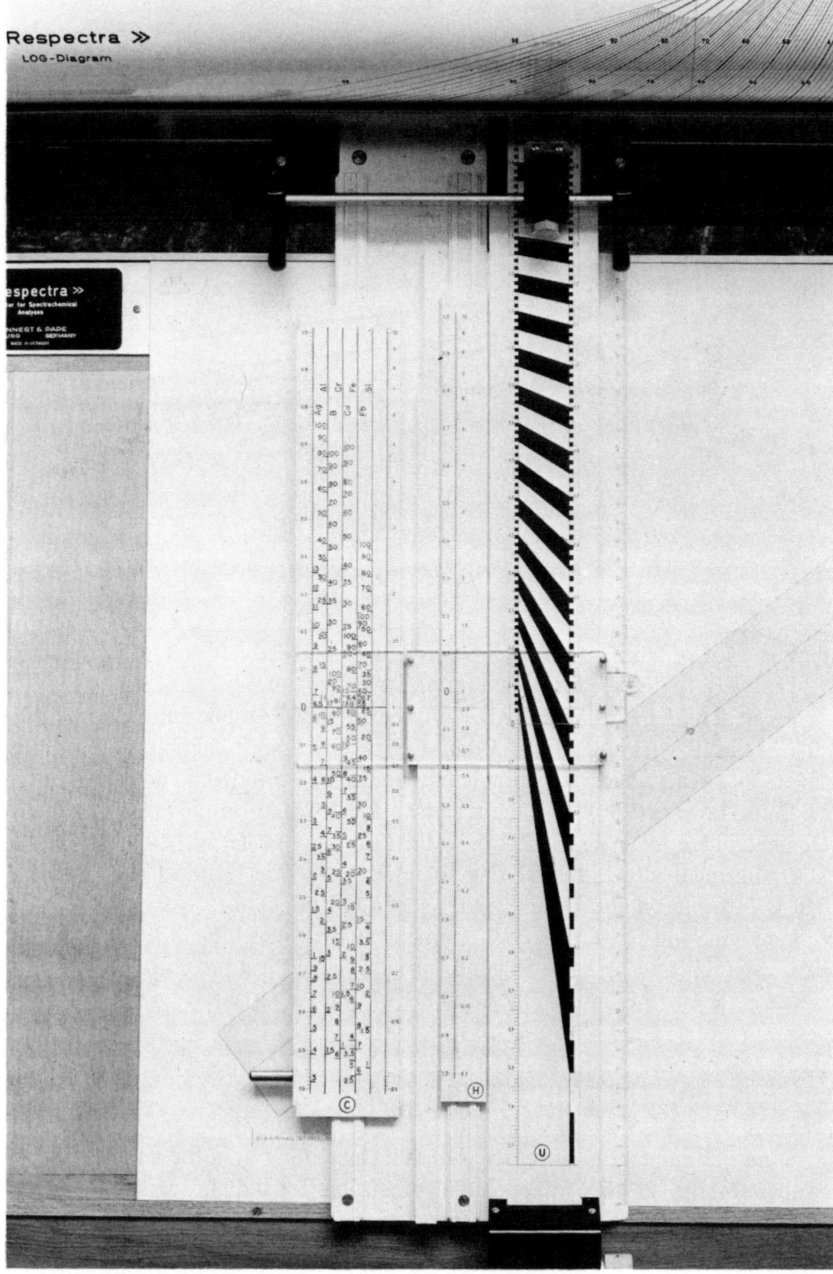

Fig. 1.30 Close-up of intensity and concentration scales and of the fan-shaped background correction scale, Respectra calculator.

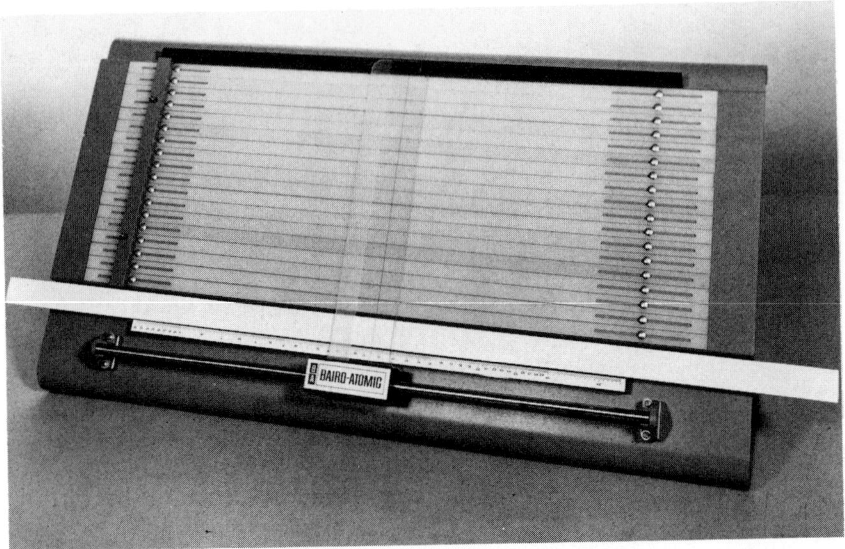

Fig. 1.31 Baird-Atomic calculating board, Model RB-1.

there is a lot of work to set up a Dunn-Lowry correctly, it does offer the most efficient way to run a series of determinations once it has been set up.

A calculating board Model RB-1 offered by Baird-Atomic and shown in Fig. 1.31 retains the feature of the Dunn-Lowry of having a calibration curve displayed on a scale which is manipulated parallel to working-curve scales. Unlike the Dunn-Lowry, it holds all the scales on a flat surface and uses a long vertical reference index to relate each scale to the calibration scale. Although this physical arrangement allows for relating to fewer analytical scales at any one time than is possible on the Dunn-Lowry, the Baird calculator does hold up to 18 working curves in position at any one time and provides for additional working-curve assemblies to be kept on hand for larger analytical programs. Calculating boards are also manufactured by a number of other companies.

1.9 SPECTRAL DATA PROCESSING

Computer control is currently promising an extension in the scope of practical spectroscopy in both direct-reading spectrometers and photographic spectrographs. In the case of the former, computer control on reading spectral information from a vidicon detector permits flexibility in selecting what lines are to be read, including background, such as has been true for photographic instruments. In the case of the latter, by use of special microphotometers having either a stepwise or continuous scanning control, coupled with computers, a large amount of data can be collected and handled almost as quickly as for direct readers, once the photographic process has been completed. Török[84] briefly reviewed progress in this area and described data treatment for both qualitative and semiquantitative analyses. Herz[85] has developed a program for the correction of emission data. Hecq and Pel[86] have proposed a computer calculation procedure for emission spectrographic analyses. Srorka and Strizh[87] reviewed spectral transformation obtained by computer and outlined methods for spectral data processing and handling large data arrays. Both Walthall[88] and Witmer *et al.*[89] have reported on systems that recorded transmittance readings from the spectrographic plates, performed the calibrations and transformations, and solved for the unknown concentrations. Thomas[90] devised an integrated-intensity procedure which improved the computer analysis of spectrographic data by the addition of a new subroutine to determine elemental concentrations.

1.9.1 Computer Program for Microphotometer Readings

At the FACSS second national meeting, Thomas[91] reported on using a minicomputer to control a scanning microphotometer to perform the determination of 64 elements in geological materials, taking only 70 s to read one spectrogram and permitting a search for up to 500 spectral lines. The procedure also includes making emulsion calibrations at 250-Å intervals on each plate and making corrections for background and spectral interferences.

Details of the equipment used appears in a report by Helz.[92] The actual microphotometer used is a special device positioned on a modified Moore No. 3 Measuring Machine* in which two light systems are used. The light for viewing the general field is passed through a yellow filter. The reading light is unfiltered in the entrance optics but goes through an entrance slit 100 μm wide. The reading light is focused down to a scanning light about 7 μm wide and 1 mm long. After passing through the spectrogram, the reading light is magnified back to its original width and passed through an exit slit and also through a blue filter before being picked up by the phototube. The use of an exit slit and the blue filter aids in eliminating light from the field illumination. The Moore measuring machine permits fast and precise movement of the spectrographic plate being read. It has been adapted to hold a special 102- by 508-mm (4- by 20-in.) plate and is capable of making a full scanning of the spectrogram in 70 s. The response on the phototube is recorded on tape at every 5-μm interval and therefore requires a rate of taking more than 1300 readings per second.

Details of the computer system and calculations used at the Geological Survey appear in a report by Walthall.[88] The program is written in Fortran IV and executed on an IBM System/360 Model 65 computer. Included in the program is a table of the spectral lines which can be recognized by the computer. Initially

*Moore Special Tool Co., Bridgeport, Conn.

an indexing is done to two cadmium lines, 2248.58 and 4415.70, and then other specific lines are located by a combination of listing them as to estimated positions in terms of the number of readings taken from the start of the scan and by correction as follows. By searching for minimum transmittance readings on 11 selected lines, corrections can be applied for locating the rest of the lines in the spectrogram. Each photographic plate includes an iron spectrum taken through a two-step filter which permits emulsion calibrations to be made in selected wavelength ranges. The computer also holds element concentration curves and may include corrections for interfering elements. With the complete information which is recorded on each spectrogram, background corrections may be readily handled. Since some 400 lines are held in memory for the 64 elements being sought, selection is made in the program with regard to what lines are used for final calculations based on priority ratings, concentration ranges, and the darkness of the lines being read.

The emulsion calibration depends upon reading a large number of pairs of iron lines in any one region, specifically 26 lines in each 250-Å range used. These are treated in a preliminary curve display similar to the density preliminary curve described in Section 1.7.2.2 and appearing as Fig. 1.17. Instead of working with density values, however, natural logarithms of the light and dark transmittance readings are used. Arbitrarily these are taken as logs of 10 times the transmittance readings. The data are fitted into a quadratic equation in which the unfiltered value is the independent variable and the filtered value the dependent variable. Furthermore, the display is made to come close to the arbitrary points of 1.0 and 1.5% T and of 99.0 and 100.0% T by weighing in the two arbitrary pairs 26 times each making a total of 78 points for the curve fitting. From this preliminary curve an H and D curve is generated. This treatment, as well as some extensive diagrams of line profiles and analytical curves, appear in an earlier article by Helz *et al.*[93] in which a simpler microphotometer was used.

An attempt has been made to apply the transmittance readings of Table 1.4 to have them fit a quadratic equation on a log preliminary plot, and it was observed that this was only an approximate match. A more detailed study was made by picking off a sequence of pairs of transmittance readings spread out over the whole range of transmittance readings from the Seidel preliminary curve, Fig. 1.18, or more specifically from Eq. (1.11), and then applied to making a quadratic fit of their log values by the method of least squares. Presumably largely because the low transmittance values did not fit well into the constraint of a quadratic configuration, poor agreement was found between the log preliminary curve and the Seidel preliminary curve when the data included a range of dark readings from 1.0 to 99.1%. Applying the weighing in of extreme points, as is done at the Geological Survey, did improve the match, except for dark line values below 7% T, and restricting the transmittances being fitted to range from 4% unfiltered to 96% filtered. An even better fit was made when the weighing in of the arbitrary low point was for 1.0/1.8 instead of for 1.0/1.5, while retaining the weighing in of the high point 99.0/100.0. These are summarized in Table 1.9. The improvement in the fit with the small change in the weighed low point shows clearly in the plots in Fig. 1.32 in which deviations in the filtered line reading from what they appeared as in the Seidel plot are plotted against unfiltered readings. Even in the most ideal match which was achieved, it would seem

Table 1.9 Log Preliminary Plots Fitting a Quadratic Equation[a]

% *T*		% *T* filtered from quadratic preliminary curve[b]		
Unfiltered	Filtered	A	B	C
1.0	2.8	2.6	1.5	1.8
1.6	4.7	4.7	3.2	3.6
2.6	7.8	8.2	6.3	6.9
3.4	10.2	10.8	9.0	9.6
4.3	12.9	13.7	12.0	12.5
5.8	17.3	18.2	16.8	17.3
7.2	21.3	22.0	21.1	21.5
9.6	27.6	28.0	27.9	28.0
11.9	33.1	33.1	33.8	33.7
15.4	40.7	39.9	41.7	41.2
19.0	47.5	46.1	48.7	47.9
24.4	56.2	53.9	57.4	56.4
29.9	63.5	60.7	64.7	63.5
41.1	74.9	71.7	75.9	74.6
52.3	83.0	80.2	83.9	82.7
62.5	88.5	86.4	89.2	88.2
71.2	92.2	90.9	92.6	92.0
78.2	94.7	94.1	94.9	94.5
83.7	96.3	96.3	96.4	96.2
87.9	97.5	97.9	97.4	97.4
91.1	98.2	99.0	98.1	98.2
93.5	98.8	99.8	98.6	98.8
95.3	99.2	100.4	98.9	99.2
96.6	99.4	100.8	99.2	99.5
97.5	99.6	101.1	99.3	99.7
98.2	99.7	101.4	99.4	99.8
98.7	99.8	101.5	99.5	99.95
99.1	99.9	101.6	99.6	100.0

[a]Basic data from linear Seidel preliminary curve, Fig. 1.18, and Eq. (1.11).

[b]A, least-squares fit of all data, $y = -0.107184x^2 + 1.7811x - 0.2621$; B, sampling from 4.0% unfiltered to 96.0% filtered, plus weighing in points for 1.0/1.5 and 99.0/100.0: $y = -0.162546x^2 + 2.4077x - 1.9704$; C, as in B, except that the low end was weighted with the point 1.0/1.8: $y = -0.145411x^2 + 2.2119x - 1.4307$.

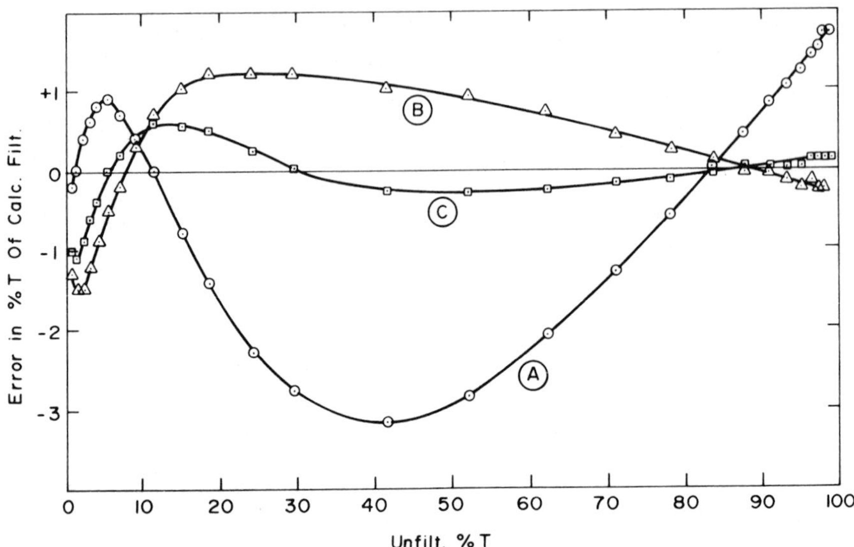

Fig. 1.32 Deviations in line pairs taken from log preliminary curves made to fit quadratic equations as detailed in Table 1.9.

questionable to go below about 5% *T*. In the program at the Geological Survey, however, the cutoff point usually is 2% *T*.

1.9.2 Use of Computer and Electronic Calculator

A combination of using a computer to determine a constant in an empirical transform for transmittance readings and then applying this to an emulsion calibration equation which can readily be handled on a programmable electronic calculator is described by Blevins and O'Neill.[94] They support the idea that if a preliminary curve from a two-line method can be represented by a straight line, then the data from scattered points can be averaged more easily. They also state that if the preliminary curve is a straight line on a log-log plot, the emulsion calibration curve will also be a straight line on a log-log plot. This is not true in the very general way it is stated but is true in the way it is applied, which is for preliminary curves that are not only linear but which also have unity slopes.

Blevins and O'Neill[94] had observed that a preliminary plot based on the Seidel transform deviated from linearity for lines of increasing darkness. Their description of the deviation is that the preliminary curve was more like two nearly straight but intersecting lines. This apparently was very similar to the curve drawn in Fig. 1.20 for theorizing on what might happen from an intensity-retardation-of-development effect as discussed in Section 1.7.2.4. They explored other ways of expressing a transform that would not only linearize the relationship but also yield a preliminary curve of unit slope. As stated, the transform is

$$f(T) = \log\left[\left(\frac{100 - T}{T}\right)\left(\frac{a + T}{T}\right)\right] \tag{1.29}$$

which may be restated as being

$$f(T) = \log\left(\frac{100 - T}{T}\right) + \log\left(\frac{a + T}{T}\right) \tag{1.30}$$

in which the main expression is seen to be the Seidel transform to which is added the modifier log $[(a + T)/T]$, where the constant a would have more effect on low transmittance readings than on high readings, in conformance with the observed divergence from linearity for lines of increasing blackness. The empirical modification appears to be a reasonable approach to those cases which do not yield linear Seidel preliminary curves. What is questionable is the necessity for having the plot be a 45° line. If the actual calibration data reported in Table 1.4 were not recognized as a forming a linear Seidel prelminary curve, some distortion would result if it were forced to yield a line of unit slope by introducing the modifier of Eq. (1.30). In order to make the data of Table 1.4 have this unity slope, the constant a would have to be 3.76. Plotting this transform on a preliminary curve would appear to give a reasonable display for a

Table 1.10 Calculations of Intensity Ratios[a]

% T line pair		Ratio of intensity A–L	Ratio of intensity B–O	Error A–L	Error B–O
81.0	95.5	1.989	2.059	−0.011	0.059
84.4	97.0	2.123	2.233	0.123	0.233
70.0	92.0	2.028	2.055	0.028	0.055
60.6	87.6	2.000	1.996	0.000	−0.004
43.5	77.8	2.053	2.004	0.053	0.004
51.2	81.6	1.958	1.930	−0.042	−0.070
31.5	65.5	2.004	1.936	0.004	−0.064
25.0	56.0	1.960	1.885	−0.040	−0.115
19.0	47.6	2.004	1.922	0.004	−0.078
24.9	56.0	1.965	1.890	−0.035	−0.110
13.5	36.0	1.971	1.896	−0.029	−0.004
10.0	28.5	1.998	1.934	−0.002	−0.066
2.9	8.5	1.974	2.048	−0.026	0.048
3.5	11.0	2.055	2.105	0.055	0.105
2.0	5.5	1.913	2.034	−0.087	0.034
1.6	4.5	1.955	2.109	−0.045	0.109

[a]A–L, Anderson and Lincoln; B–O, Blevins and O'Neill.

Table 1.11 Transmittance–Seidel Conversion Table[a]

% T	0.0	0.1/0.9	0.2/0.8	0.3/0.7	0.4/0.6	0.5/0.5	0.6/0.4	0.7/0.3	0.8/0.2	0.9/0.1	0.0	% T
											−3.892	98
2	3.892	3.842	3.794	3.749	3.705	3.664	3.623	3.585	3.547	3.511	−3.476	97
3	3.476	3.442	3.409	3.378	3.347	3.317	3.288	3.259	3.231	3.204	−3.178	96
4	3.178	3.152	3.127	3.103	3.079	3.055	3.032	3.009	2.987	2.966	−2.944	95
5	2.944	2.924	2.903	2.883	2.863	2.844	2.825	2.806	2.788	2.769	−2.752	94
6	2.752	2.734	2.717	2.700	2.683	2.666	2.650	2.634	2.618	2.602	−2.587	93
7	2.587	2.571	2.556	2.541	2.527	2.512	2.498	2.484	2.470	2.456	−2.442	92
8	2.442	2.429	2.415	2.402	2.389	2.376	2.363	2.351	2.338	2.326	−2.314	91
9	2.314	2.301	2.289	2.278	2.266	2.254	2.242	2.231	2.220	2.208	−2.197	90
10	2.197	2.186	2.175	2.164	2.154	2.143	2.132	2.122	2.111	2.101	−2.091	89
11	2.091	2.081	2.070	2.060	2.051	2.041	2.031	2.021	2.012	2.002	−1.992	88
12	1.992	1.983	1.974	1.964	1.955	1.946	1.937	1.928	1.919	1.910	−1.901	87
13	1.901	1.892	1.883	1.875	1.866	1.857	1.849	1.840	1.832	1.824	−1.815	86
14	1.815	1.807	1.799	1.791	1.782	1.774	1.766	1.758	1.750	1.742	−1.735	85
15	1.735	1.727	1.719	1.711	1.704	1.696	1.688	1.681	1.673	1.666	−1.658	84
16	1.658	1.651	1.643	1.636	1.629	1.621	1.614	1.607	1.600	1.593	−1.586	83
17	1.586	1.579	1.572	1.565	1.558	1.551	1.544	1.537	1.530	1.523	−1.516	82
18	1.516	1.510	1.503	1.496	1.489	1.483	1.476	1.470	1.463	1.457	−1.450	81
19	1.450	1.444	1.437	1.431	1.424	1.418	1.411	1.405	1.399	1.393	−1.386	80
20	1.386	1.380	1.374	1.368	1.361	1.355	1.349	1.343	1.337	1.331	−1.325	79
21	1.325	1.319	1.313	1.307	1.301	1.295	1.289	1.283	1.277	1.272	−1.266	78
22	1.266	1.260	1.254	1.248	1.243	1.237	1.231	1.225	1.220	1.214	−1.208	77
23	1.208	1.203	1.197	1.191	1.186	1.180	1.175	1.169	1.164	1.158	−1.153	76
24	1.153	1.147	1.142	1.136	1.131	1.125	1.120	1.115	1.109	1.104	−1.099	75

25	1.099	1.093	1.088	1.083	1.077	1.072	1.067	1.062	1.056	1.051	–1.046	74
26	1.046	1.041	1.036	1.030	1.025	1.020	1.015	1.010	1.005	1.000	–0.995	73
27	0.995	0.990	0.984	0.979	0.974	0.969	0.964	0.959	0.954	0.949	–0.944	72
28	0.944	0.940	0.935	0.930	0.925	0.920	0.915	0.910	0.905	0.900	–0.895	71
29	0.895	0.891	0.886	0.881	0.876	0.871	0.866	0.862	0.857	0.852	–0.847	70
30	0.847	0.843	0.838	0.833	0.828	0.824	0.819	0.814	0.809	0.805	–0.800	69
31	0.800	0.795	0.791	0.786	0.781	0.777	0.772	0.768	0.763	0.758	–0.754	68
32	0.754	0.749	0.745	0.740	0.735	0.731	0.726	0.722	0.717	0.713	–0.708	67
33	0.708	0.704	0.699	0.695	0.690	0.686	0.681	0.677	0.672	0.668	–0.663	66
34	0.663	0.659	0.654	0.650	0.646	0.641	0.637	0.632	0.628	0.623	–0.619	65
35	0.619	0.615	0.610	0.606	0.602	0.597	0.593	0.588	0.584	0.580	–0.575	64
36	0.575	0.571	0.567	0.562	0.558	0.554	0.549	0.545	0.541	0.537	–0.532	63
37	0.532	0.528	0.524	0.519	0.515	0.511	0.507	0.502	0.498	0.494	–0.490	62
38	0.490	0.485	0.481	0.477	0.473	0.468	0.464	0.460	0.456	0.452	–0.447	61
39	0.447	0.443	0.439	0.435	0.431	0.426	0.422	0.418	0.414	0.410	–0.405	60
40	0.405	0.401	0.397	0.393	0.389	0.385	0.381	0.376	0.372	0.368	–0.364	59
41	0.364	0.360	0.356	0.352	0.347	0.343	0.339	0.335	0.331	0.327	–0.323	58
42	0.323	0.319	0.315	0.310	0.306	0.302	0.298	0.294	0.290	0.286	–0.282	57
43	0.282	0.278	0.274	0.270	0.266	0.261	0.257	0.253	0.249	0.245	–0.241	56
44	0.241	0.237	0.233	0.229	0.225	0.221	0.217	0.213	0.209	0.205	–0.201	55
45	0.201	0.197	0.193	0.189	0.185	0.180	0.176	0.172	0.168	0.164	–0.160	54
46	0.160	0.156	0.152	0.148	0.144	0.140	0.136	0.132	0.128	0.124	–0.120	53
47	0.120	0.116	0.112	0.108	0.104	0.100	0.096	0.092	0.088	0.084	–0.080	52
48	0.080	0.076	0.072	0.068	0.064	0.060	0.056	0.052	0.048	0.044	–0.040	51
49	0.040	0.036	0.032	0.028	0.024	0.020	0.016	0.012	0.008	0.004	–0.000	50

[a]Table based on the natural logarithm Seidel function of ln $[(100/\%\ T) - 1]$. For T below 50%, Seidel values are read from left to right and are all positive. For T above 50%, Seidel values are read from right to left and are all negative. Thus, 23.4 would read as +1.186, whereas 76.6 would read as –1.186. To convert to Seidel values based on the common log base 10, divide the values of the table by 2.303.

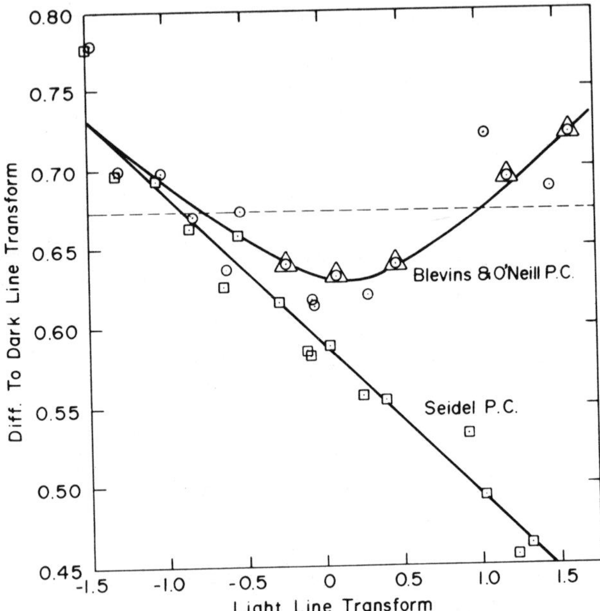

Fig. 1.33 Differences in the transform values of preliminary curve line pairs as determined by the Seidel transform and by the modified Seidel transform of Blevins and O'Neill,[94] which yields a preliminary curve having a slope of unity.

straight line. On closer inspection, however, it can be seen to be warped. In calculating errors between observed and calculated line pairs in the preliminary curve, the sum of squares of the errors is about five times greater on the Blevins-O'Neill transform than it is for the Seidel plot.

This can be seen more graphically in Fig. 1.33, in which the difference between the dark line transform and the light line transform is plotted against the light line transform. This is done for both the Seidel plot and the modified Seidel. In the case of the former, a systematic, linear change in the difference shows in support of the observation that this plot did not have a unity slope, even though it was straight. In the case of the latter, the average difference agrees with the intercept of the straight-line preliminary curve, which was calculated to be 0.6720. It is shown as a dashed line in the figure. The bowing of the actual difference curve accounts for the increases in the errors. The effect on computation can be seen if ratios of intensity are calculated by each method on the line pairs of Table 1.4. Table 1.10 summarizes these results, which carry through to the emulsion equation of Anderson and Lincoln, specifically using Eq. (1.19), and for the intensity ratio equation of Blevins and O'Neill:

$$\log\left(\frac{I_D}{I_L}\right) = m(D - L) \tag{1.31}$$

in which m is the ratio of the log of the filter factor and the intercept of the preliminary curve and D and L are the transform values of the dark and light lines. For the data under consideration, the filter factor is 2.0 and the intercept is 0.6720, making $m = 0.4480$.

Applying the data to Eq. (1.19) yields a value of S_s of 6.9461 as described in Section 1.7.4.3. The value for k then follows readily from noting in Eq. (1.11) that either the light line has a Seidel value of -0.6446 when the dark line is 0.0 or that the dark line is 0.5899 when the light line is 0.0. These fit Eq. (1.19), to give either

$$\log 2 = k \log \left(\frac{6.9461 + 0.6446}{6.9461} \right) \tag{1.32}$$

or

$$\log 2 = k \log \left(\frac{6.9461}{6.9461 - 0.5899} \right)$$

from which $k = 7.811$. The intensity-ratio calculations are shown in Table 1.10, and the errors are summarized showing the generally larger error from using the modified Seidel transform.

ACKNOWLEDGMENT

The author wishes to thank Arno Arrak for his proposals on material for this chapter, particularly with respect to historical background and theory.

1.10 REFERENCES

1. J. W. Ritter, *Ann. Physik 12*, 409 (1803).
2. H. Gernsheim, *The History of Photography*, Oxford University Press, London (1955), pp. 20–84, 261.
3. J. M. Eder, *History of Photography* (Edward Epstein, trans.), Columbia University Press, New York (1945).
4. E. Becquerel, *Biblio. Univ. Geneve 40*, 341 (1842).
5. J. W. Draper, *Phil Mag. 21*(3), 348 (1842); *22*(3), 360 (1943).
6. H. A. Rowland, *Amer. J. Sci. 33*, 182 (1887).
7. H. A. Rowland, *Astron. Astrophys. 12*, 321 (1893).
8. G. R. Harrison, R. C. Lord, and J. R. Loofbourow, *Practical Spectroscopy*, Prentice-Hall, Inc., Englewood Cliffs, N.J. (1948), p. 331.
9. Eastman Kodak Co., Kodak Plates and Films for Scientific Photography, Kodak Publication, P-315, 1st ed. (1973), p. 28.
10. Ref. 8, p. 146.
11. Ref. 9, pp. 3–4.
12. W. R. Brode, *Chemical Spectroscopy*, John Wiley & Sons, Inc., New York (1943), p. 358.
13. Photographic Processing in Spectrochemical Analysis, Recommended Practice E115-71,

Methods for Emission Spectrochemical Analysis, 6th ed., American Society for Testing and Materials, Philadelphia (1971).
14. F. Hurter and J. C. Driffield, *J. Soc. Chem. Ind. 9*, 455 (1890).
15. Ref. 9, pp. 7–8.
16. K. Schwarzschild and W. Villiger, *Astrophys. J. 23*, 287 (1912).
17. G. Eberhard, *Physik. Z. 13*, 288 (1912).
18. S. Kostinsky, *Mitteilungen der Nikolaihauptsternwarte zu Pulkowo 1*(11) and *2*(14) (1906).
19. N. H. Nachtrieb, *Principles and Practice of Spectrochemical Analysis*, McGraw-Hill Book Company, New York (1950), p. 116.
20. L. Strock, *Proceedings of the Seventh Summer Conference on Spectroscopy*, John Wiley & Sons, Inc., New York (1940), p. 134.
21. A. Arrak, *Appl. Spectrosc. 16*, 124 (1962).
22. M. Slavin, *Appl. Spectrosc. 16*, 127 (1962).
23. C. E. Harvey, Spectrochemical Procedures, Applied Research Laboratories, Glendale, Calif. (1950), p. 58.
24. Ref. 9, p. 10.
25. T. Török, G. Heltai, and I. Moharos, *Spectrochim. Acta 27B*, 215 (1972).
26. T. Török, G. Heltai, and I. Moharos, *Magy. Kem. Folyoirat 78*, 594 (1972).
27. T. Török, G. Heltai, and I. Moharos, *Magy. Kem. Folyoirat 78*, 598 (1972).
28. T. Török, G. Heltai, and I. Moharos, *Acta Chim. (Budapest) 77*, 11 (1973).
29. T. Török, G. Heltai, and I. Moharos, *Acta Chim. (Budapest) 77*, 117 (1973).
30. Ref. 9, p. 15.
31. M. Slavin, *Emission Spectrochemical Analysis*, vol. 36 of *Chemical Analysis* (P. J. Elving and I. M. Kolthoff, eds.), John Wiley & Sons, Inc. (Interscience Division), New York (1971), p. 146.
32. R. W. Bunsen and H. E. Roscoe, *Ann. Physik 108*(2), 193 (1859).
33. K. Schwarzschild, *Astrophys. J. 11*, 89 (1900).
34. Ref. 19, p. 108.
35. Eastman Kodak Co., *Tech Bits* (1964), No. 4, p. 9.
36. Ref. 19, p. 107.
37. Eastman Kodak Co., *Tech Bits* (1967), No. 1, p. 2.
38. H. Green, *Appl. Spectrosc. 7*, 24 (1953).
39. B. O'Brien and V. L. Parks, *Phys. Rev. 41*, 387 (1932).
40. J. H. Webb, *J. Opt. Soc. Amer. 23*, 157, 316 (1933).
41. R. A. Sawyer and H. B. Vincent, *J. Opt. Soc. Amer. 33*, 185 (1943).
42. W. C. Pierce and N. H. Nachtrieb, *Ind. Eng. Chem. Anal. Ed. 13*, 774 (1941).
43. M. Malpica, *Gen. Elec. Rev. 73*, 288 (1940).
44. Ref. 9, p. 92.
45. G. R. Harrison, *J. Opt. Soc. Amer. 24*, 60 (1934).
46. J. R. Churchill, *Ind. Eng. Chem. Anal. Ed., 16*, 653 (1944).
47. G. H. Dieke and H. M. Crosswhite, *J. Opt. Soc. Amer. 33*, 425 (1943).
48. L. H. Ahrens and S. R. Taylor, *Spectrochemical Analysis*, 2nd ed., Addison-Wesley Publishing Company, Inc., Reading, Mass. (1961), pp. 152–153.
49. J. W. Anderson and A. J. Lincoln, *Appl. Spectrosc. 22*, 753 (1968).
50. Ref. 23, p. 70.
51. Ref. 48, pp. 149–152.
52. Ref. 19, p. 39.
53. I. H. Bond, International Nickel Ltd., London, private communication (1972)
54. Photographic Photometry in Spectrochemical Analysis, Recommended Practice E116-70a, Ref. 13.
55. G. V. Wheeler, *Appl. Spectrosc. 10*, 11 (1956).

56. Ref. 23, p. 80.
57. E. A. Baker, *Proc. Roy. Soc. Edinburgh 45*, 166 (1925); *47*, 34 (1927); *48*, 106 (1928).
58. R. A. Sampson, *Monthly Notices Roy. Astron. Soc. 85*, 212 (1925).
59. J. M. McCrea, *Develop. Appl. Spectrosc. 4*, 501 (1965).
60. P. F. Verhulst, *Correspondance mathématique et physicque*, A. Quételet, Bruxelles (1838), vol. 10, pp. 113–121, in: The Determination and Consequence of Population Trends, Population Studies 17 (ST/SOA/Ser. A/17), Department of Social Affairs, United Nations Publication (1953), pp. 41–42.
61. J. Noar, *Phot. J. 91B*, 64, 99 (1965).
62. N. Trandafir, *Appl. Spectrosc. 18*, 175 (1964).
63. H. Kaiser, *Spectrochim. Acta 2*, 1 (1941); *3*, 159 (1948); *4*, 351 (1950–1952).
64. E. N. Severin and V. S. Rossikhin, *Zh. Prikl. Spekstrosk. 14*, 454 (1971).
65. R. Schmidt, *Rec. Trav. Chim. 67*, 737 (1948).
66. J. W. Anderson, *Appl. Spectrosc. 10*, 195 (1956).
67. A. Arrak, Quantitative Theory of Photographic Action and the Behavior of Linear Density Transformations, Report ADR 09-02-65.1 (Feb. 1965), Grumman Aircraft Development Corp., Bethpage, N.Y.
68. S. D. Rasberry, M. Margoshes, and B. F. Scribner, Applications of a Time-Sharing Computer in a Spectrochemistry Laboratory; Optical Emission and X-Ray Fluorescence, *Natl. Bur. Std. (U.S.) Tech. Note 407* (Feb. 1968).
69. M. Green, *Appl. Spectrosc. 12*, 149 (1958).
70. H. Kaiser, *Spectrochim. Acta 3*, 159 (1951).
71. W. Seith, *Spectrochim. Acta 3*, 188 (1951).
72. F. J. Haftka, *Spectrochim. Acta 7*, 242 (1955).
73. T. Török and K. Zimmer, *Ann. Univ. Sci. Budapest. Rolando Eötvös Nominatae Sect. Chim. 8*, 11–22 (1966).
74. T. Török and K. Zimmer, *Acta Chim. Acad. Sci. Hung. 41*, 97–104 (1964).
75. T. Török, *Acta Chim. Acad. Sci. Hung. 41*, 155–160 (1964).
76. K. Zimmer, T. Török, and I. Asztalos, *Chem. Anal. (Warsaw) 11*, 1065 (1966).
77. T. Török and K. Zimmer, *Quantitative Evaluation of Spectrograms by Means of Ϙ-Transformation*, Heyden and Sons, Ltd., London (1972).
78. G. T. Chamberlain, *Appl. Spectrosc. 21*, 32 (1967).
79. Ref. 19, pp. 139–140.
80. O. S. Duffendack and R. A. Wolfe, *Ind. Eng. Chem. Anal. Ed. 10*, 61 (1938).
81. Ref. 48, pp. 158–159.
82. M. Slavin, *Appl. Spectrosc. 19*, 28 (1965).
83. C. R. Boswell, S. S. Berman, and D. S. Russell, *Appl. Spectrosc. 23*, 268 (1969).
84. T. Török, *Kem Kozlem 37*, 167 (1972).
85. M. L. Herz, U.S. National Technical Information Service Report AD 745126 (1972), 46 pp.
86. W. E. Hecq and D. J. Pel, U.S. National Technical Information Service Report EUR 4780 (1972), 56 pp.
87. L. M. Srorka and T. A. Strizh, Report JINR-P10-6702 (1972), 137 pp.; Lab. Nucl. Probl., Jt. Inst. Nucl. Res., from *Nucl. Sci. Abstr. 28*, 1518 (1973).
88. F. G. Walthall, *J. Res. U.S. Geol. Surv. 2*, 61 (1974).
89. A. W. Witmer, V. A. J. Jansen, and G. H. van Gool, *Philips Tech. Rev. 34*, 322 (1974).
90. C. P. Thomas, *J. Res. U.S. Geol. Surv. 3*, 181 (1975).
91. C. P. Thomas, A Minicomputer Based Emission Spectrographic Analysis System, presentation at FACSS Second National Meeting, Indianapolis, Ind. (Oct. 7, 1975).
92. A. W. Helz, *J. Res. U.S. Geol. Surv. 1*, 475 (1973).
93. A. W. Helz, F. G. Walthall, and S. Berman, *Appl. Spectrosc. 23*, 508 (1969).
94. D. R. Blevins and W. R. O'Neill, *Appl. Spectrosc. 30*, 190 (1976).

Laser Emission Excitation and Spectroscopy

2

R. H. Scott and A. Strasheim

2.1 INTRODUCTION

Soon after the first report of laser action in ruby in 1960[1] it was generally recognized that the intense output beam can be used to excite material into a state of optical emission. In this chapter we shall devote our attention to the interaction which occurs when pulsed laser radiation is focused onto the surface of a solid, causing material to be vaporized by the process of thermal absorption and resulting in the formation of a plasma above the focal spot.

Since laser radiation can be produced in a number of diverse modes, it is apparent that the physical properties of the interaction will vary according to the characteristics of the radiation used. Pulsed laser radiation can essentially be produced in two modes: conventional and *Q*-switched. Conventional radiation has a relatively low power (kilowatt range), while that produced by *Q*-switching has power in the megawatt range. In each case the observed surface phenomena created by the interaction of the laser beam with the surface is a result of the close interaction of a number of different mechanisms, including thermal absorption by electron excitation, thermionic emission, surface vaporization, space-charge effects, and other plasma processes.

When conventional radiation is used, a relatively deep crater is formed in the surface of the solid, accompanied by the ejection of luminous vapor and molten matter. Because the heating rate is relatively slow, the vapor temperature is close to the phase-transition temperature. The expansion rate of the plume is generally subsonic and, owing to the low particle density, little absorption of incident radiation occurs in the vapor. In the higher-power mode, the vaporization mech-

---

R. H. Scott and A. Strasheim • National Physical Research Laboratory, Council for Scientific and Industrial Research, Pretoria, South Africa.

anism is very rapid and the crater depth is only of the order of a few micrometers. The dense high-temperature plasma expands at a supersonic rate, and strong ionization is in evidence. Furthermore, a significant fraction of the incident laser energy is absorbed in the plasma.

Both conventional- and *Q*-switched-mode laser plasmas emit characteristic optical radiation which can be spectrographically analyzed to yield qualitative and quantitative data for a small amount of sample material. This principle led to the development of the laser as an analytical tool by Brech.[2,3] This laser energy source consisted of a ruby laser head combined with a metallurgical microscope equipped with a special objective of uncemented glass. First, the sample surface is brought into focus and the spot to be analyzed is positioned under the microscope crosshairs. By rotating a mirror within the microscope, the laser is brought into alignment. The laser is then fired and the high-power output pulse, focused exactly onto the preselected spot by the microscope objective lens, vaporizes the surface, resulting in the formation of a plasma containing the emitting analytical species.

This new spectrochemical light source has several unique capabilities. It is primarily a microprobe because of the ease with which it may be focused onto small areas of sample. It can be used to analyze a variety of samples: electrically conductive and nonconductive, organic and inorganic, absorbing or transparent. Useful fields include mineralogy, biology, archaeology, medicine, metallography, and crime detection.

A laser may be used as a spectrochemical source in two ways: first, part of the energy of the laser pulse is used to vaporize the sample and part to heat the vapor to temperatures at which characteristic emission will occur. One of the advantages of this single-step laser microspectroscopy is that optical power is used rather than electrical power to produce excitation, and therefore no contamination of the sample can occur. The laser-excited spectrum will contain only lines of the elements constituting the sample and surrounding atmosphere. In the second method (developed by Brech[3]), the laser radiation is primarily used to vaporize the sample material, whereas an auxiliary discharge is used to enhance the radiation from the vaporized species. In this two-step process, the vaporization and excitation processes are separated and can be controlled independently.

Most laser microprobes are provided with a choice of operating parameters: laser output energy can usually be selected by adjusting the electrical energy into the flashtube, and the power density and spot size can be selected by means of a diaphragm setting within the laser resonator as well as by changing the microscope objective lens. The optional auxiliary excitation normally consists of an electrical discharge between two graphite electrodes mounted close to the sample surface on either side of the focal spot. The electrode gap is triggered by the ionized vapor ejected by the laser pulse. The discharge parameters (voltage, inductance, resistance, and capacitance) can be adjusted to provide arclike or sparklike excitation conditions.

By virtue of the small spot size to which a laser pulse can be focused under certain conditions, the most important area of laser emission excitation is undoubtedly single-shot *in situ* microanalysis. Because of the fact that only a single shot is used to perform an analysis (as in the case of a minute inclusion), the problem of insufficient spectral intensity is sometimes present, particularly when no auxiliary excitation is used. The exposure duration is seldom more than 5 μs for the *Q*-switched mode. A fast spectrograph is therefore an essential requirement. Furthermore, fast spectrographic plates are necessary to increase the overall sensitivity of the technique. Recently, with the development of fast pulsing lasers, attention has been paid to their use for macroanalysis of large homogeneous samples. The main advantages of this method above conventional techniques are the spatial stability of the laser source and the fact that samples need not be electrically conductive. Time resolution has been used with some success in decreasing the contribution of the spectral continuum to the analytical signal. More details of this and other techniques to improve the analytical method are given in Section 2.5.

In the following section, the origin and properties of pulsed laser radiation are discussed briefly. This is followed by a section dealing with physical aspects of the interaction between the laser beam and the surface. Thereafter, attention will be focused on the spectroscopy of laser-induced plasmas and the application of these plasmas as spectrochemical light sources.

2.2 PULSED LASER RADIATION

2.2.1 Characteristics

Pulsed laser radiation is derived from solid laser material in an optically resonant cavity consisting of two parallel reflectors, one of which is partially transmissive. The material is optically pumped by a suitable flashlamp. Typical solid laser materials are ruby and neodymium-doped glass rods. The radiation produced by such lasers is not continuous, but occurs in pulses during the flashlamp discharge after the threshold pump energy has been exceeded.

When the laser is operated in the conventional mode, the output light beam consists of a train of pulses of irregular intensity for approximately the duration of the flashlamp discharge, usually about 1 ms. The light pulsations are a direct consequence of the process of amplication by stimulated emission. As the laser material is being excited, the population inversion between the levels of the laser transition reaches the oscillation threshold and the system becomes unstable. A photon, randomly produced by fluorescence, will stimulate the emission of more photons, and when this occurs along the optical axis of the resonance cavity the electromagnetic field is amplified, resulting in an output pulse and a net decrease in the number of atoms in the excited (metastable) state. When the threshold value is reached, the field amplitude dies out and the process is repeated for the duration of the flashlamp discharge.

When operated with the *Q*-switch (see Section 2.2.4), the output usually consists of one or a few "giant" pulses of radiation. Exceedingly high power is obtainable because the radiated energy is emitted in a very short time, typically less than 0.1 μs. The irregular light pulsations described above do not occur with fast *Q*-switching, because the resonant cavity is only introduced after maximum population inversion has been achieved.

Owing to the mechanism by which stimulated emission adds energy to the wave of radiation in the laser crystal, the emitted beam is coherent for durations of the order of individual pulse lengths. The term "coherence" is often used rather vaguely, as its implications are difficult to grasp. Coherence implies both spatial and temporal coherence. The classical interpretation is as follows. If the phase difference between any two points on a wavefront of the emitted beam remains constant as the wavefront propagates, this wavefront possesses spatial coherence. If the phase difference between two points situated along an axial path of the beam remains constant, the radiation along this path possesses temporal coherence. The maximum distance between these points at which the phase difference is still constant is termed the coherence length. Coherence time is equal to the coherence length divided by the speed of light. For the beam to possess both spatial and temporal coherence, the phase difference must remain constant between any point on a wavefront and any other point on another wavefront within the coherence length.

The output beam of a laser is essentially monochromatic because of the well-defined upper and terminal energy levels of the lasing element. The exact degree of monochromaticity will become evident only after studying the mode structure (see Section 2.2.3). Furthermore, by virtue of the construction of the lasing element in the form of a rod, the device will emit a highly collimated beam. Waves propagating in off-axial directions will, after a number of oblique reflections, be reflected away from the system, so strong emission cannot occur except in the direction closely parallel to the rod axis.

The degree of collimation, or the angular divergence of the output beam, is theoretically of the order of λ/D, where λ is the laser wavelength and D the aperture of the device. It is, however, difficult in practice to attain this small value ($\sim$0.1 mrad for a 1-cm-diameter neodymium rod) and divergences in the order of 5 mrad are usually achieved. Off-axis oscillations coupled with imperfect resonator reflectors are usually responsible for this increase in the divergence of the output beam above the theoretical value.

One of the most striking features of pulsed laser radiation, apart from those mentioned, is its very high intensity. By using a simple lens to focus the beam, it is possible to attain extremely high radiation densities. This important feature forms the basis of investigations concerned with the interaction between the laser beam and materials and the application of this interaction in analytical spectroscopy.

2.2.2 The Laser as a Radiation Source

2.2.2.1 Ruby

The processes of pumping and stimulated emission in a ruby crystal (Al_2O_3 doped with approximately 0.05 wt% Cr_2O_3) are shown schematically in Fig. 2.1. Ruby can be regarded as a three-level fluorescent solid, consisting of the ground state E_0, the broad pump bands E_2 and E_3, and the metastable state E_1, which has a lifetime of approximately 3 ms. The ruby is pumped to levels E_2 and E_3 by optical excitation using wavelengths within the bands 5000–6000 Å and 3200–4400 Å to produce absorption. A spontaneous nonradiative transition then takes place to the intermediate level E_1 (a process known as phonon relaxation into internal energy of the crystal in the form of heat). Ordinary fluorescence occurs when radiation is emitted, which is associated with the spontaneous transition $E_1 \rightarrow E_0$. When this fluorescence travels through the crystal, additional $E_1 \rightarrow E_0$ transitions are stimulated.

When the crystal is placed within a resonant cavity, the radiation is reflected repeatedly in both directions through the crystal, inducing more $E_1 \rightarrow E_0$ transitions. Thus, every passage through the crystal is accompanied by an amplification of the radiation intensity, provided that the population of the E_0 level does not exceed that of the E_1 level. This is ensured by continuous repumping of ground-state ions. Laser output is obtained by partial transmittance through one of the resonator reflectors. The laser may thus be described as a high-gain amplifier with positive feedback. The wavelength of the ruby laser line is 6943 Å.

2.2.2.2 Neodymium-Doped Glass

The four-level system of neodymium-doped glass (glass doped with approximately 3 wt% Nd_2O_3) is shown schematically in Fig. 2.2. In the neodymium system fluorescing does not occur directly to the ground state from the upper laser level as in the ruby system. The terminal level E_1 is an intermediate one located about 200 cm^{-1} from the ground state E_0. The transition from this laser terminal level to the ground state occurs by phonon relaxation. Because the terminal level is essentially empty at room temperature, the population of the

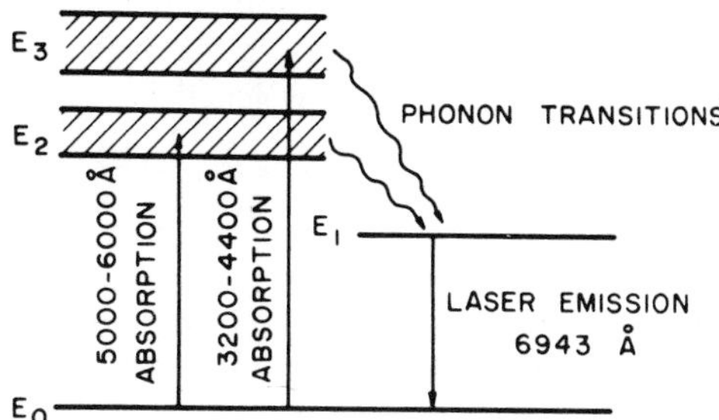

Fig. 2.1 Three-level energy system of ruby crystal.

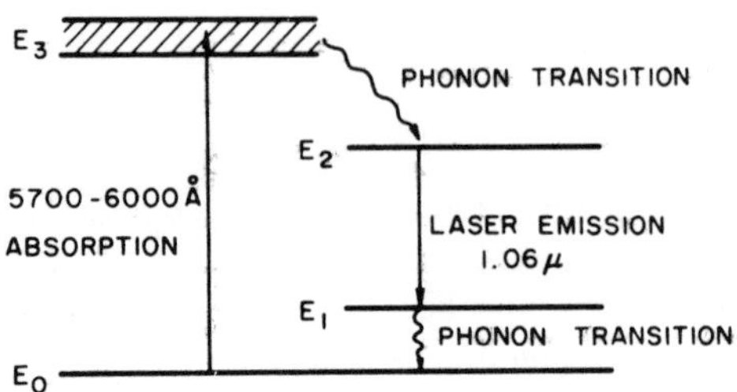

Fig. 2.2 Four-level energy system of neodymium-doped glass.

E_2 level can be increased by a relatively small amount of pump power above that of the E_1 level. This is unlike the ruby system, where more than half the chromium atoms must be raised to the fluorescence level before population inversion is achieved, and is the reason why the threshold pump power of the three-level ruby system is so much higher than that of the four-level neodymium system. The lifetime of the E_2 state of neodymium is approximately 0.3 ms, that is, about $\frac{1}{10}$ that of the upper level of ruby. The wavelength of the laser transition of neodymium is 10,600 Å.

2.2.3 Mode Structure

An optical resonator with plane-parallel reflectors at each end of the laser rod and perpendicular to the optical axis may be assumed. An oscillation between the reflectors along any one particular path through the laser crystal is characterized by the quality factor Q of the path. This factor is defined by the formula

$$Q = 2\pi \nu U P_l^{-1} \tag{2.1}$$

where ν is the resonant frequency, U the energy of the oscillation, and P_l the power loss along the path. An oscillation along an off-axis path will have a high rate of energy dissipation due to off-normal reflections. Another factor which limits off-axis oscillations in ruby crystals is the high refractive index of ruby ($\eta = 1.76$). The quality factors of off-axis oscillations are therefore low compared to those of axial oscillations, which will contain most of the energy.

The frequency of oscillation along any one path is determined by the resonance condition; that is, twice the distance between the reflectors is equal to an integral multiple of the wavelength in the crystal. This condition is given by the formula

$$n\lambda = 2\eta L \tag{2.2}$$

where n is an integer, η the refractive index of the laser crystal, and L the distance between the reflectors. The wavelength separation between two adjacent resonances is given by

$$\Delta\lambda = \frac{\lambda^2}{2\eta L} \tag{2.3}$$

For a ruby laser with a cavity length of 10 cm, this wavelength separation is approximately 0.01 Å. Therefore, more than one resonance may occur within the atomic line width (usually greater than 0.01 Å). These resonant frequencies are referred to as the resonance or longitudinal modes.

A practical resonant cavity has dimensions which are many orders of magnitude greater than the wavelength of the oscillating light, and it is possible that a large number of cavity resonances are involved, owing to the irregular reflecting surfaces of the resonator reflectors. Thus, the resonance conditions for the many paths between the reflectors are satisfied by the longitudinal modes related to the individual path lengths. The field pattern in a plane perpendicular to the output beam will therefore reveal a spatial distribution of discrete frequencies within the atomic line width. This spatial distribution is termed a transverse mode.

The rate of excitation of longitudinal modes is highest for paths having the highest Q factors and for the resonance frequencies near the center of the atomic line. As all the modes are driven by the same supply of excited atoms, these paths will contain almost all the radiative energy. Thus, a transverse-mode pattern consists of a spatial distribution of energy at discrete frequencies close to the center of the atomic line.

The ideal laser produces an output wavefront having a unique phase across the surface, with an approximately Gaussian intensity falloff at the edges. It is said to operate in the lowest-order transverse mode. It is difficult to achieve this mode of operation because of imperfections in the resonator reflectors, inhomogeneously doped crystals, inclusions, thermally induced strains, and doping concentration gradients. These features usually cause multimode operation in which higher-order modes predominate. This can be prevented by inserting a mode-locking device such as a Fabry-Perot etalon within the laser resonator or by reducing the aperture of the system. Mode-locking usually results in a comparatively large loss of output energy.

2.2.4 *Q*-Switching

Emission of laser radiation generally begins when population inversion occurs. However, when one of the reflectors of the resonator is removed, the radiation cannot oscillate even though inversion has been achieved. Because there is no feedback, amplification of stimulated emission (apart from fluorescent emission) does not occur. Thus, a very high peak inversion is reached if the pumping rate is sufficiently high. If the reflector is replaced suddenly, the high peak inversion causes a very high growth rate of energy in the ensuing wave of radiation through the crystal. An intense pulse of radiation is therefore emitted in a very short time.

This technique of obtaining a high power output is known as Q-switching. The one reflector of the resonator can be rotated into optical alignment as a means of Q-switching. Other methods of doing this are by inserting an electro-optical shutter in the cavity in the form of a Kerr cell or Pockels cell, or a cell filled with a bleachable dye. The dye becomes transparent due to laser irradiation only after a high population inversion is achieved. This type of Q-switch, known as a passive Q-switch, can be combined with one of the other types if a single pulse of extremely short duration is required.

2.2.5 Properties of the Focused Beam

The principal purpose of focusing the laser beam is to obtain the maximum intensity at the focal point. The condition that the spot size approaches the diffraction limit (i.e., the Airy disc) can only be achieved if the wavefront has the same phase across its entire surface. Ideally, therefore, the laser must be operated in the lowest-order transverse mode. This results in a far-field diffraction pattern which consists of the Airy central disc containing about 84% of the energy. The rest of the energy falls into a series of concentric diffraction rings.

Classical optical theory predicts the spot diameter at the focus of a simple lens, for uniform plane wavefronts, to be

$$d_0 \approx 2.44 \frac{\lambda f}{D} \tag{2.4}$$

where λ is the wavelength of the laser radiation, f the focal length of the lens, and D the aperture of the laser device.

Imperfections in the focusing optics will result in a spot size larger than the Airy disc. The important factor that contributes to a very much larger spot size than that established theoretically is that the laser does not usually produce a uniform-plane wavefront. Although the output of a ruby or neodymium laser can be approximated to a plane wave by mode locking or by reducing the aperature of the system, this leads to an energy loss which cannot always be tolerated. The output beam normally has a divergence of over 1 mrad, resulting in a spot size one or two orders of magnitude greater than that predicted for plane waves.

A good approximation of the spot diameter can be derived from a geometrical construction. Consider Fig. 2.3: if the divergence angle of the laser beam is θ, the spot diameter is given by

$$d \approx 2f \tan \frac{\theta}{2} \tag{2.5}$$

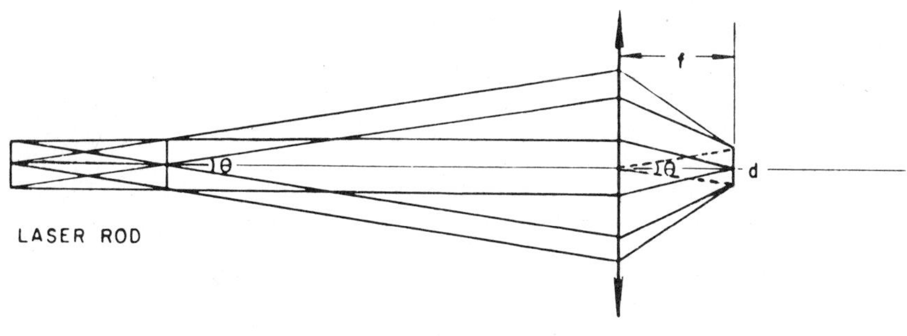

Fig. 2.3 Geometrical construction to determine spot size.

where f is the focal length of the lens at the laser wavelength. Angle θ is generally very small, so $\tan \theta/2 \approx \theta/2$, giving

$$d \approx \theta f \tag{2.6}$$

The diameters of craters formed in metal samples by focused laser pulses are invariably greater than this value, owing to processes such as lateral heat diffusion or transfer of energy from the laser plasma to the sample.

Important parameters influencing the interaction between the laser beam and the surface are the energy and power density of the radiation falling onto the sample surface. To determine these parameters, the laser-beam energy must first be measured calorimetrically. The power can be calculated once the time profile is known. In the case of a single pulse (Q-switched mode) the time profile must be recorded oscillographically. The half-width of the pulse is then measured and the mean power calculated. This is equal to the total pulse energy divided by the half-width. The mean power of a multipulse discharge such as a conventional-mode laser-pulse train is determined by measuring the total energy of the pulse train and dividing this by the total duration of the train.

The mean energy and power densities produced by focusing the laser beam are averages not only in time but also in space, because of the transverse-mode structure of the beam. Thus, the mean energy density is defined as the total energy deposited on the focal spot divided by the area of the spot, $\pi\theta^2 f^2/4$. The The mean power density of a focused Q-switched laser pulse of energy E and half-width τ is given by

$$\bar{\phi} = \frac{4E}{\pi\theta^2 f^2 \tau} \tag{2.7}$$

2.3 PHYSICAL ASPECTS OF THE LASER BEAM: SURFACE INTERACTION

Heating by conventional-mode laser pulses may be treated using ordinary thermodynamic theory, because the heating rate is relatively slow,[4-8] but as the power density is increased, the validity of treating the problem in this manner eventually becomes questionable, owing to the short time period of the interaction. A distinction has therefore been made in the case of the high-power *Q*-switched mode. A model has been proposed[4] in which it is assumed that as the surface of the target material rises to the vaporization temperature and beings to vaporize, a high recoil pressure pulse is produced[9] which would raise the boiling point of the underlying material as heat continues to be conducted into the interior. When the temperature rises high enough, a critical point is reached at which the heat of vaporization falls to zero, allowing the emission of high-pressure material to proceed. This phenomenological model was supported by photographic evidence of a time delay which occurred between the initial irradiation of a metal surface and the bulk vaporization of material.[10] However, no time delay for the appearance of spectral data relative to the laser pulse could be measured by other investigators.[11,12] In the model above the fusion process was neglected in order to reasonably simplify the calculations of total energy requirements for complete vaporization of the displaced material. Baldwin, however, has shown empirically that the process responsible for removal of most of the crater material is ablation of molten metal (a Cu-Zu alloy sample was used) and that the fusion process occurs as a physical phenomenon which cannot be ignored.[13] The sequence of events visualized is that a thin layer of the metal is melted, the surface begins to vaporize, and the recoil pressure of the escaping vapor causes ablation of the underlying melt, thereby establishing a situation in which undisturbed target material is continually made available for the removal process.

During ejection, the material is ionized, presumably by the laser pulse through the mechanism of gas breakdown at optical frequencies,[14] resulting in very rapid energy absorption in the plasma.[15-18] The laser flux threshold (minimum power density) for plasma formation has been determined for a number of target materials.[19] If sufficient laser power is deposited in a very small amount of material, very high plasma temperatures are attained, leading to high expansion velocities and ion energies.

The quantity of laser energy absorbed during the interaction can be empirically determined by measuring the difference between the incident and reflected energies.[20-22] Results have shown that for a conventional-mode laser pulse incident on copper-zinc alloys, approximately 20% of the energy is expended in the heating, liberation, and excitation of atoms, while the remaining portion is lost in reflection from the metal surface. This fraction corresponds to the reflection coefficient of the metal surface. Therefore, because the reflection coefficient of an alloy is dependent on its chemical composition, the laser energy available for heating and excitation will vary for various alloys in the case of the low power mode. At power densities in excess of 10^{10} W cm^{-2}, however, the reflection coefficient drops below 10% of the normal value.[22] It is therefore important from an analytical viewpoint to note that reflection losses are less significant for Q-switched laser interactions than for conventional-mode interactions.

Energy balance measurements have shown that apart from the influence of the reflection coefficient, the thermal properties of the target are important parameters in determining crater shapes and sizes. Indications are that energy is lost by thermal conduction and that in some cases a quantity of material is ejected in the liquid phase, because the energy necessary for vaporizing the total mass of ejected material exceeds the laser-beam energy.[23]

Detailed investigations of the conventional-mode interaction using high-speed photography have revealed a complex hydrodynamic system.[24,25] It has been shown using metal targets that a significant amount of material is ejected in the liquid phase as the result of a melting-flushing mechanism which sets in after an initial relatively pure vaporization process. This mechanism is responsible for the formation of a wall of resolidified material which builds up around the crater above the surface of the sample. During its formation, while in a partially solidified state, this sheath of material was observed to contract and to eventually interrupt the incident laser beam, resulting in elevated vaporization and scattering by the blow-off material. The phenomenon was thought to be due to the low pressure accompanying the high flow velocity of vapor from within the crater. This melting-flushing mechanism did not appear to occur to the same degree in the case of the high-power Q-switched pulse interaction, which was found to be more of a vaporization process accompanied by little transportation of molten material.

Studies of momentum transfer effects[26-28] have revealed that the target momentum due to incidence of a Q-switched pulse is greater than that due to a conventional pulse, even when the total energy of the latter is much higher. The lost material is therefore ejected with a considerably higher velocity for Q-switched pulse impacts. An optimum laser pulse power density was found to exist at which a maximum momentum transfer per joule of laser energy occurred. The existence of this optimum means that all the momentum is delivered by material vaporized from the surface, and that at lower power densities a large amount of energy is conducted away and does not contribute toward vaporization. At power levels above the optimum, a significant fraction goes into increasing the degree of ionization and the ion energies in the plume. This optimum power density is of the order of 10^9 W cm^{-2}. The expansion velocity of a laser plasma has been found to be dependent on the power density of the radiation as well as on the target material.[25,29] Supersonic velocities associated with Q-switched laser plasmas result in shock-wave formation, which has been the subject of further investigation.[30-34]

An interesting aspect of laser interaction is the emission of electrons and ions from an irradiated metal or semiconductor target.[35-43] These particles can be detected using a suitably biased collector near the target in vacuum. The electrons are observed as a sharp pulse the instant the laser pulse impinges on the surface, while the ions are observed as a broad pulse of opposite polarity a short time afterward. The broadness of the ion pulse is due to the ion-charge distribution, and the time delay is due to the lower velocity of the ions compared to that of the electrons. Field emission can be ruled out, since at optical frequencies the laser radiation field does not act in one direction long enough to free electrons from the target. The possibility of a multiphoton mechanism is highly unlikely, since a three-photon photoelectric effect would be required. Thus, thermionic emission is considered to be the most favorable mechanism. If the Richardson equation for thermionic emission is applied to measured values of the electron current density, the calculated surface temperatures range between the melting and boiling points of the samples.

The energies of the ions in laser-induced plasmas have been measured for a number of power densities and experimental conditions.[41,44-52] Generally, the ion energies are directly proportional to the power density of *Q*-switched laser pulses, and are in some cases independent of the target element and far too large to be attributed to thermal effects. The production mechanism has been considered to be more directional than thermal, owing to the existence of an accelerating electric field formed by space-charge separation.[48] Electrons, being the more mobile species, are ejected with a higher velocity than ions, causing a buildup of space charge and associated field. After plasma formation, it is no longer possible to distinguish thermionic processes from other ionization mechanisms in the plasma which yield high-energy ions with multiple charges.[53,54]

In conclusion, it is evident from the discussion above that the interaction of focused laser radiation with solids is a highly complex phenomenon. Variations in the characteristics of the laser radiation used in various experiments, and in some cases the absence of fundamental data, make it difficult to describe explicitly the mechanisms that occur. Generally, however, it is evident that vaporization and excitation processes characterize the submicrosecond high-power *Q*-switched mode interaction, while melting and flushing processes dominate that of the lower-power, longer-duration conventional mode. The *Q*-switched mode is more suitable for analytical purposes because of this, as well as because of the reduced influence of physical parameters of the target material on the energy transfer.

2.4 SPECTROSCOPY OF *Q*-SWITCHED LASER PLASMAS

2.4.1 Time Resolution

Because of the transient nature of laser plasmas it is necessary to study the plasma emission using a time-resolution technique. For time-resolved spectros-

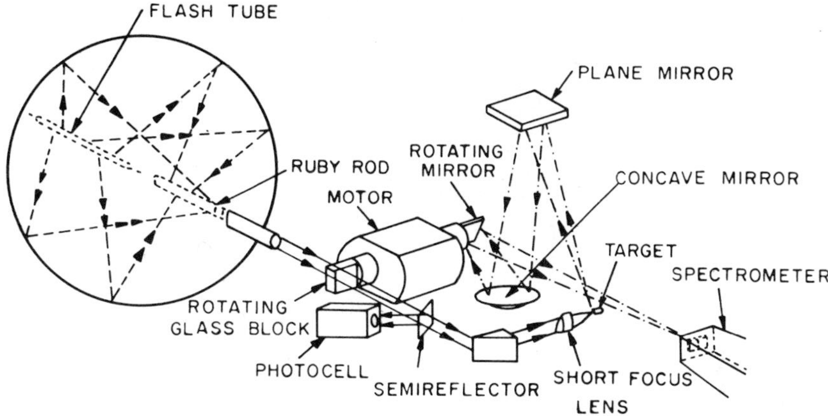

Fig. 2.4 Experimental setup used for time-resolved studies of laser plasmas. (From Archbold *et al.*[55])

copy a system is required whereby the spectrograph entrance slit is scanned with an image of the plasma, such as is shown in Fig. 2.4. This system was used by Archbold *et al.*[55] to study laser plasmas produced by 1-MW pulses. The *Q*-switch, a rotating glass block, was mounted on one end of the armature of a motor and the scanning mirror on the other. In this way proper synchronization could be achieved between the timing of the laser pulse and the position of the plasma image relative to the entrance slit of the spectrograph. A sweep speed of 0.47 mm/μs was achieved with a *Q*-switch motor speed of 6000 rpm.

Another system has been described by Piepmeier and Malmstadt.[17] In their system the scanning spectrograph mirror was electronically synchronized with the rotating prism laser *Q*-switch by means of a sophisticated coincidence detection system, shown in Fig. 2.5. A sweep speed of approximately 2 mm/μs was accomplished, leading to a time-resolving capability of 0.3 μs. This resolution is generally dependent on the quality of the optical system as well as on the time dispersion. Spatial resolution was incorporated in this study by observing the plasma at different heights above the sample surface.

Time-resolved spectroscopy, using a direct-reading spectrometer, offers the advantage that no scanning mechanism is required.[25,56,57] The method is simpler, in that a fast oscilloscope is the only additional requirement. However, it is not possible to record the plasma spectrum over a wide wavelength range as in photographic recording. For photoelectric recording of highly transient phenomena, the time constant of the electronic readout system should be sufficiently short so that a true representation of the time profiles of the various spectral lines is recorded. It is also necessary to bridge the resistors of the dynode chains of the spectrometer photomultipliers with capacitors to obtain a satisfactorily fast response to the intense light pulses.

Time-resolved studies incorporating spatial resolution have revealed the following properties of *Q*-switched laser plasmas: Initially, a plasma is formed in

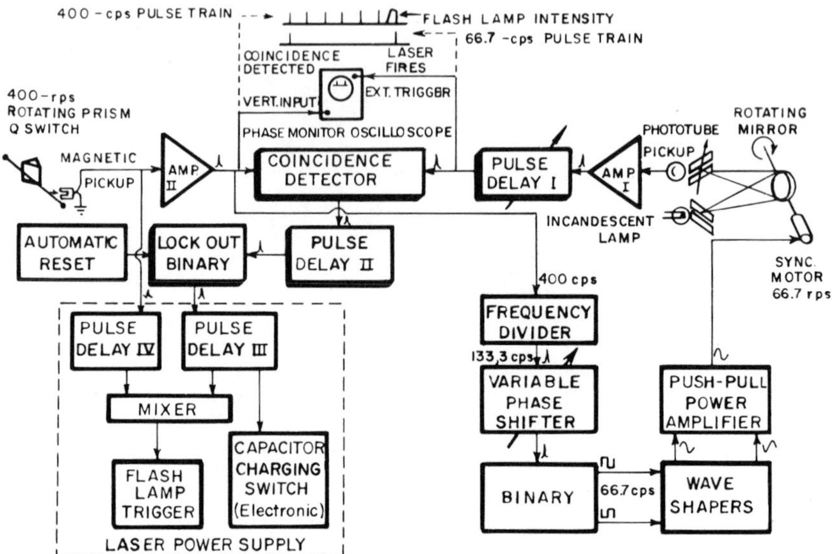

Fig. 2.5 Functional diagram of system to synchronize the rotating mirror of the spectrograph with the rotating prism laser Q-switch. (From Piepmeier and Malmstadt.[17])

the atmosphere above the target, initiated by the ejection of high-energy electrons and ions from the sample.[17] This plasma may provide the means of inverse Bremsstrahlung absorption of laser energy in the atmosphere. At this stage, the emitting atmospheric plasma contains relatively few sample species. Nitrogen emission was observed by Piepmeier and Malmstadt[17] using an 8-MW laser pulse focused onto an aluminum alloy target in air, whereas Archbold *et al.*,[55] using a 1-MW pulse, reported no spectral lines of nitrogen. Atmospheric plasma formation therefore tends to occur more readily at higher laser powers. The atmospheric plasma is rapidly replaced by sample species as target material is vaporized.

During the laser pulse, strong continuous emission occurs. This continuum decays rapidly after the laser pulse has ended. An example is given in Fig. 2.6, which shows the variation in the signal-to-background ratio of Ca II (3159 Å) for three laser pulse energies.[58] During the early stages of the laser pulse, spectra of multiple ionized atoms originating from the target are observed. The spectra of these highly ionized species are of shorter duration than those of atomic or singly ionized species.

Spectral line reversal for transitions terminating in the ground state or a low metastable level of the neutral atom occurs strongly throughout the plasma emission period, indicating a mantle of cool atoms surrounding the radiating core of the plasma. This reversal is asymmetrical with a shift of both emission and absorption peaks toward longer wavelengths.[25]

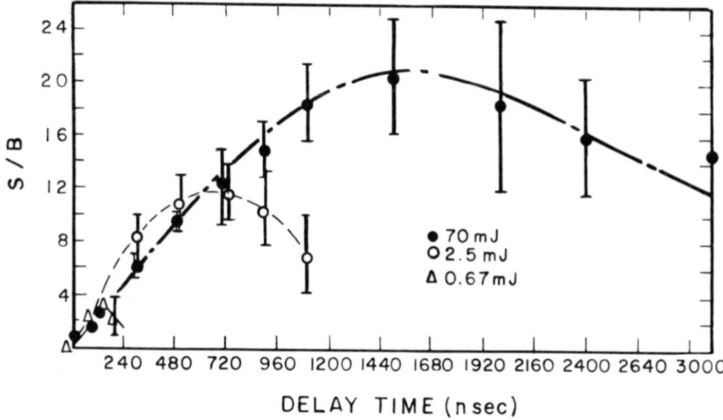

Fig. 2.6 Signal-to-background ratio of the line Ca II (3159 Å) as a function of the time delay after the laser pulse, for three laser energies. (From Treytl *et al.*[58])

Piepmeier and Malmstadt[17] reported an apparent separation of sample atoms and ions by using a second laser pulse to sample the plasma composition 0.5 μs after the initial pulse. A probable explanation can be derived from the proposed existence of an electronic space charge, causing the ions to be accelerated at a higher rate than the atoms.

Results of time-resolved studies suggest that a significant reduction in the contribution of the intense continuum to the analytical signal can be accomplished by using a time-resolution technique in the case of laser microprobe analysis. The temporal separation between the continuum and the peak spectral intensity of most atom and singularly charged ion lines is sufficient (of the order of 1 μs) to permit a practical analytical system to be developed.[25,58] The use of such a system in analysis is discussed in Section 2.5.

2.4.2 Characteristics of Spectra

The time-integrated emission spectra contain lines of the major constituents which are resonance-broadened to half-widths as great as 5 Å.[59] The asymmetrical self-absorption can be explained in terms of a Doppler shift. The absorbing atoms form part of the radially expanding outer zone of the plasma and the high-velocity atoms between the central radiating portion and the spectrograph entrance slit will absorb radiation at a longer wavelength than if they were stationary. Scott and Strasheim[25] have measured shifts of both absorption and emission lines. The method consisted of superimposing the spectrum of a hollow cathode lamp on the spectrum produced by the laser plasma. The absorption line shifts were found to be of the order of 0.3 Å for the aluminum 2568-Å and 2575-Å lines. Photometer scans of various aluminum resonance lines

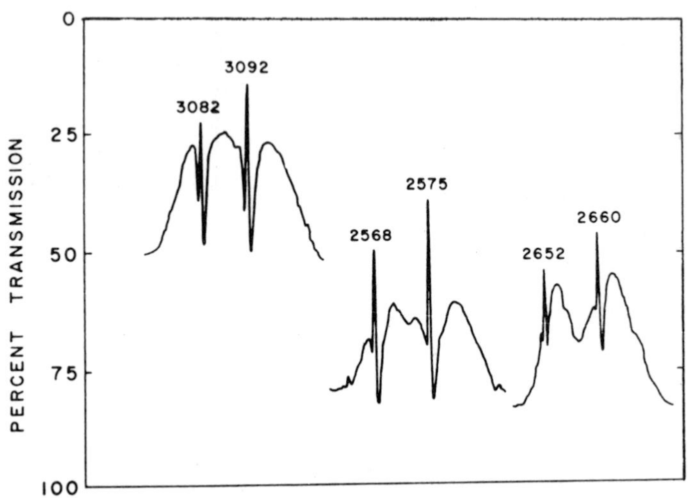

Fig. 2.7 Densitometer tracings of the aluminum resonant lines 3082, 3092, 2568, 2575, 2652, and 2660 Å for the semi-Q-switched-mode plasma, superimposed with the resonant lines of an aluminum hollow cathode lamp to show the laser plasma line shifts.

are shown in Fig. 2.7 to illustrate the relative shifts. Similar red shifts were measured by Allemand.(57)

The radial expansion velocity of the plasma may be calculated from known values of the absorption line shift, or vice versa. Assuming a Doppler shift, we have

$$\frac{\Delta\lambda}{\lambda} = \frac{v}{c} \tag{2.8}$$

where $\Delta\lambda$ is the line shift, λ the wavelength for a stationary source, v the velocity of the absorbing atoms, and c the velocity of light. For $\Delta\lambda = 0.3$ Å, $\lambda = 2568$ Å and $c = 3.0 \times 10^{10}$ cm s^{-1}, the velocity of the absorbing atoms equates to 3.5×10^{6} cm s^{-1}, which is in the range of measured plasma expansion velocities.

In addition to the Doppler shift, it is also possible that the line shifts are due to pressure and/or the quadratic Stark effect. The latter effect has been assumed to be the major cause of line broadening in Q-switched laser plasmas.(56)

The degree of ionization is dependent on the available laser power. Spectra of Al VIII have been observed as sharp emission lines in the extreme ultraviolet region using a peak laser power of 100 MW.(60) In fact, ion charge multiplicities as high as 25 have been created for a number of elements using gigawatt laser pulses.(53,61) Investigations of the extreme ultraviolet spectra have also revealed sharp absorption lines corresponding to the interaction of continuum radiation from the plasma core with atoms at various stages of ionization present in the outer regions.[62-64] The laser powers used in the experiments above are greater by at least an order of magnitude than those commonly used for analytical applications and, in general, ion charge multiplicities of greater than 4 are not found in plasmas created by laser pulses of less than 10-MW peak power, corresponding to approximately 10^9 to 10^{10} W cm^{-2}.

The origin of the continuum emission is predominantly the lower central zone of the plasma.[25,57] This emission is a direct result of partial absorption of the laser pulse (the trailing edge of the initial pulse or the second pulse if a double pulse is used) in the lower core region by the mechanism of inverse Bremsstrahlung. The emission spectra of highly ionized species also predominate in this region, which is bounded somewhat by the physical dimensions of the laser beam.

Increased analytical line-to-background ratios of up to 40 have been reported when observing the plume just next to the central zone rather than in its center,[17] mainly as a result of the central continuum. This result suggests an improvement in the analytical method by spatially selecting a region for analysis. A selection in the vertical direction may also be employed.[15]

2.4.3 Influence of Atmospheric Pressure and Composition

The pressure of the atmosphere surrounding the target material is a parameter that has a marked effect on the characteristics of the plasma. As the pressure is reduced from 760 to 1 Torr, the height of the emitting region of the target atoms increases.[65] This effect is partly due to increased sampling at low pressures, as illustrated in Fig. 2.8. The amount of target material (copper) which is vaporized is seen to be dependent on the laser energy for an ambient pressure of 1 Torr. This dependency decreases with increasing pressure, and at a pressure of 760 Torr the sampling is relatively independent of laser energy. A

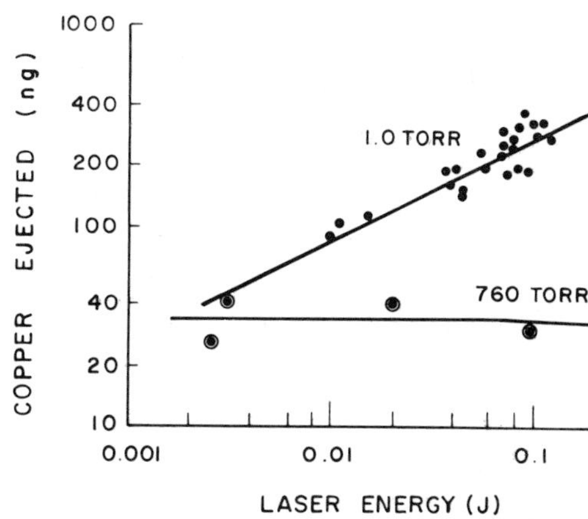

Fig. 2.8 Amount of copper vaporized for various laser energies at 1 and 760 Torr ambient pressures in air. (From Piepmeier and Osten.[65])

possible explanation for this phenomenon is that the atmospheric plasma which is initially formed absorbs the laser energy and allows only a given amount to reach the sample. This process has been treated theoretically by examining a mathematical model for an atmospheric radiation-supported shock wave which propagates in the atmosphere toward the laser beam.[65] These calculations have shown that isolation of the laser beam is theoretically possible for the experimental conditions used.

The relative intensity of the continuum versus laser energy at atmospheric pressure is plotted in Fig. 2.9 for two heights above the target surface. It is clear that the continuum intensity of the local region at the surface is relatively constant with laser energy (curve A) while the continuum increases in the extensive region (0.05–0.5 cm) linearly with increase in laser energy (curve B). This seems to indicate that the local region receives a small constant amount of laser energy, while all the other additional energy is absorbed by the extensive region.

Possible atmospheric influences are emission quenching, molecular association, and the obstruction of the expanding target vapors by atmospheric species.

The most important indications which stem from studies of pressure effects are the following:

1. Once a shock wave is initiated in the atmosphere above the target, the wave becomes opaque, thus limiting sample vaporization.

2. The use of a longer single laser pulse should allow more energy to be used to vaporize more material.

3. At low ambient pressure, the analytical line-to-background ratio in the upper region of the plasma is significantly improved.

The effects of atmospheric composition, that is, argon, air, oxygen, nitrogen, and helium and also a vacuum (5 Torr) on the optical emission of laser plasmas from metallic, nonmetallic, and biological samples containing iron or magnesium

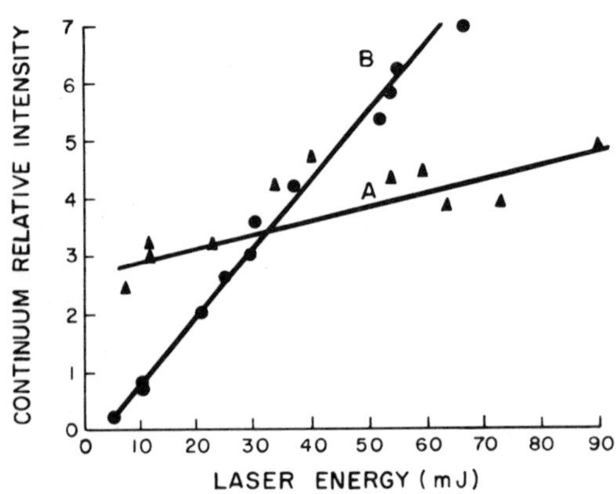

Fig. 2.9 Influence of a change in laser energy on the relative continuum intensities at two positions above a copper sample at atmospheric pressure. Curve A is for the region up to 0.2 mm above the surface and curve B for the region 0.5–5 mm above the surface. (From Piepmeier and Osten.[65])

(as test elements), have been studied from an analytical viewpoint.[66] From oscillographic recordings of 3020-Å Fe line and spectral background at 3015 Å, it was found that the amplitude of the Fe signal in helium was approximately 100% greater than the corresponding pulses in other atmospheres.

The signal pulse durations were found to be longer in argon and shorter in helium than those obtained in air, oxygen, and nitrogen. No systematic differences were observed in the latter three atmospheres. The plasmas produced in vacuum were generally of short duration and less intense than others. Although it may be advantageous in certain cases to select an appropriate atmosphere in order to optimize signal-to-background ratios, the atmospheric composition does not appear to affect the ratios sufficiently to warrant changing to atmospheres other than air for time-integrated analysis.

However, Demitrov and Petrakiev[67] substituted nitrogen, oxygen, and argon for air for metal analyses in alloys with a neodymium laser. They reported marked increases for intensities of atomic and ionic lines for Si, Mg, Mn, Cr, Ni, and Fe; the Mn II 2610-Å line intensity using nitrogen, oxygen, and argon was 1.4- to 4.4-fold higher than in air. The signal-to-noise ratios also improved; for example, for air, nitrogen, oxygen, and argon, the ratios were 6.9, 72.0, 90.6, and 26.0, respectively. The slopes of the analytical curves were affected. Kubota and Ishida[68] also investigated spectral line intensities of various elements in air, oxygen, nitrogen, argon, and helium–oxygen and observed that the greatest intensities were with the argon atmosphere. A small amount of oxygen in argon, 0.1 volume to 2 volumes, improved the sensitivity of magnesium and zinc in aluminum alloys. The oxygen in argon was necessary to maintain the high spark voltages when auxiliary spark excitation was used. These authors[69] also demonstrated that self-absorption is reduced by blowing argon across the spark gap. Self-absorption was also reduced by increasing the spark voltage and condenser capacity and by decreasing the resistance of the spark circuit and the laser energy output. Another study[70] simply reported that the use of argon as the protective gas improved the signal for the major metal components while pressures greater than 1 atm were needed to improve the signal for trace elements.

2.4.4 Temperature

Spectroscopic temperature measurements of laser plasmas present difficult problems as a result of the inhomogeneous spatial characteristics and highly transient nature of these plasmas. A number of methods have been used by various investigators to determine laser plasma temperatures.[56,71,72] Bogershausen and Honle,[72] for instance, used the Ornstein two-line method and obtained a maximum temperature of 29,000°K for a 1-MW Q-switched laser plasma. Because of high temperature gradients in laser plasmas, temperature values obtained by this method could depend on the excitation potentials of the spectral lines chosen. This is because the lines do not necessarily radiate from the same region in the plasma. Ideally, a time- and space-resolved readout of the spectral line intensities is required, followed by the necessary Abel transformations. However, the dynamic nature of laser plasmas, the extreme temperature gradients, and the time period in which they exist are factors which would certainly limit the accuracy of such a difficult study.

David and Weichel[71] have shown by interferometric measurements that the maximum electron temperature attained for a laser-heated carbon plasma, using an energy density of the order of 100 J cm^{-2}, was approximately 116,000°K. At this relatively low energy density, the plasma was not in local thermodynamic equilibrium, with the electron temperature nearly an order of magnitude higher than the plasma temperature. A decrease in the electron temperature to that of the plasma (Saha) temperature was observed for increasing laser energy density. This decrease implied that the electrons transferred their energy to the more massive particles in the plasma. At 1,000 J cm^{-2} the plasma was found to be in thermodynamic equilibrium, with a maximum temperature of approximately 24,000°K.

Studies of the high-temperature effects associated with extremely powerful (gigawatt) pulse interactions have been published.[31,73-80] The use of ultrahigh powers results in characteristic x-ray emission from the plasma, which can be measured and used to calculate the electron temperature.[79]

The generation of these ultra-high-temperature plasmas using solid hydrogen or deuterium targets to produce thermonuclear fusion reactions is a relatively new development.[81,82] Technological problems associated with laser and optical design to achieve the high powers necessary are formidable,[83] but power exceeding 15 GW has already been achieved. This produced more than 10,000 neutrons per pulse.[82] This field is, however, beyond the scope of this chapter.

2.5 SPECTROCHEMICAL ANALYSIS

The reader is referred to the book *Laser Micro-Spectrochemical Analysis* by H. Moenke and L. Moenke-Blankenburg, recently translated into English by R. Auerback,[84] which deals more at length with the topic than is possible in this section.

2.5.1 Influence of Power Density

The power density and duration of the laser radiation are determined principally by the design of the laser source and the focal length of the focusing lens, Eq. (2.7). In comparing the effects of power densities in the range 10^5-10^{10} W cm^{-2} with pulse durations in the range 10^{-3}-10^{-8} s, it is evident that low-power-density radiation of the order 10^5 W cm^{-2} and of long pulse length, typically greater than 10^{-3} s, causes predominantly melting. As the power density is increased to approximately 10^7 W cm^{-2}, a flushing of molten material occurs, provided the pulse duration is suitably long, typically greater than 10^{-4} s. Vaporization is in evidence at this power density but is confined to the initial stage of the interaction.[25] A melting-flushing mechanism may well occur at power densities of 10^8 W cm^{-2}, provided that the irradiation period is sufficently long.[85] Vaporization, however, appears to dominate the interaction for pulse durations in the *Q*-switched range, even at low power densities. Detailed observations of craters produced by single *Q*-switched pulses with power density of 10^9 W cm^{-2} and duration less than 10^{-7} s, as well as study of high-speed photographs, have shown less evidence of a melting-flushing mechanism. Melting is in evidence in the craters, which are of a pitted nature, and the general appearance indicates a rapid resolidification of the molten material, suggesting very localized heating, with little heat flow into the deeper and surrounding regions.

It has been established that the interaction becomes less dependent on the physical properties of the sample as the power density is increased. For instance, it is not possible to vaporize soda-glass with a *Q*-switched pulse of power density of 10^8 W cm^{-2}, but at 10^{10} W cm^{-2} a crater and intense plasma are formed at the surface, signifying a successful sampling of the normally transparent material.

For analytical purposes it is obvious that a power density (by *Q*-switching) of at least 10^9 W cm^{-2} is essential if the laser sampling technique is to be applicable to all known solids. The use of relatively low power (usually conventional mode) lasers in spectrochemistry[20,56,86–89] has two major disadvantages: the quantity of material liberated is dependent on the chemical composition of the sample, such as the silicon concentration in silicon brasses[87] and preferential vaporization tends to occur, such as the expulsion of zinc in preference to copper from a brass sample.[87] These effects are due, at least in part, to a dependence of the mass of liberated material on the amount of laser energy utilized in the interaction, which is related to the reflection coefficient of the sample (or, in the case of transparent material, the transmission coefficient) at the laser wavelength during the prebreakdown period.[20] A comparison of reflection coefficients of various metal targets with the quantity of reflected laser energy (as measured calorimetrically) and the mass of liberated material is shown in Table 2.1. These data indicate a satisfactory agreement between the fraction of reflected energy and the reflection coefficient for most samples. The thermal properties of the samples also play a significant role, as can be seen, for example, in the case of tin and lead. Both these metals have low cohesive energies, resulting in a more efficient utilization of laser energy after surface breakdown, during which the reflectivity is diminished.[21]

Other factors that support the contention that the *Q*-switched mode is superior to the conventional mode with respect to the variety of samples that can be analyzed are: *Q*-switched pulses are more reproducible than conventional pulses in terms of both power and intensity/time profiles; the plume produced

Table 2.1 Comparison of the Reflection Coefficients of Metals with the Reflected Laser Energy and Mass of Liberated Material[a]

Element	Reflection coefficient (%)	Amount of energy reflected (%)	Amount of material liberated (mg)
Copper	91	83	0.6
Aluminum	74	66	2.1
Nickel	72	38	1.4
Zinc	67	45	4.8
Iron	65	40	1.0
Tungsten	60	43	1.7
Tin	54	29	10.4
Lead	54	13	15.1
Bismuth		12	21.6
Graphite	27	23	0.4

[a]From Panteleev and Yankovski.[20]

by a Q-switched pulse has a more reproducible geometry from one sample to the next; high-speed photography of laser plumes has revealed more random effects, which occur as a result of the absorption of conventional-mode radiation,[25] and this suggests that a single Q-switched pulse may have analytical advantages; less material is liberated by a Q-switched pulse—usually less than 1 μg—resulting in improved microsample capabilities and an increase in the absolute sensitivity of the method.

2.5.2 Influence of Laser Wavelength

The laser wavelength naturally depends on the type of laser used. Neodymium glass, with its low threshold energy and high efficiency, is preferred to ruby in these respects. The 6943-Å ruby radiation, on the other hand, has a lower reflection coefficient for most solids than the 10,600-Å neodymium radiation. The difference in the reflection coefficients are mostly small, however, and for Q-switched pulses they are negligible. A further characteristic in comparing the two wavelengths is the theoretically determined spot size produced by focusing a uniphase wavefront. Optical theory shows that the Airy disc diameter is proportional to the wavelength, Eq. (2.4). Therefore, ruby radiation, under this ideal condition, can be focused to a smaller spot than neodymium radiation. Unfortunately, the means for producing a uniphase wavefront (mode selection) inherently decreases the laser energy output and this is a disadvantage. The spot size produced by a multimode laser depends primarily on the divergence of the laser beam and the focal length of the focusing lens, Eq. (2.6). Therefore, the difference in spot size between ruby and neodymium radiation can be regarded as close to negligible for multimode laser operation.

The absorption coefficient for laser radiation in a fully ionized plasma, due to free-free transitions, is proportional to the square of the laser wavelength if the plasma frequency is much lower than the frequency of the radiation.[90] Thus, if this mechanism dominates the absorption processes, it would appear that neodymium radiation, with its longer wavelength, is more effective in heating the plasma than ruby radiation by a factor of 2.33 (the square of the ratio of the respective wavelengths). Whether or not free-free processes (inverse Bremsstrahlung) occur to a significant extent can only be speculated. The free-free absorption coefficient may be approximated[91] as

$$K_\nu \approx \frac{Zn_e \cdot 10^{-7}}{3\nu^2 (kT_e)^{3/2}} \tag{2.9}$$

where Z is the ionic charge, n_e the electronic density, k Boltzmann's constant, T_e the electron temperature, and ν the frequency of the laser radiation. Absorption of radiation occurs according to the relationship

$$E_{\text{ab}} = E_0 [1 - \exp(-K_\nu l)] \tag{2.10}$$

where E_{ab} is the absorbed energy, E_0 the initial energy, and l the path length. For a path length of 1 mm, it can be easily calculated that for appreciable absorption, the electron density would have to be of the order of 10^{19} cm^{-3}. Measurements of the electron density of laser plasmas from spectral line widths[56] have yielded values of Q-switched laser plasmas (2×10^8 W cm^{-2}) of 2.5×10^{18} and 1.8×10^{18} cm^{-3} at heights of 1.1 and 1.75 mm, at atmospheric pressure. (At reduced pressure of 0.5 Torr, the electron densities are somewhat lower, owing to an increase in the expansion rate.) It is therefore quite possible that in a region close to the target surface, the electron density is sufficiently high to permit substantial absorption by inverse Bremsstrahlung. The thermal contact between this region of the plasma and the surface of the target presumably results in a transfer of energy to the solid. This transfer mechanism was postulated by Piepmeier and Malmstadt[17] as counteracting the decrease in sampling that is due to partial screening of the laser beam from the target by the plasma. Fully effective screening will occur only if the plasma frequency is higher than the frequency of the laser radiation. The plasma frequency is given by[92]

$$\nu_p = 8.9 \times 10^4 n_e^{1/2} \tag{2.11}$$

The frequencies of neodymium and ruby radiation are 2.829×10^{14} s^{-1} and 4.321×10^{14} s^{-1}, respectively. Equating these two frequencies to the plasma frequency gives the electron density required for nonpenetration in the two cases:

$$n_e \geqslant 1.01 \times 10^{21}\ \text{cm}^{-3} \quad \text{for neodymium radiation}$$

$$\geqslant 2.35 \times 10^{21}\ \text{cm}^{-3} \quad \text{for ruby radiation}$$

As the atomic density of most metals is higher than 10^{22} cm^{-3}, it is possible that a highly ionized plasma formed from the solid may initially reflect the incident energy before expansion lowers the electron density and thus the plasma frequency. More than double the electron density that would be required for reflection of the neodymium radiation is necessary for reflection of the ruby radiation.

To conclude, it is obvious from the discussion above that the advantages of using neodymium radiation as far as the absorption processes are concerned are not highly significant. Nevertheless, it does seem that if a laser pulse is used for both vaporization and excitation, some advantage may be gained in using neodymium radiation, whereas if auxiliary excitation is employed, the 6943-Å ruby radiation, with its higher plasma penetration ability, may be superior. Factors such as low threshold energy and high efficiency are the more obvious advantages of using the 10,600-Å radiation for spectrochemical applications.

2.5.3 Single-Step (Laser Excitation) Analysis

In single-step analysis the laser pulse is used both to vaporize and to excite the analytical species. The risk of contaminating the sample is therefore negligible. This method may be used when highly localized sampling is required, typically for less than 10 μm crater diameter.[93] The use of auxiliary spark excitation results in spark burning of the area surrounding the sample, and thus the addition of extraneous material to the sampling area. The elimination of the auxiliary excitation places severe demands on the spectrographic and detection systems in order to maintain a favorable ratio of signal to background. The demands on the laser source and focusing optics to maintain a sufficiently large power density at these extremely small spot sizes are also severe. Instrumental improvements for a laser microprobe system have been developed by Peppers *et al.*,[94] and resulted in a laser beam of extremely low divergence and also fine control of its energy. Both these factors are necessary to extend the sampling resolution to below 10 μm. The laser source was mode-locked using a Fabry–Perot etalon, and a double *Q*-switch was used in the form of a rotating prism and bleachable dye combination to shorten the laser pulse. Transverse-mode selection was incorporated, and control of the laser energy could be accomplished by a liquid attenuator in the form of a cell filled with aqueous copper sulfate solution and mounted in the path of the laser beam. Using a high-quality 8-mm-focal-length lens and a 1-mm-diameter diffraction-limited beam, the authors were able to produce symmetrical craters less than 14 μm in diameter. With fine control of the copper sulfate concentration, craters of less than 1 μm diameter could be formed. This made possible the sampling of single blood cells with production of spectral lines of sufficient intensity to be useful for analysis.

The single-step technique can also be used for macroanalysis. Quantitative macroanalysis using solely a *Q*-switched laser pulse for vaporization and excitation was first reported by Runge *et al.*[95] Using stainless steel samples and a relatively large spot diameter of 400 μm they were successful in obtaining calibration curves for the major constituents: nickel and chromium. Similar curves were obtained for molten metal samples.[96] The calibration curves are shown in Fig. 2.10. Taganov and Fainberg[97] also studied the parameters for spectrographic analyses with direct sample excitation with a focused laser beam. Relatively high energy pulses, 40–60 J, were used with metals, glass, and alloys. Good reproducibility was reported.

Another method of laser macroanalysis consists of moving the sample by means of a scanning device such as that designed by Felske *et al.*[98,99] to permit a rapid sequence of laser pulses to sample an extended region so as to obtain a representation of the mean content or a topographical map giving the distribution of the elements of interest. Figure 2.11 shows the craters produced in steel using the scanning laser microprobe. The analytical curves obtained for various elements in homogeneous steel and cast iron samples are shown in Fig. 2.12. Rapid-firing laser systems are necessary for these studies. Felske *et al.* used a ruby laser capable of producing 1 pulse per second.[98] Technological improvements in the manufacture of laser rods has since progressed and has led to the development of a laser microprobe with a firing rate of up to 25 pulses per second.[100] At the present time many lasers are commercially available with *Q*-switched pulse repetition rates of up to 50 pulses per second while still maintaining substantial individual pulse power. Pulse repetition rates of a few thousand per second are also available using continuous wave lasers which are *Q*-switched at the required frequency, but generally these pulses have insufficient power for spectrochemical applications.

The main disadvantages of using a laser pulse to simultaneously vaporize and excite the sample are the following: The spectrum contains an intense background continuum, which makes it difficult to identify weak lines; the spectral lines are wide and in some cases diffuse; the lines of major constituents are strongly self-reversed; and the spectral intensity is poor, so that it is neces-

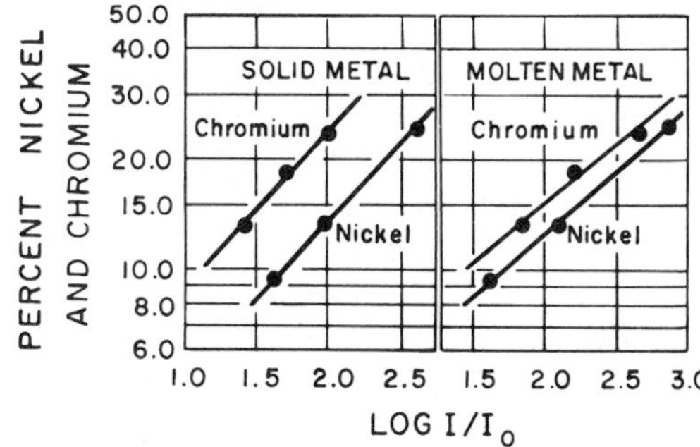

Fig. 2.10 Analytical curves for nickel and chromium based on laser-excited spectra of solid and molten steel samples. (From Runge *et al.*[96])

Fig. 2.11 Laser craters in a steel sample, produced by a scanning laser microprobe. (From Felske *et al.*[98])

sary to use a fast spectrograph (preferably less than F12) and fast spectrographic plates, such as Ilford Selochrome or Kodak SA1 or B10. The large grain size, which is an inherent characteristic of fast plates, results in imprecise line photometry for quantitative determinations.

Spatial resolution may be employed to increase the line-to-background ratios by forming an enlarged image of the source on the spectrograph entrance slit. This also allows maximum illumination of the spectrograph. A quantitative analysis is difficult under this condition, as the spectral lines are nonuniform, and plate calibration using a step filter at the entrance slit is inaccurate. A

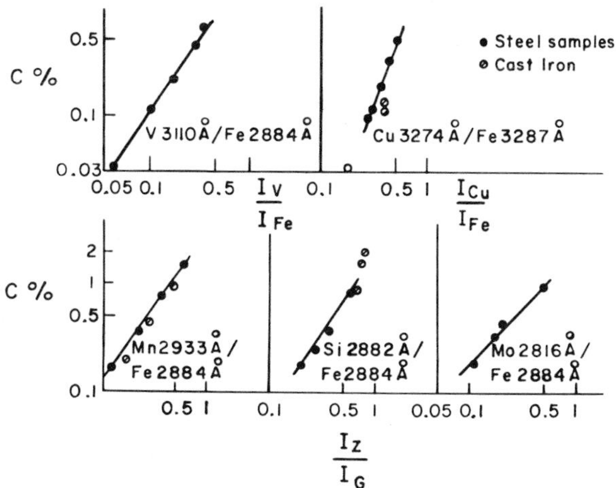

Fig. 2.12 Analytical curves for steels using a scanning laser microprobe. (From Felske *et al.*[98])

rotating sector for plate calibration cannot be used because of the transient nature of the laser plasma. Thus, in order to gain any meaningful quantitative information, an external optical system must be employed which produces uniform illumination of the entrance slit and step filter. This causes a decrease in spectrographic illumination and produces relatively weak spectra. To spatially resolve the laser plasma using this system, an enlarged image of the source should be formed on a diaphragm containing an adjustable slit which is set to pass light from a selected region of the plasma. The diaphragm aperture should be fully imaged onto the collimating lens of the spectrograph to obtain the maximum possible illumination for the case in question.

Spectrometric (direct-reading) techniques employing time resolution have been developed for laser analysis to reduce the contribution of the spectral background to the analytical signals.[25,58,101] The high spectral background appears in the initial 0.5 μs of a single-pulse laser plasma, whereas the atomic and low-energy ion lines reach maximum intensities only after 1 to 3 μs, depending on the excitation energy of the lines. By appropriate electronic gating, the analytical lines can be separated from the spectral continuum. This results in output signals with improved line-to-background ratios. Electronic gating at the collectors of the photomultipliers offers the further advantage of being able to reduce the contribution to the integrated signals by photomultiplier dark current during the "off" period.[25] This is favorable because the duty cycle of single-shot laser analysis is very low, and the analytical signals are very small. It is necessary to use small integrating capacitors to accumulate sufficient voltage over the short integration time (typically 5 μs) to be accurately measured using a digital voltmeter.

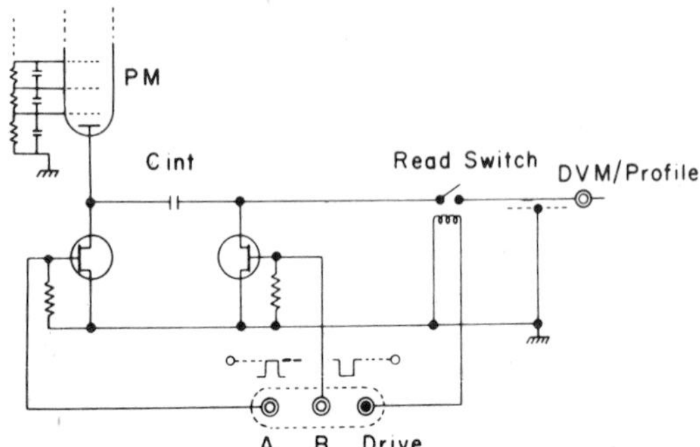

Fig. 2.13 Circuit diagram of the photomultiplier gating device. (From Schroeder *et al.*[101])

The operating principle of a suitable gating device[101] is as follows. Each photomultiplier tube is gated at its collector using a pair of field effect transistors as the switching elements (Fig. 2.13). These transistors act as switches which are controlled by appropriate signals on their bases. The output current from the photomultiplier is gated either to ground or to the integrating capacitor. This is achieved by controlling the switching transistors with complementary signals from a pulse generator which is triggered by the laser pulse. The on/off times of these complementary signals as well as the delay time between the laser pulse and the integration period are controlled by the pulse generator. Figure 2.14 shows the improvement in analytical sensitivity achieved by the use of this gating device.

Conventional-mode radiation has been used by some authors for qualitative[84,87] and quantitative analyses.[25,87,89] The dependence of the characteristics of the laser interaction on the physical properties of the sample is

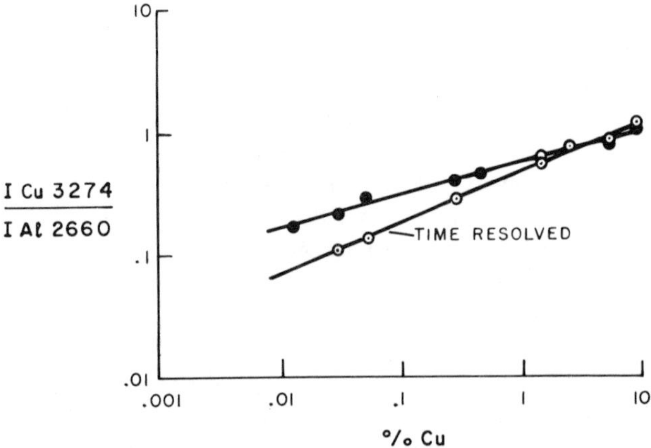

Fig. 2.14 Analytical curves for copper in aluminum alloys, showing the improved sensitivity obtained by time resolution.

unfavorable for quantitative analysis.[87] Nevertheless, the line spectra obtained by this method are superimposed on a less intense background continuum than in the case of the Q-switched mode.

A comparative analytical study between conventional- and Q-switched-mode laser plasmas[25] has revealed that the sensitivity, that is, the relative increase in analytical signal with an increase in the concentration of the analyte in the sample, is higher in the former case. However, by using electronic time resolution to reduce the spectral background contribution, the sensitivity of the Q-switched mode can be increased significantly, as seen in Fig. 2.14. The reproducibilities of the intensities of the spectral lines for replicate (single-shot) analyses were found to be about 10%, compared to about 25% for the conventional mode. The poor gradients of the calibration curves in both cases resulted in analytical precisions of the order of 20% and 35%, respectively. These reported values are single-shot variations, and it is possible to predict the coefficient of variation for the case where a number of samplings are integrated, using statistical theory. For an integration of N samplings, these variation coefficients should decrease by a factor $N^{-1/2}$. Generally, it has been found that somewhat higher variation coefficients are observed for multiple samplings than those calculated theoretically.[98,100] This is probably due to an abnormal distribution function of the intensities of the individual samplings over the time taken to make the measurements, such as may be caused by prelasing effects.

Repeated laser samplings on one spot on the sample surface using a fast-pulsing Q-switched laser result in a variation in intensity of the emitted spectral

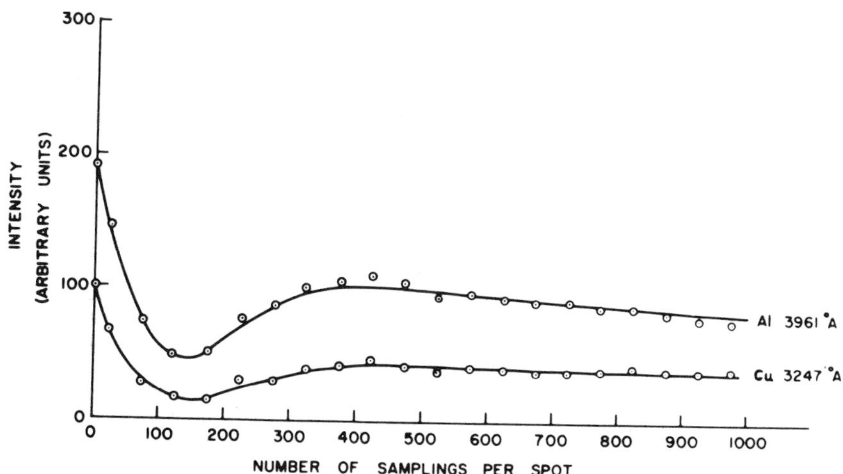

Fig. 2.15 "Spark-off" curves of relative intensities of the spectral lines Cu 3247 and Al 3961 Å for an aluminum alloy.

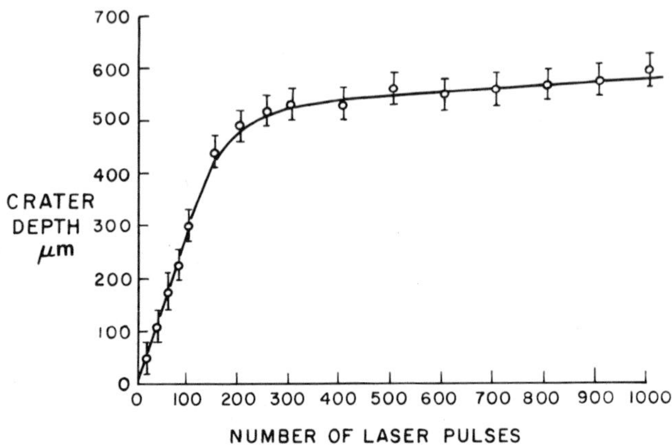

Fig. 2.16 Crater depth as a function of the number of laser samplings (aluminum sample).

lines similar to that observed in normal spark sources, termed "spark-off."[(87,88)] The variation in the intensities of two spectral lines as a function of the number of previous laser samplings is shown in Fig. 2.15. The growth of the crater in the sample is shown in Fig. 2.16.

A number of factors could be responsible for these observations. These are the masking action of the crater walls with respect to the aperture of the laser light bundle entering the crater, the defocusing of the laser radiation due to the receding surface, and possibly the screening effect of the plasma within the crater. In this latter possibility it is assumed that the plasma, partly trapped within the crater, absorbs a significant fraction of the laser energy. An energy transfer may then occur to the crater walls, resulting in a second vaporization mechanism, which could influence the intensities of the spectral lines emitted during the expansion of the plasma above the surface.

Table 2.2 Spectral Line Intensities for Various Sample Surface Conditions[a]

Mode	Surface condition	I_{Cu3247} (arbitrary units)	I_{Al3961} (arbitrary units)	I_{Cu3247}/I_{Al3961}
No prelasing	Polished	328	414	0.793
	Dull	282	375	0.753
	Rough	228	322	0.688
With prelasing	Polished	108	223	0.487
	Dull	110	228	0.484
	Rough	105	215	0.489

[a]From Scott and Strasheim.[(100)]

It is clear from the observations above that an integration of a few hundred samplings should be attempted only after a prelasing period of approximately 400 shots, when "stable" conditions have been attained. The influence of the surface condition (reflectivity, roughness, presence of oxide layer) can be eliminated by prelasing as shown in Table 2.2. (The foregoing observations and conclusions are for the case in question and do not necessarily apply to all experimental configurations.)

With gated integration of multiple samplings and prelasing, the analytical precision for some elements can be improved to ±2%.(100) Thus the combination of a high-repetition-rate laser source and a gated integration detection system can be used to obtain satisfactory analytical precisions if analytical samplings are integrated.

2.5.4 Double-Step (Laser Plus Spark) Analysis

This technique is most commonly used in laser microprobe analysis. The primary function of the laser pulse is to vaporize material. The vapor plume is then heated by a spark discharge between two graphite electrodes mounted above the sample surface as shown in Fig. 2.17. The distance between the electrode tips and the height above the sample surface are normally set at 1–2 mm. When conducting samples are analyzed, the distance between the electrode tips should be less than twice the height or else the spark discharge will occur via the sample surface and not via the vapor plume.

The spark circuit has the usual adjustable parameters of voltage, capacitance, inductance, and resistance. The useful ranges of the parameters are: 1500–2000 V, 2–20 μF, 0–150 μH, and 0–10 Ω. The effects of these parameters on the spectral intensity of laser plasmas are qualitatively similar to those observed for normal spark excitation, except that the spark has no part in the sampling

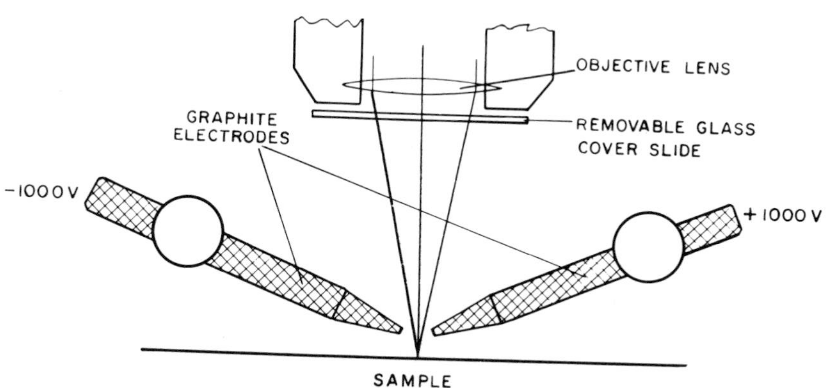

Fig. 2.17 Electrode assembly used for auxiliary spark excitation of the laser plasma.

mechanism. Generally, an increase in the electrical power delivered to the laser plume results in an increase in the relative intensity of the ion to the atom spectra.[59]

Choice of the spark-operating parameters, including the interelectrode distance and geometry, is dependent on the analytical application. For instance, in a more specialized application, the analysis of cytological samples in the crater diameter range 10–25 μm,[102] it is necessary to modify the shape of the graphite electrode tips and their position, and to reduce the spark energy to less than 10 J in order to prevent spark burning of the peripheral zone of the small laser-vaporized area. Spark vaporization extends the sampling region.[103] This could be undesirable in microprobe studies.

A critically damped spark discharge is the best suited for obtaining high spectral intensities, reduced background, and also a reduction in the coefficient of variation in replicate analyses.[102] The total impedance of a discharge circuit for critical damping is given by

$$R = 2\left(\frac{L}{C}\right)^{1/2} \tag{2.12}$$

where L and C are the series inductance and capacitance, respectively.

The occurrence of an atmospheric plasma above the sample surface prior to the arrival of sample species, as observed by Piepmeier and Malmstadt,[17] leads to the supposition that this plasma could cause premature breakdown of the electrical gap, resulting in undesired spectra and wasted electrical energy. Furthermore, the expansion of the laser plume containing the species of analytical interest could be hindered by the presence of the expanding spark channel. However, external control of the ignition of the spark has shown that these are not serious problems. For low-melting-point compounds, best results are obtained when the laser and spark are triggered simultaneously. For high-melting-point compounds, a short delay in the spark ignition is preferable.[104]

A problem associated with auxiliary spark excitation, the staining of the sample surface surrounding the area of analytical interest due to the deposition of a thin oxide layer, can be overcome in one of two ways. A heavy paper shield may be used to cover the surface around the sample site.[59] Shields can be made by drilling a 1-mm-diameter hole through a stack of ordinary paper file cards, which can then be trimmed to size. A card is placed on the sample surface with the aperture over the area under investigation. A second method consists of coating the sample surface with collodion prior to analysis.[105] The collodion (nitrocellulose/alcohol/ether solution) dries rapidly after application, leaving a thin layer. No staining occurs in either of the two methods.

The spectral characteristics of a laser plasma with auxiliary spark excitation are preferred to the case where no additional excitation is employed. The addition of the spark usually results in an increase in the peak intensity of spectral lines by more than an order of magnitude. The spectral line widths are decreased and less self-reversal occurs. Cyanogen bands are formed if the surrounding atmosphere is air. (These bands are absent in pure laser spectra.) Flushing with an inert atmosphere such as argon may be necessary to eliminate these bands, should they cause spectral interferences.

2.5.5 Qualitative and Quantative Analyses of Various Samples

2.5.5.1 Minerals

The identification of microgram quantities of minerals in polished rock sections using a laser microprobe was first described by Maxwell.(106) A number of subsequent studies have been reported in the literature.(84, 107–112) Preparation of the rock samples is not essential but it does aid in the selection of the target area. The allowable degree of roughness is limited by the need to maintain a suitable distance between the surface and the cross-excitation electrodes. More difficulty is experienced in sampling silicate minerals compared to ore minerals, presumably as a result of one or more of the physical properties of silicates, such as brittleness, tenacity, and cleavage. The tendency of some minerals to shatter or fracture due to the nature of the crystal bonding is another factor which influences the magnitude of laser action. Semiquantitative results

Table 2.3 Analysis of Ore Minerals by Laser[a]

Mineral	Major	Minor	Trace	Faint trace
Bornite	Fe, Cu	Cr	–	Mg, Si, Ca, Ag
Sphalerite	Zn	Fe	Pb, Cd(?)	Cu, Si, Mg, Ca
Magnetite	Fe	–	Mg	Si, Mn, Ca
Sphalerite[b]	Zn	Fe	Si	Pb, Mg, Ca
Chalcopyrite[b]	Fe	Cu	–	Mg, Ca
Pyrite	Fe	Ni	–	Mg, Ca, Cu, Co, Ag
Pyrrhotite	Fe	Ni	–	Mg, Cr, Ca, Ag
Galena[b]	Pb	–	Fe, Ag	Si, Mg, Bi, Ca
Chalcopyrite[b]	Fe, Cu	Al	Cr, Mg	–
Siegenite[c]	Ni, Fe	–	Co, Mg, Cu, Zn	–
	Ni, Fe	Cu, Zn	Co, Mg	–
	Ni, Fe	Cu, Zn	Co, Mg	–
Niccolite	As	Ni	Mg, Fe	–
Skutterudite	As, Co	Ca, Fe	Ni, Cu, Cr	–
Arsenopyrite	Fe, As	Ni, Cr	Mg	–
Chromite	Fe, Cr	Ti	Mg, Al	Mn, Cu, V, Ca
Chromite	Cr	Fe, Al	Mg	–
Chromite	Fe, Cr	Al, Mn	Mg	V, Ca
Chromite	Fe, Cr	Ti, V	–	Al

[a] From Maxwell.(106)
[b] Two minerals in the same polished section.
[c] Three shots were taken.

Table 2.4 Analysis of Silicate Minerals[a]

Mineral	Major	Minor	Trace	Faint trace
Clinopyroxene[b]	Si	Ca, Mg, Fe	Al	Ti
Orthopyroxene[b]	Si	Mg, Ca, Fe	Mn	Ti, Al, Cr
Orthopyroxene[b]	Si, Ca	Fe, Mg	–	Ti, Al, Cr
Plagioclase[c]	–	Mg, Ca	Si, Al	B, Fe
	–	Ca	Mg, Si, Al	B, Fe
Plagioclase[c]	–	Ca	Mg, Si, Al	B, Fe
	–	Ca	Mg, Si, Al	B, Fe
K feldspar[d]	Si, Na	K, Al	Mg, Ca	Fe
	Si, Na	K, Al	Ti, Mg, Ca	Mn, Fe
	Si, Na	K, Al	Ti, Mg, Ca	Mn, Fe
K feldspar[e]	–	Si	Ca, Mg	Al
	–	Si	Ca, Mg	Al
Kamacite[f]	Fe	Ni, Si	Cr	Mg, Ca
Enstatite[f]	Si, Ti	Mg	Fe, Al	–

[a] From Maxwell.[(106)]
[b] In pyroxenite.
[c] In anorthosite; two shots taken at ends of crystal.
[d] In metamorphic rock; three shots taken across crystal.
[e] In granophyre; two shots taken at edge and center of crystal.
[f] In polished section of meteorite.

for the identification of a number of ore and silicate minerals are given in Tables 2.3 and 2.4, respectively.[(106)]

It has been demonstrated that a laser microprobe is effective in analyzing meteorites.[(110)] The distributions of basic components and chondrules (for example) and the various mineral formations can be investigated *in situ*. It is also possible to obtain semiquantitative information characterizing different parts of a single chondrule. Nonhomogeneity in composition in terms of both the qualitative representation of the separate elements as well and the quantity of individual elements has been found to exist.[(110)] In this and most other mineralogical studies, the diameter of the sample area and therefore the resolution of the technique is 50–100 μm.

A neodymium glass laser beam analyzer[(113)] incorporating a device with a variable refractive index has been described for powdered minerals in which there is no fractional distillation of the sample. This device consists of a hollow cylinder placed between the laser rod and the reflecting mirror. Piezoelectric vibrations of the cylinder permit only the transmission of the important emission peaks. The unit is totally automatic.

2.5.5.2 Biological Samples

The technique of laser microanalysis is applicable to both *in vivo* and resected material. It is of interest to note that due to the short duration of a *Q*-switched laser pulse (typically 50 ns), no sensation is felt if the pulse is focused onto living tissue, even though vaporization occurs and a plasma is initiated. This factor plays an important role in the sampling of live tissue such as rat liver. The high temperature produced by the laser pulse causes immediate coagulation of the lesion which is so small that little or no functional impairment of the organ occurs. Cell destruction by this method has been investigated as a possible treatment for certain types of cancer.

The analysis of resected tissue has been amply demonstrated in the literature.[(114-122)] Histochemists have commented that until the advent of the laser microprobe, no practical method for *in situ* spectroscopic analysis existed, because selection from intact sections was not possible without considerable tedious manipulation.

Up to the present time this technique has been essentially a qualitative or at most a semiquantitative technique. One of the problems that is inherent in quantitative biological microanalysis is known to be sample heterogeneity. Other problems are the preparation of suitably homogeneous standards, and the difficulty with internal standardization. Semiquantitative determinations of the concentrations of calcium and magnesium in frozen dried sections of stomach[(118)] and the analysis of calcified tissue[(121)] have yielded estimates of concentrations with errors between 10 and 50%.

2.5.5.2a Absolute Detection Limits. Values equal to or less than 10^{-12} g have been reported for calcium and magnesium in stomach tissues,[(115)] and magnesium, phosphorus, silicon, aluminum, copper, and zinc in microsites of tooth, calculus, and bone.[(121)] Their redesigned laser microprobe system,[(123)] with an improved laser cavity for greater stability and more precise optical alignment of sample plume with the spectrometer, enabled Glick and Marich[(124)] to extend detection limits down to 10^{-12}-10^{-15} g for a sample size in the region 10^{-8}-10^{-10} g of tissue.

2.5.5.3 Briquetted Samples

In order to assess the quantitative capabilities and detection sensitivity of the laser microprobe technique, Whitehead and Heady[(125)] analyzed briquetted samples which were prepared from pure aluminum oxide in the form of a finely divided powder with a particle size of about 5 μm, doped with known traces of various metal oxides having particle sizes of 1–3 μm. Homogeneity was obtained by extremely thorough grinding and mixing of the powders in an ethanol slurry, followed by dry grinding and mixing. The specimens were then hand-packed into shallow trays and analyzed spectrographically.

The average variation in the intensities of the individual spectral lines for replicate analyses was 22%. The use of internal standardization resulted in an improvement in the precision capability as the average variation coefficient of the intensity ratios was 11%. This improvement shows that the major cause of

Table 2.5 Assessed Detection Limits in Briquetted Samples[a]

Spectral line	Å	Detection level (ppm) Single exposure	Double exposure	Triple exposure
B 1	2497.7	11	3	0.8
Be 11	3130.4	0.8	0.4	0.4
Co 1	3453.5	23	12	6
Cr 1	3578.7	11	3	3
Fe 11	2599.4	45	12	3
Mo 1	3132.6	21	11	11
Ni 1	3414.8	25	6	2
Pb 1	4057.8	30	15	8
Sn 1	2840.0	50	25	12
Sr 11	4077.7	10	5	5
Ti 22	3349.4	20	10	5
W 1	4008.8	400	200	200

[a] From Whitehead and Heady.(125)

the poor variation coefficients obtained for the relative intensities of the spectral lines is a variation in the sampling rather than sample heterogeneity. Yoshida and Murota(126) fused rock samples and geochemical standards with lithium tetraborate, then pressed small discs for laser microprobe analyses. With silica as the internal standard, linear calibration curves were obtained for the minor components.

2.5.5.3a Detection Limits. The detection limits of a number of elements present as trace oxide impurities in Al_2O_3 are given in Table 2.5. It is noted that in most cases the detection limit (expressed as parts per million) decreases for superimposed spectra from successive laser shots. This is due to a lack of spectral background in the single-shot (and in some cases the double-shot) spectra.

In an earlier study,(107) Snetšinger and Keil determined the detectability of a number of elements in a briquetted sample consisting of a spectrochemically pure graphite matrix into which was mixed a standard powder containing 42 elements, each at a concentration of 1000 ppm. Unlike the former investigation, a pressure of 40,000 psi was used to compact the sample *in vacuo*. The authors calculated that approximately 20 μg of material was liberated by one laser shot. The results of this investigation are given in Table 2.6.

Table 2.6 Elements Detected[a] in 20 μg at 1000 ppm for Briquetted Sample[b]

Si, Al, Li, Na, Be, Mg, Ca, Sr, Ba, Ti, V, Cr, Mn, Fe, Co, Ni, Cu, Ag, Cd, Hg, Sn, Pb, Bi, Sb, Te, Zr, Cb, Mo, W, Ce, Th, U, B, P, Ga, Ge, In, Tl

[a] Elements not detected: Zn, As, Hf, and Ta.
[b] From Snetšinger and Keil.(107)

2.5.5.4 Glass and Ceramics

The laser analysis of glass and ceramics, specifically the analysis and identification of glass defects such as stones, cords, and knots, is a valuable supplement to petrographic studies.

For the identification of an inclusion, it is necessary to grind away the host glass between the inclusion and the surface, leaving the inclusion within about 10 μm of the surface. Final careful grinding and polishing is required to expose the inclusion. Laser pulses may also be used to drill into the glass, thus exposing the inclusion, which can then be sampled by subsequent pulses. The inclusion depth should not be more than about 20 μm in this case

Successful identifications of aluminous stones and knots, chrome spinel stones, and a stone resulting from a siliceous clay contact refractory, as well as nephelite worms, have been reported.[(127)] Inclusions which defy positive petrographic identification, such as very finely divided zirconia and also cassiterite as an impurity of a feldspar, are easily identified by laser analysis. The analysis of siliceous cords has, however, been reported to be unsuccessful.

In a study of the application of laser analysis to nonmetallic samples, Felske *et al.*[(99)] have reported the following:

1. Opaque samples with high boiling points (e.g., ceramic, silicon, etc.) behave in the same manner as metallic samples. Auxiliary excitation is not necessary, although it results in improved power of detection at the expense of poorer precision.

2. Transparent samples with high boiling points (e.g., glass) need a considerably higher power density on the sample surface for reliable vaporization of material.

3. Samples with low boiling points (transparent as well as opaque) can only be analyzed if auxiliary excitation is used, the reason being that in this case the laser pulse produces mainly small droplets and dust particles and only a few free atoms.

2.5.5.5 Metals

The most homogeneous samples available to accurately assess the capabilities of the laser sampling technique are in the form of certified (standard) alloys. The reproducibility of the method will, of course, vary with the quality of the laser used. Values reported in the literature generally lie within the ranges given in Table 2.7.

In addition to the analyses of metals, Vasil'eva *et al.*[(128)] reported on the chemical composition of the oxide phase layers in Si–Ti–Cr oxide coatings with the use of laser volatilization and x-ray emission microanalysis; 1×10^{-7} to 5×10^{-8} g of the coating material was volatilized with a single laser pulse from a crater depth of approximately 5 μm. This laser microanalyzer was especially

Table 2.7 Reproducibility of Laser Output, Crater Dimensions, and Spectral Intensities for Single-Shot Double-Step Analysis (%)

Laser energy	4–10
Crater diameter	2– 3
Crater depth	5–10
Spectral intensity	10–50
Intensity ratios	5–20

suitable for examining coating films with no crystal structure. See Table 2.8 for detection limits.

2.5.6 Dependence of Crater Dimensions on Sample Material

It has been well established that although the characteristics of a *Q*-switched mode interaction are less dependent on the physical properties of the sample

Table 2.8 Detection Limits for Metal Samples[a]

Element	Assessed detection limits[b] for 3 -μg sample (ppm)
Al	200
Be	12
Ti	280
V	830
Cr	60
Mg	40
Co	>700
Ni	210
Ag	10
Sn	480
Pb	>200
Bi	180
Sb	>500
Te	>400
W	>480
B	>32
P	>1100
Ga	80
Si	270
Fe	240

[a] From Snetšinger and Keil.(107)
[b] These limits may be lowered by superimposing spectra from more than one laser shot(125) and further by spatial resolution, until a spectral background is obtained at the wavelength of the spectral line used for the determination.

Table 2.9 Crater Dimensions for Various Samples[a]

Sample	Crater diameter (μm)	Crater depth (μm)
Aluminum	70	70
Lead–antimony alloy	70	100
Steel	65	60
Ceramic	60	50
Glass	60	30
Tooth	65	60

[a]Courtesy of F. Brech, Jarrell-Ash Company.

than those of the conventional-mode interaction, they are not independent, at least at the power levels which are normally used for analytical applications (10^9–10^{10} W cm^{-2}).

Comparative crater dimensions for various samples for a power density of 1.3×10^{10} W cm^{-2} are given in Table 2.9. Although the crater diameter remains fairly constant for the various samples, the depth varies considerably from low-melting-point low-cohesive-energy materials to those with high melting point and high cohesive energy.

Crater dimensions for a number of metallic samples for two laser power densities are given in Table 2.10. The actual values given in the table cannot be compared directly with those given in Table 2.9 because of differences in the lasers and focusing lenses used. However, Table 2.10 shows a far greater depen-

Table 2.10 Crater Dimensions and Mass of Vaporized Material for Various Samples and Power Densities[a]

Power density:	5.8×10^{10} W cm^{-2}			1.8×10^{10} W cm^{-2}		
Focal spot diameter:	33 μm			60 μm		
Sample material	Diameter (μm)	Depth (μm)	Vaporized material (μg)	Diameter (μm)	Depth (μm)	Vaporized material (μg)
Mo	160	14	1.43	192	20	2.85
Cu	160	28	2.5	1.92	21	2.67
Ag	160	32	3.35	187	21	2.94
Al	198	35	1.43	257	27	1.9
Steel	128	16	0.8	192	13	1.4
Mg	192	45	1.08	200	37	1.0
Cd	160	30	2.6	225	24	4.0
Sn	192	43	4.42	380	26	10.3
Pb	320	32	14.2	470	21	20.3

[a]From Felske *et al.*[(99)]

dence of the crater diameters on the sample material than is indicated in Table 2.9. A dependence of crater size on sample material was also found by Allemand using Q-switched laser pulses of 3.4 mJ.[57]

The reasons for the variation in crater dimensions are obvious. Each material is characterized by a number of thermal and other physical properties, such as its melting point, boiling point, specific heat, thermal conductivity, latent heat of vaporization, cohesive energy, and reflectivity. It is thus a combination of these properties as well as the laser pulse characteristics which finally determines the crater dimensions.

2.5.7 Matrix Effects

A comparison between the laser technique and the dc arc technique for the analysis of briquetted powder samples of different matrices showed that matrix effects were relatively less severe for the laser technique.[91,125] In another study,[129] matrix effects were found to be principally physical rather than chemical in nature. The size and shape of the laser-induced plasmas depended on the nature and concentration of the matrix. Thus the amount of emitting material in the plasma and the fraction of light entering the spectrograph vary with the composition of the sample.

Cerrai and Trucco[130] have shown that there are other factors which influence the analytical line intensities besides the concentration of the analytical element in metal or ceramic pellet samples. These factors may be summarized as internal metallurgical conditions, and include grain size, mechanical strain, and the degree of sintering in ceramic samples. Since similar matrix effects were not observed in an electron microprobe analysis, the authors concluded that the effects originated in the sample evaporation process. The amount of vaporized material depends on the heating efficiency or absorbed laser energy in the sample, which is dependent on metallurgical properties.

Matrix effects have also been reported in the laser spectral analysis of steels[131] and rare earth elements in several host materials.[132] However, Buravlev *et al.*[133] reported that the use of laser excitation in place of arc or spark excitation considerably reduced the third element effect in the analyses for Si, Mn, Cr, Ti, and Al in low-alloy steels but did not suppress this effect for the determination of Al and Ti in heat-resistant alloys. Similar findings were reported by Krevchikova.[134] In their careful study of matrix effects, Kirchheim *et al.*[135] used high-purity single-crystal metals and alloys with different orientations. Crystal orientation was found to be an important factor in density and reproducibility of characteristic lines. In general, the density of lines increased with the density of atoms within the crystal plane. In the quantitative analysis of a single crystal, Cu–Al alloys, the anisotropic effect was canceled by the use of a matrix line as the internal standard line.

2.5.8 Some Other Applications

In the field of air pollution, Uchida *et al.*[136] have demonstrated the use of the laser microprobe for the multielement analysis of airborne particulates collected on Mylar film with a cascade impactor. Particulates 4, 6, 8, and 10 μm in diameter were individually analyzed for the seven elements Mg, Al, Si, Ca, Cu, Fe, and Zn. Yamane[137] detected halogens and sulfur. The absorption of laser energy has been used to measure atmospheric dust concentrations.[138] Roth *et al.*[139] have developed an optical particle counter. The laser light scattering of single aerosol particles is recorded photoelectrically. Particle size is determined by pulse heights of the electronic signals referenced to a calibration curve.

The possibilities of atomic absorption spectrometry have been studied.[140] Laser atomization of samples were performed with different targets as graphite and atmospheres as argon.

Bingham[141] demonstrated the use of the laser for the analyses of trace elements in solid materials by mass spectrometry. High-powered CO_2, ruby, and NdYAG laser probes were used for ionization of the sample. The spectra were less complex than those produced by the spark source. Surface analyses were performed by defocusing the laser beam. Moenke and Moenke-Blankenburg[142] also used the laser with mass spectrometry.

With a pulsed CO_2 laser with an average power of 8 W pulsed at 50 Hz, Cali *et al.*[143] obtained significant evaporation rates from alumina and other highly refractive materials for the production of thin films. Vacuum was 10^{-5} Torr.

With a narrow-band dye laser, Tallant and Wright[144] developed a technique for selective laser excitation of charged compensated sights in erbium (3+)-doped calcium fluoride crystals. Sheveleva and Kropŏtkin[145] determined the concentration of aqueous solutions of various strong bases, strong acids, and salts by measuring the intensity of the laser radiation reflected from the solution. The dependence of the relative coefficient of reflection on concentration was distinct, especially for the sulfates.

The tunable dye laser provides high-intensity monochromatic light at the maximum absorbance wavelength for many solutions. Thus it is possible to increase the measurable concentration range of a solution and to measure lower concentration ranges.[146]

2.6 CONCLUSION

In less than two decades, the laser developed from no more than a theoretician's dream into a light source that has revolutionized science. The parallel development of the laser microprobe as a new and unique spectrochemical source has been facilitated both by the rapid technological advances in the field of laser research and a better understanding, through fundamental research, of the interlinked mechanisms associated with laser plasma production.

Trends are toward the development of laser microprobes having greater stability and the development of more sophisticated excitation, focus, and detection systems,[57,123,147,148] as well as methods to improve the excitation of the laser-produced vapor cloud. As examples of the latter, mention can be made of the use of an electrodeless high-frequency gas discharge for additional excitation[149] and the use of an external magnetic field with spark cross-excitation, resulting in improved sensitivity and reproducibility.[150]

Although it has been realized that the spectrochemical laser probe is not a "universal source," it has nevertheless found its place among the few micro-analytical sources on the stockpile of useful tools available to spectroanalysts, and will continue to be used in cases where analysis by any other means is difficult or impossible.

2.7 REFERENCES

1. T. H. Maiman, *Phys. Rev. Letters 4*, 564 (1960).
2. Anonymous, *Chem. Eng. News 40*, 52 (1962).
3. F. Brech and L. Cross, *Appl. Spectrosc. 16*, 59 (1962).
4. J. F. Ready, *J. Appl. Phys. 36*, 462 (1965).
5. P. I. Ulyakov, *Soviet Phys. JETP (English Transl.) 25*, 537 (1967).
6. G. M. Rubanova and A. P. Sokolov, *Soviet Phys.-Tech. Phys. (English Transl.) 12*, 1226 (1968).
7. F. A. Richards and D. Walsh, *Brit. J. Appl. Phys. (J. Phys. D) 2*, 663 (1969).
8. J. P. Babnel-Peyrissac, C. Fauquignon, and F. Floux, *Phys. Letters 30A*, 290 (1969).
9. G. A. Askar'yan and E. M. Moroz, *Soviet Phys. JETP 43*, 2319 (1962).
10. J. F. Ready, *Appl. Phys. Letters 3*, 11 (1963).
11. S. I. Andreyev, Yu. I. Dymshits, L. N. Kaporskii, and G. S. Musatova, *Soviet Phys.-Tech. Phys. (English Transl.) 13*, 657 (1968).
12. A. Felske, W. D. Hagenah, and K. Laqua, *Spectrochim. Acta 27B*, 295 (1972).
13. J. M. Baldwin, *Appl. Spectrosc. 24*, 429 (1970).
14. R. G. Meyerand and A. F. Haught, *Phys. Rev. Letters 11*, 401 (1963).
15. W. I. Linlor, *Phys. Rev. Letters 12*, 383 (1964).
16. A. W. Ehler, *J. Appl. Phys. 37*, 4962 (1966).
17. E. H. Piepmeier and H. V. Malmstadt, *Anal. Chem. 41*, 700 (1969).
18. I. V. Nemichinov and S. P. Popov, *Soviet Phys. JEPT Letters (English Transl.) 11*, 312 (1970).
19. S. I. Andreyev, I. V. Vershikovskiy, and Yu. I. Dymshits, *Soviet Phys.-Tech. Phys. (English Transl.) 15*, 1109 (1971).
20. V. V. Panteleev and A. A. Yankovskii, *Zh. Prikl. Spektrosk. 3*, 350 (1965).
21. A. M. Bonch-Bruevich, Ya. Y. Imas, G. S. Romanov, and M. N. Libenson, *Soviet Phys.-Tech. Phys. (English Transl.) 13*, 640 (1968).
22. N. G. Vasov, V. A. Boiko, O. N. Krokhin, O. G. Semenov, and G. V. Skilizkov, *Soviet Phys.-Tech. Phys. (English Transl.) 13*, 1581 (1969).
23. M. K. Chun, *J. Appl. Phys. 41*, 2 (1970).
24. T. J. Harris, *IBM J. Res. Develop. 7*, 342 (1963).

25. R. H. Scott and A. Strasheim, *Spectrochim. Acta 25B*, 311 (1970).
26. F. Neuman, *Appl. Phys. Letters 4*, 167 (1964).
27. D. W. Gregg and S. J. Thomas, *J. Appl. Phys. 37*, 2787 (1966).
28. A. M. Bonch-Bruevich and Ya. A. Imas, *Soviet Phys.-Tech. Phys.* (*English Transl.*) *12*, 1407 (1968).
29. L. I. Grechikhin and L. Ya. Min'ko, *Soviet Phys. JETP* (*English Transl.*) *12*, 846 (1967).
30. N. G. Basov, O. N. Krokhin, and G. V. Sklizkov, *ZhETF Pis'ma 6*, 683 (1967).
31. N. G. Basov, V. A. Gribkov, O. N. Krokhin, and G. V. Sklizkov, *Soviet Phys. JETP* (*English Transl.*) *27*, 575 (1968).
32. D. C. Emmony and J. Irving, *Brit. J. Appl. Phys.* (*J. Phys. D*) *2*, 1186 (1969).
33. V. A. Batonov, F. V. Bunkin, A. M. Prokhorov, and V. B. Fedrorv, *Soviet Phys. JETP Letters* (*English Transl.*) *11*, 69 (1970).
34. N. G. Basov, O. N. Krokhin, and G. V. Skilzkov, *Trudy Fiz. Inst. Akad. Nauk SSSR 52*, 171 (1970).
35. D. Lichtman and J. F. Ready, *Phys. Rev. Letters 10*, 342 (1963).
36. C. M. Verber and A. H. Adelman, *Appl. Phys. Letters 2*, 220 (1963).
37. J. J. Muray, *Bull. Amer. Phys. Soc. 8*, 77 (1963).
38. E. Honig, *Appl. Phys. Letters 3*, 8 (1963).
39. F. Giori, L. A. Mackenzie, and E. J. McKinney, *Appl. Phys. Letters 3*, 25 (1963).
40. D. Lichtman and J. F. Ready, *Appl. Phys. Letters 3*, 115 (1963).
41. W. I. Linlor, *Appl. Phys. Letters 3*, 210 (1963).
42. J. K. Cobb and J. J. Muray, *Brit. J. Appl. Phys. 16*, 271 (1965).
43. L. G. Pittaway, J. Smith, E. D. Fletcher, and B. W. Nicholls, *Brit. J. Appl. Phys.* (*J. Phys. D.*) *1*, 711 (1968).
44. N. R. Isenor, *Appl. Phys. Letters 4*, 152 (1964).
45. N. R. Isenor, *Can. J. Phys. 42*, 1413 (1964).
46. E. Bernal, J. F. Ready, and L. P. Levine, *Phys. Letters 19*, 645 (1966).
47. E. Bernal, J. F. Ready, and L. P. Levine, *IEEE J. Quant. Electron. QE-2*, 480 (1966).
48. P. Langer, G. Tonon, F. Floux, and A. Ducanze, *IEEE J. Quant. Electron. QE-2*, 493 (1966).
49. Yu A. Bykovskiy, N. N. Degtyarenko, V. I. Dymovich, V. F. Elesin, Yu P. Rozyrev, B. I. Nikolayev, S. V. Ryzhikh, and S. M. Sil'nov, *Soviet Phys.-Tech. Phys.* (*English Transl.*) *14*, 1269 (1971).
50. Yu A. Bykovskiy, N. N. Degtyarenko, V. F. Elesin, Yu P. Kozyrev, and S. M. Sil'nov, *Soviet Phys.-Tech. Phys.* (*English Transl.*) *15*, 2020 (1971).
51. F. J. Allen, *J. Appl. Phys. 41*, 3048 (1970).
52. N. G. Basov, V. A. Boiko, U. A. Drozhbin, S. M. Zakharov, O. N. Krokhin, G. V. Sklizkov, and V. A. Yakovlev, *Soviet Phys. "Doklady"* (*English Transl.*) *15*, 576 (1970).
53. V. V. Apollonev, Yu A. Bykovskiy, N. N. Degtyrenko, V. F. Elesin, Yu P. Kosyrev, and S. M. Sil'nov, *Soviet Phys. JETP Letters* (*English Transl.*) *11*, 252 (1970).
54. T. S. Green, *Phys. Letters 32A*, 530 (1970).
55. E. Archbold, D. W. Harper, and T. P. Hughes, *Brit. J. Appl. Phys. 15*, 1321 (1964).
56. W. Borgershausen and R. Vesper, *Spectrochim. Acta 24B*, 103 (1969).
57. C. D. Allemand, *Spectrochim. Acta 27B*, 185 (1972).
58. W. J. Treytl, J. B. Orenberg, K. W. Marich, and D. Glick, *Appl. Spectrosc. 25*, 376 (1971).
59. S. D. Rasberry, B. F. Scribner, and M. Margoshes, *Appl. Opt. 7*, 81 (1967).
60. F. P. J. Valero, D. Goorvitch, B. S. Fraenkl, and B. Ragent, *J. Opt. Soc. Amer. 59*, 1380 (1969).
61. C. R. Stumpfel, J. L. Robitaille, and J. J. Kunze, *J. Appl. Phys. 43*, 902 (1972).

62. P. Dhez, P. Jaegel, S. Leach, and M. Velghe, *J. Appl. Phys. 40*, 2545 (1969).
63. A. W. Ehler, *Appl. Phys. 8*, 89 (1966).
64. A. Carillon, P. Jaelge, and P. Dhez, *Phys. Rev. Letters 25*, 140 (1970).
65. E. H. Piepmeier and D. E. Osten, *Appl. Spectrosc. 25*, 642 (1971).
66. W. J. Treytl, L. W. Marich, J. B. Orenberg, P. W. Carr, D. Craig-Miller, and D. Glick, *Anal. Chem. 43*, 1452 (1971).
67. G. Demitrov and A. Petrakiev, *Fiz. Fak 1973*, 531–552; *Chem. Abstr. 83*, 90299 (1975).
68. M. Kubota and R. Ishida, *Bunko Kenkyu 24*, 89 (1975).
69. M. Kubota and R. Ishida, *Bunko Kenkyu 23*, 74 (1974).
70. J. Mohr, *Exptl. Tech. Physik 22*, 327 (1974).
71. C. D. David, Jr., and H. Weichel, *J. Appl. Phys. 40*, 3674 (1969).
72. W. Bogershausen and K. Honle, *Spectrochim. Acta 24B*, 71 (1969).
73. N. G. Basov, O. N. Krokhin, and G. V. Sklizkov, *IEEE J. Quant. Mech. QE-4*, 988 (1968).
74. N. G. Basov, V. A. Boiko, V. A. Gribkov, S. M. Zakharov, O. N. Krokhin, and G. V. Slizkov, *Soviet Phys. JETP Letters (English Transl.) 9*, 315 (1969).
75. J. Dawson, P. Kaw, and B. Green, *Phys. Fluids 12*, 875 (1969).
76. H. Puell, H. J. Neusser, and W. Kaiser, *Z. Naturforsch. 25A*, 1815 (1970).
77. W. Seka, C. Breton, J. L. Schwob, and C. Minier, *Plasma Phys. 12*, 73 (1970).
78. B. Wolterbeek Muller and T. S. Green, *Plasma Phys. 13*, 73 (1971).
79. A. J. Alcock, P. P. Pashinin, and S. A. Ramsden, *Phys. Rev. Letters 17*, 528 (1966).
80. M. J. Bernstein and G. G. Comisar, *J. Appl. Phys. 41*, 729 (1970).
81. F. Floux, D. Cognard, J. L. Bobin, F. Delobeau, and C. Fauquingnon, *Compt. Rend. 269B*, 697 (1969).
82. F. Floux, D. Cognard, L. G. Denoeud, G. Pair, D. Parisot, J. L. Bobin, F. Delobeau, and C. Fauquingnon, *Phys. Rev. 1A*, 821 (1970).
83. J. de Metz, *Appl. Opt. 10*, 1609 (1971).
84. H. Moenke and L. Moenke-Blandenburg, *Laser Micro-Spectrochemical Analysis*, Adam Hilger Ltd., London (1973).
85. V. P. Veiko, Ya A. Imas, A. N. Kokova, and M. N. Libenson, *Soviet Phys.-Tech. Phys. (English Transl.) 12*, 1410 (1968).
86. A. V. Karyakin, M. V. Akhamanova, and V. A. Kaigorodov, *J. Anal. Chem. USSR 20*, 133 (1965).
87. V. V. Panteleev and A. A. Yankovskii, *Zh. Prikl. Spektrosk. 3*, 96 (1965).
88. H. Klocke, *Spectrochim. Acta 24B*, 263 (1969).
89. A. V. Karyakin and V. A. Kaigorodov, *J. Anal. Chem. 22*, 444 (1967).
90. J. M. Dawson and C. R. Oberman, *Phys. Fluids 5*, 517 (1962).
91. J. M. Dawson, *Phys. Fluids 7*, 281 (1964).
92. L. Spitzer, *Physics of Fully Ionised Gases*, John Wiley & Sons, Inc. (Interscience Division), New York (1956).
93. I. Harding-Brown, E. S. Beatrice, and D. Glick, *Federation Proc. 26*, 780 (1967).
94. N. A. Peppers, E. J. Scribner, L. E. Alterton, and R. C. Honey, *Anal. Chem. 40*, 1178 (1968).
95. E. F. Runge, R. W. Minck, and F. R. Bryan, *Spectrochim. Acta 20*, 733 (1964).
96. E. F. Runge, S. Bonfiglio, and F. R. Bryan, *Spectrochim. Acta 22*, 1678 (1966).
97. K. I. Taganov and L. M. Fainberg, *Zh. Prikl. Spektrosk. 20*, 571 (1974).
98. A. Felske, W. D. Hagenah, and K. Laqua, *Z. Anal. Chem. 216*, 50 (1966).
99. A. Felske, W. D. Hagenah, and K. Laqua, *Spectrochim. Acta 27B*, 1 (1972).
100. R. H. Scott and A. Strasheim, *Spectrochim. Acta 26B*, 707 (1971).
101. W. W. Schroeder, J. J. van Niekerk, L. Dicks, A. Strasheim, and H. vd Piepen, *Spectrochim. Acta 26B*, 331 (1970).

102. E. S. Beatrice, I. Harding-Brown, and D. Glick, *Appl. Spectrosc. 23*, 257 (1969).
103. T. Yamane and S. Matsushita, *Spectrochim Acta 27B*, 27 (1972).
104. W. Quillfeldt, *Exptl. Tech. Physik 17*, 415 (1969).
105. H. N. Barton and J. Benallo, *Appl. Spectrosc. 24*, 614 (1970).
106. J. A. Maxwell, *Can. Mineralogist 7*, 727 (1963).
107. K. G. Snetsinger and K. Keil, *Am. Mineralogist 52*, 1842 (1967).
108. W. Blackburn, Y. Pelletier, and W. Deunen, *Appl. Spectrosc. 22*, 278 (1968).
109. L. Georgiere and A. Petrakiev, *Bol. Geol. Mineral. 80*, 491 (1969).
110. N. Nikolov, A. P. Petkov, G. Dimitrov, and D. Dimov, *Bol. Geol. Mineral. 80*(4), 485 (1969).
111. N. Eremin and V. Kel'kh, *Dokl. Akad. Nauk SSSR 191*, 166 (1970).
112. H. Blandenburg and L. Moenke, *Krist. Tech. 1*, 351 (1966).
113. J. P. Meric, Ger. Offen. 2,336,635 (Jan. 31, 1974).
114. R. C. Rosan, M. K. Healy, and W. F. McNary, Jr., *Science 142*, 236 (1963).
115. R. C. Rosan, D. Glick, and F. Brech, *Federation Proc. 24*, 542 (1965).
116. D. Glick and R. C. Rosan, *Microchem. J 10*, 393 (1966).
117. D. Glick, *Clin. Chem. 15*, 753 (1969).
118. M. Kozik, J. Warchol, and B. Archimowicz, *Histochemie 26*, 212 (1971).
119. D. Glick, K. W. Marich, P. W. Carr, and E. S. Beatrice, *Ann. N.Y. Acad. Sci. 168*, 507 (1970).
120. P. Haggerty, *Peridontol-Peridontics 40*, 330 (1969).
121. D. B. Sherman, M. P. Ruben, H. M. Goldman, and F. Breck, *Ann. N.Y. Acad. Sci. 122*, 767 (1965).
122. S. F. Brokeshoulder and F. R. Robinson, *Appl. Spectrosc. 22*, 758 (1968).
123. K. W. Marich, W. J. Treytl, J. G. Hawley, N. A. Peppers, R. E. Meyers, and D. Glick, *J. Phys. E Sci. Instr. 7*, 830 (1974).
124. D. Glick and K. W. Marich, *Clin. Chem. 21*, 1238 (1975).
125. A. B. Whitehead and H. H. Heady, *Appl. Spectrosc. 22*, 7 (1968).
126. M. Yoshida and M. Murota, *J. Geosci. 19*, 81 (1975).
127. J. R. Ryan, C. B. Clark, and E. Ruth, *Bull. Amer. Ceram. Soc. 45*, 260 (1966).
128. V. V. Vasil'eva, V. S. Terent'eva, J. Ugoste, and M. V. Niketina, *Neorg. Organosilikatnye Pokrytiya 6*, 63 (1975); *Chem. Abstr. 85*, 71634a (1976).
129. K. W. Marich, P. W. Carr, W. J. Treytl, and D. Glick, *Anal. Chem. 42*, 1775 (1970).
130. E. Cerrai and R. Trucco, *Energia Nucl. (Milan) 15*, 581 (1968).
131. Yu. M. Buravlev, I. I. Morokhovskaya, V. N. Muraviev, and D. A. Nadezh, *Zh. Anal. Khim. 27*, 2337 (1972).
132. T. Ishizuka, *Anal. Chem. 45*, 538 (1973).
133. Y. M. Buravlev, B. P. Nadezhda, and L. N. Babanskaya, *Zavodsk Lab. 40*, 165 (1974).
134. E. P. Krevchikŏva, *Zh. Anal. Khim. 30*, 187 (1974).
135. R. Kirčhheim, U. Nagormy, L. Maier, and G. Tölg, *Anal. Chem. 48*, 1505 (1976).
136. H. Uchida, F. Adachi, O. More, and R. Negishi, *Bunseki Kagaku 24*, 325 (1975).
137. T. Yamane, *Bunko Kenkyu 22*, 321 (1973).
138. S. A. Schleusener, J. D. Lindberg, and K. O. White, *Appl. Opt. 14*, 2564 (1975).
139. C. Roth, J. Gebhart, and G. Heigwer, *J. Colloid Interface Sci. 54*, 265 (1976).
140. E. K. Vul'son, A. V. Karyakin, and A. F. Yanushkevich, *Zh. Prikl. Spektrosk. 24*, 13 (1976).
141. R. A. Bingham, *Proc. Anal. Div. Chem. Soc. 13*, 93 (1976).
142. H. Moenke and L. Moenke-Blankenburg, *Mikrochim. Acta Suppl. 5*, 377 (1973).
143. C. Cali, V. Daneau, A. Oriole, and S. Riva-Sanseverine, *Appl. Opt. 15*, 1327 (1976).
144. D. R. Tallant and J. C. Wright, *J. Chem. Phys. 63*, 2074 (1975).
145. T. Y. Sheveleva and M. A. Kropotkin, *Kvantovaya Elektron. (Moscow) 2*, 198 (1975).
146. W. E. Baucum, *Nucl. Sci. Abstr. 31*, 28916 (1975).

147. W. J. Treytl, W. K. Marich, and D. Glick, *Anal. Chem. 47*, 1275 (1975).
148. L. Moenke-Blankenburg, H. Moenke, J. Mohr, W. Quillfeldt, K. Wiegand, W. Grassme, and W. Schrön, *Spectrochim. Acta 30*, 227 (1975).
149. W. D. Hagenah, K. Laqua, and F. Leis, *XVII Colloq. Spectrosc. Intern. Lecture Preprints 2*, 491 (1973).
150. A. Petrakiev and L. Georgieva, *XVII Colloq. Spectrosc. Intern. Lecture Preprints 2*, 508 (1973).

Electrode Material and Design for Emission Spectroscopy

3

J. W. Mellichamp

3.1 INTRODUCTION

An important consideration in atomic excitation by an electrical discharge is the electrode material and design. Electrodes are an integral part of the excitation conditions and are essential for the effectiveness and utilization of emission spectroscopy. With excitation by the arc (low voltage-high current), the electrodes, anode and cathode, are the conductive medium and terminals as well as the support for nonconductive sample material. The electrode material is consumed with the sample material in varying degrees according to the procedure. The entry rate of the sample into the arc column is determined by the electrode material and design along with arc current, thus affecting both sensitivity and reproducibility of analytical results.

In excitation by a spark discharge (high voltage-low current) the material to be analyzed is usually a metallic conductor which can serve as one or both electrodes. Another material often is used as the counter electrode. The design or shape of the electrodes, whether as self-electrodes or counter electrodes, plays a major part in determining discharge stability as well as repeatability of results. The large choice of excitation conditions, besides the arc and the spark, in combination with different electrode materials in various shapes and forms has given flexibility to spectroscopy and makes it one of the most effective methods for inorganic analysis.

J. W. Mellichamp • U.S. Army Electronics Command (DRSEL-TL-EC), Fort Monmouth, New Jersey 07703

3.2 BACKGROUND AND HISTORY

An electrical discharge in the form of a spark as a natural phenomenon has always been known to man and is observed as lightning or other static discharges. Conditions exist in nature whereby an electrical potential can be created to cause an ionized path in air, thus the spark. However, natural conditions do not exist for the arc. It was not until the development of the galvanic cell that man first produced an arc. Long before the galvanic cell sparks had been produced in the laboratory using devices such as the Leyden jar. As early as 1800 Sir Humphry Davy employed carbon in the electrical arc.(1) His electrodes were charcoal points or rods cut from carbonized wood. He employed the battery of the Royal Institution, constructed after Wollaston's design, consisting of 2000 pairs of plates, from which he obtained an arc of about 4 in. Later his electrodes were made from a mixture of powdered wood charcoal and a thick tar molded under pressure and baked. It was the great Michael Faraday (1791-1867) who demonstrated that electricity from static and galvanic sources was identical.(2) He also introduced the terms "electrode," "cathode," "anode," and others now in common use. Other scientists of his period began to recognize the significance of the electrical discharge for atomic excitation.(3) Talbot, Wheatstone, Ångström, Alter, Willigen, and others applied the spark for atomic excitation. Around 1849, Foucault used an arc between charcoal electrodes to compare its emission with that of the sun. The practicability of the carbon arc for illumination, originally for lighthouses, was realized in 1857 with the construction of a suitable generator by Holms.(4) Thus the carbon arc became available for spectroscopy.

3.3 PREPARATION AND MANUFACTURING OF ELECTRODE MATERIAL

Historically, carbon in one of its several forms is the preferred material as electrodes in spectroscopy. Discovery of carbon dates from prehistoric times and is now utilized for many purposes. Until comparatively recently the spectroscopist adapted available forms to his specialized needs. The development of carbon for spectroscopy parallels that of demands for many purposes, of which spectroscopy represents but a small fraction. The spectroscopist is indeed fortunate to have the high-quality material now in use because his demands alone would not have justified the efforts for its perfection. Spectroscopy has benefited from the development of graphite for nuclear reactors and for many other modern-day technologies.

The preference for carbon as electrode material is determined by its many properties that are matched by no other material. Some of these properties are the following:

1. Its electrical conductivity assists the production of the arc or spark, and makes it possible for nonconductive materials to be analyzed.

2. It is nonfusible and sublimes in the arc, in contrast to the effect of metal electrodes.

3. It is readily machinable to close tolerances and to the variety of electrode shapes required for analysis.

4. It can be purified to a high order.

5. It has a simple emission spectrum. In air, cyanogen bands are emitted that can be eliminated by operating in an atmosphere free from nitrogen.

6. It has an excitation potential of a sufficiently high order (10 eV) to ensure that the low-potential elements are not obscured.

7. Its high sublimation point (>3500°C) makes most refractory analyses possible.

8. It is resistant to most corrosive agents, nontoxic either cold or in the arc, absorbs little moisture, and is highly resistant to thermal shock.

A study of the properties of pure carbon in all its various forms is beyond the scope of this chapter. Some concept of the diversity of this element and the complexity in defining its physical properties can be found in the literature.[5] Forms of carbon and graphite have been developed that meet the desired properties required for spectroscopy. These forms are manufactured throughout the world, usually by large companies as an adjunct to their major production of carbon for other industrial applications. The manufacturing procedures differ among companies; however, the fundamentals are essentially the same. It is these variations in techniques that give different characteristics to the product of each company.

The desired properties of the finished electrode are dependent on the selection of raw materials and the steps taken in the processing. These steps are controlled by the manufacturing techniques and significantly affect final properties. Carbon or graphite can be produced from almost any organic material that gives a high carbon residue when thermally decomposed. Petroleum coke, a refinery by-product, is by far the major source of graphite today, because it is in large supply, relatively free from impurities, and reasonably inexpensive.[6] Other materials, such as coal, lampblack, and pitch coke, are employed in special circumstances.

The production of spectrographic electrode material requires close attention to detail and strict control at all stages. The raw materials are inspected and tested to ensure that their characteristics are uniform. These materials, when in lump form, are crushed and ground to graded powders, mixed in definite proportions, and then blended with tar as an agglomerant for a number of hours. This long mixing is carried out at a consistent temperature throughout the whole operation. The paste thus produced is compressed into the form of cylinders and is subsequently extruded in powerful hydraulic presses of several tons capacity into rods whose diameter is somewhat larger than the dimensions of the desired finished products. While still soft the rods are cut into convenient lengths and loaded into containers to be baked at approximately 1200°C to complete the carbonization of the raw materials. The stable carbonized product is then inspected for uniformity, straightness, roundness, and other possible defects. After special packing, the rods are graphitized by a process based on that patented by Acheson in 1896 whereby the carbon is heated by electrical resistance in association with one or more oxides to a temperature sufficiently high to cause chemical reaction between the constituents, and with further heating, about 3000°C, the carbon is separated out in the free state.[1] This process imparts to the carbons the required mechanical, electrical, and high refractory properties. Great care is exercised at all stages in this operation, which may take several weeks. The rods are removed from the kilns, examined for mechanical and electrical properties, then cut to size and trued.

The purity of the electrodes is of prime concern with respect to low-level and trace analysis, and in the manufacturing procedure all steps are taken to ensure this purity. The normal graphitizing process eliminates the lower-boiling-point impurities such as Si, Ca, Al, and Mg, but to remove the more tenacious refractory metals which form high-temperature carbides requires further purification with halogen gases subjected to temperatures between 2400 and 3000°C. The impurities are volatilized as the halides. The most tenacious impurity, boron, requires fluorine. The stock rods are cut and machined to specific dimensions on precision tools to shapes desired by the spectroscopist. The "performed" electrodes are then subjected to further purification. The electrodes are selectively analyzed to determine the remaining impurities. The very sensitive cathode-layer spectrographic method is used for the detection and evaluation of more than 15 possible elements, among which are Al, Fe, Mg, Si, Ag, B, Ca, Cu, Mn, Ti, and V. In the best-quality electrodes, not more than 2 ppm of any one element or greater than a total of 6-ppm spot impurities for all elements is permitted. Ash content is kept below 1 ppm. The preformed electrodes are packed in clean rooms and shipped to the customer. More information about purification and testing of quality is given by Mitteldorf[7] and by Weinard.[8] Since then there has been steady improvement in the manufacturing procedures.

In their studies of pollutants, Robinson and Hindman[9] found the metallic impurities As, Cd, Cu, Fe, Hg, Pb, Se, and Zn on/in spectroscopic carbon electrodes. The zinc and copper contamination was high. However, no quantitative data were reported. Of greater significance was the fact that the level of contamination showed wide variation from batch to batch. A unique method to remove metallic surface contaminants for ultra-trace analyses had been reported by Mannkoff and Fesser.[10] The electrode to be purified is the anode and is in close contact with the cathode during the passage of a dc current. Traces of Mn, Mg, Si, Fe, Ti, Pb, Cu, and Ca initially found on the anode was not found after purification.

3.4 ELECTRODE DESIGN

All analysts will agree that the geometry of the electrode is significant in the attainment of analytical results. The consistency of maintaining electrode dimensions as well as the quality of the electrode material are important to the reproducibility of results. Thanks to the suppliers of carbon electrodes, most laboratories are not faced with these problems because preformed electrodes can be purchased that are not only consistent in quality and dimensions but are also shaped to meet the requirements of the user. It remains for the spectroscopist to select the proper carbon grade in the correct shape to obtain optimum analytical results. Most books on analytical spectroscopy discuss electrode designs for specific analytical problems. For example, Ahrens and Taylor(11) discussed a wide variety of electrode shapes for dc arc analysis of geological and related materials. Harvey(12) has designed a series of electrodes for semiquantitative analysis that is based on sample weights. Most laboratories select the electrodes that best suit their analytical problems and over a period of time compile data that give both the reproducibility of results and the limitations of the electrodes chosen. Once the procedures are established, laboratories are reluctant to make changes unless they are reasonably certain that they will be beneficial.

The number of electrode designs is large: hundreds if not thousands. This was because each spectroscopist tended to develop his special design for each spectrochemical procedure. A study of the literature shows that from this large variety there has evolved a relatively few that serve most purposes. This is largely due to extensive evaluation by the American Society for Testing and Materials,(13) which has a recommended practice for the designation of shapes and sizes of graphite electrodes. The five classes of graphite electrodes covered are the following:

1. Class C: electrodes used opposite the sample or opposite the electrodes that hold the sample.
2. Class S: electrodes used with solution residues (dried in the cup) or particulate samples. The thinner-wall cups allow quicker burnoff of carbon and faster consumption of the sample. Undercut pins cause greater heating of the sample.
3. Class P: electrodes used with samples that require a center post, which tends to stabilize the arc.
4. Class PC: electrodes used with solution samples only.
5. Class D: electrodes of disc shape.

Each of the classes above represents a number of designs. Examples, given in Fig. 3.1, are as follows: No. 1 is the C-5A, which can be used either as the cathode for dc arc excitation or as the counterelectrode in spark excitation; No. 2, the S-13, is a design by Harvey(12) for use in his "burned-to-completion" semiquantitative procedure; No. 3, the P-2, has the center post which is an aid in arc

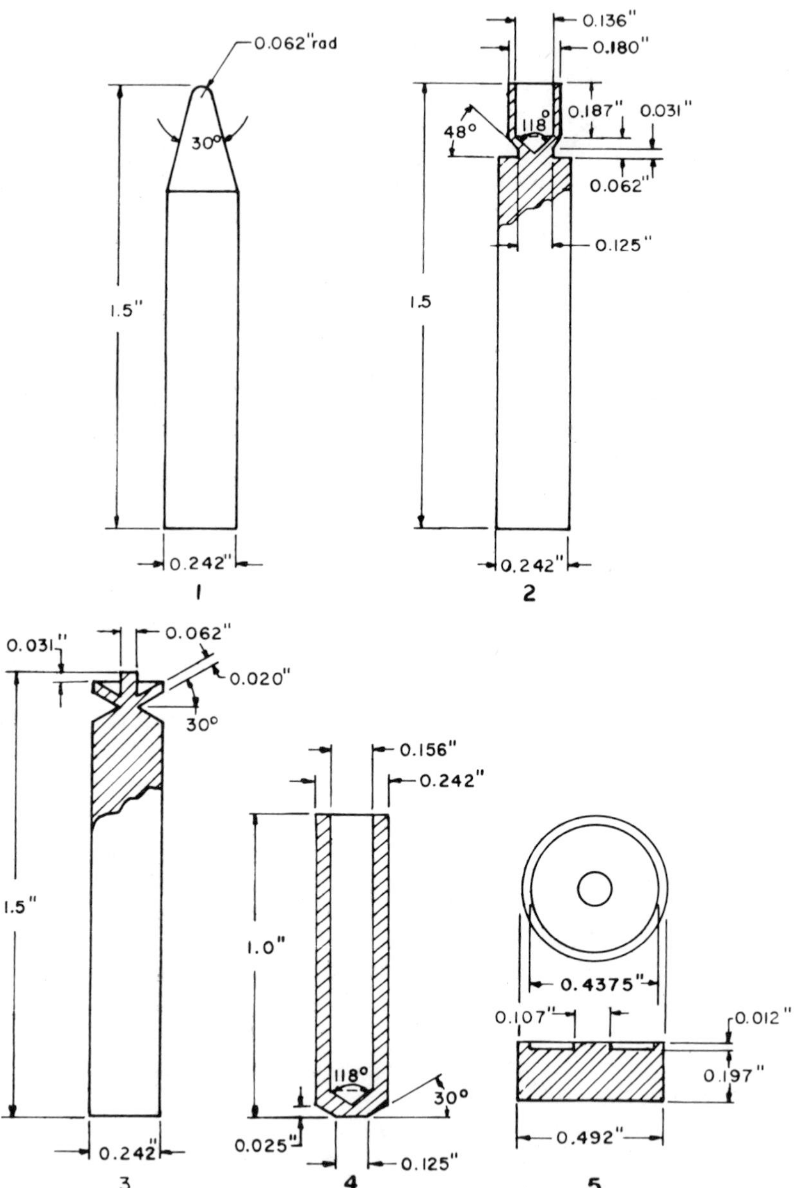

Fig. 3.1 Examples of the five classes of electrodes as designated by the American Society for Testing and Materials: (1) C-5A counter electrode for spark or as a cathode for dc arc analysis; (2) S-13 anode for dc arc for burned-to-completion semiquantitative method; (3) P-2 electrode with center post for arc stabilization; (4) PC-1 electrode for solution analysis by the porous-cup method; (5) D-3 electrode for solution analysis by rotating-platform method. (From *Methods for Emission Spectrochemical Analysis.*[13])

stabilization; No. 4 is the PC-1, which is used in solution analyses by the porous cup method; and No. 5, the D-3, is the rotating platform or "platrode," which is widely used for solution and oil analyses. The complete list of basic designs is given in *Methods for Emission Spectrochemical Analysis*, the ASTM publication.[13] In addition to the basic designs recommended by the ASTM additional variations and designs are supplied as preformed electrodes by the manufacturers. Special needs can be fulfilled through the use of high-purity rods from which the desired design can be shaped in the laboratory. The ASTM[13] also has a table with permissible tolerances for the various dimensions of a given electrode. Variation in precision resulting from random dimensional variations[7] indicate that these tolerance are important. In spite of the important work by Weinard[8] on random dimensional variation and the extensive form and classification evaluation work by the ASTM,[13] reports continue to be published on electrode forms. For example, a cup dimension most suitable for biological samples has been reported[14] but with no data to justify the design. Voronov[15] studied the effect of electrode shape and diameter on reproducibility using both the high-voltage and the low-voltage spark. A one- to three-fold improvement in reproducibility is claimed, but he found that the optimum shape varied for different analytical lines. For powdered samples, Tesarik[16] controlled cup temperature by drilling a hole through the electrode 1 mm below the crater. It is claimed that volatilization of sample from a 6-mm electrode with a 4-mm hole is as good as other electrodes but is stronger. Sample volatilization can be changed by filling the hole with a brass plug or by drilling a smaller hole.

Grikit and Galushko,[17] Peter,[18] and Raikhbaum *et al.*[19] have reported on electrode designs for fractional vaporization of volatile elements. These studies and designs are but slight modifications of designs tested and reported in the literature two or three decades earlier. Electrode design should combine inherent simplicity with versatility, speed of analyses, and acceptable sensitivity. Papp *et al.*[20] described a new counter-electrode shape with four peaks that decreases the interelement effect in the analyses of steels. For example, almost identical analytical curves for silicon are obtained when tested with either high- or low-alloy steels.

3.5 PHYSICAL PROPERTIES OF GRAPHITE AND CARBON

Manufactured carbon or graphite should not be viewed as a single material but as a family of materials with each member essentially pure carbon but varying in such characteristics as grain size, orientation of crystallites, degree of graphitization, apparent density, pore size, and porosity.[6] While the terms "carbon" and "graphite" may be somewhat confusing when used interchangeably, this need not be the case. In general, "carbon" refers to material with small crystallites and low orientation, called, questionably, amorphous; "graphite" refers to

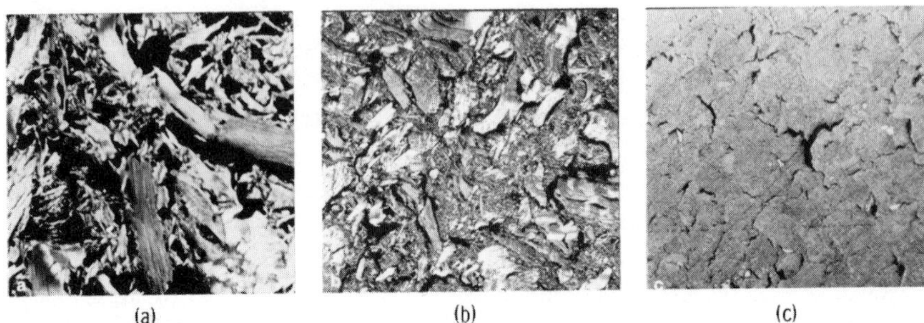

Fig. 3.2 Comparison of three grades of spectroscopic electrodes showing differences in porosity and particle size: (a) graphite grade AGKSP–relatively good electrical and thermal conductivity; (b) graphite grade SPK–intermediate electrical and thermal conductivity; (c) carbon grade L113SP–poor electrical and thermal conductivity. (From Coulter.[6])

material with a highly ordered structure. Figure 3.2 shows photomicrographs of three grades of spectroscopic carbons. Differences in texture such as porosity and particle size can be seen.

An indication of the degree of graphitization of carbon can be obtained from x-ray diffraction measurements of the average linear diameter of crystallites, L_a, and the interlayer spacing of the graphitic planes, d spacing. Table 3.1 is a comparison of grades of commercial spectroscopic rods with amorphous carbon and calculated values for pure graphite. The L_a values for the rods are obtained from their d spacing by comparing with d spacing and L_a measurements made by

Table 3.1 Comparison of Carbon and Graphite Rods with Carbon Black and Pure Graphite as an Indication of the Degree of Graphitization[a]

	d Spacing of graphitic planes (Å)	Average linear diameter of crystallites (L_a) (Å)	Surface area ($m^2\ g^{-1}$)
Carbon black G. L. Cabot Co.	3.59	33	200
Carbon rod Union Carbide Corp. carbon grade L113SP	3.41	200	4.6
Graphite rod Union Carbide Corp. graphite grade AGKSP	3.37	600	1.6
Pure graphite calculated values	3.354	1000	1.0

[a]From Castle.[21]

others. A decrease in particle surface areas is another indication of an increase in graphitization, which is also shown in Table 3.1.

Although interested in the many properties of the electrode material, the spectroscopist's major concern is performance in spectrochemical analysis. The vaporization of the electrode material into the arc is based on properties such as particle size and shape and crystallization which are controlled by manufacturing techniques. Properties of greatest interest are thermal conductivity, electrical resistivity, and ease of machining into the desired shapes.

Thermal conductivity through polycrystalline carbons is mostly along the graphitic planes of the crystallites with lower conductivity at the intercrystalline boundaries. Graphite rods with larger crystallites have higher thermal conductivity at lower temperatures than carbon rods with smaller crystallites and greater intercrystalline resistance. Figure 3.3, taken from Castle,[21] shows the temperature dependency of thermal conductivity of polycrystalline carbons as a function of crystallite size and assuming intercrystalline resistance at high temperatures. Polycrystalline carbons with crystallites less than 100 Å in linear diameter and crystallites greater than 100 Å are compared with completely graphitized carbon and baked carbon with little or no graphitization.

At room temperature the electrical resistivity of carbon is much greater than that of graphite, but at 2000°C it is about the same. This is shown in Fig. 3.4. Carbon responds to temperature changes as a semiconductor material, whereas graphite responds as a metal. In graphite, resistance rises with thermal scattering of electrons caused by increased thermal agitation of atoms. In carbon, the major source of resistance is thermal scattering along boundaries of

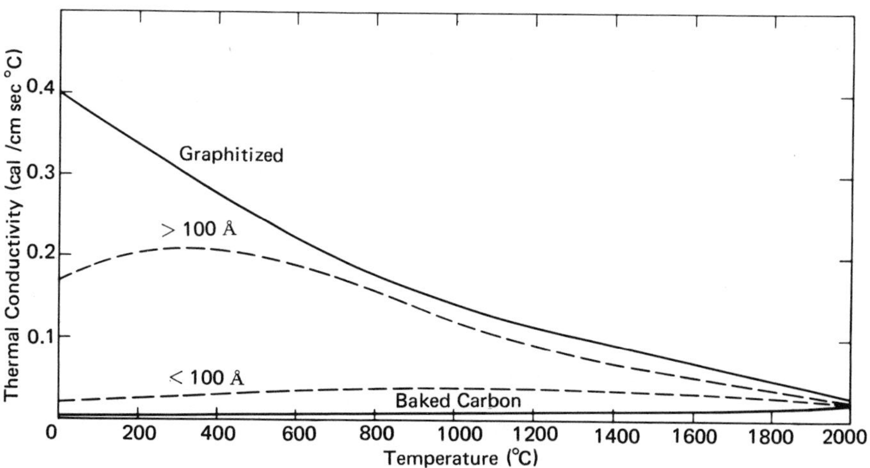

Fig. 3.3 Thermal conductivity of polycrystalline carbons of different sizes as a function of temperature. (From Castle.[21])

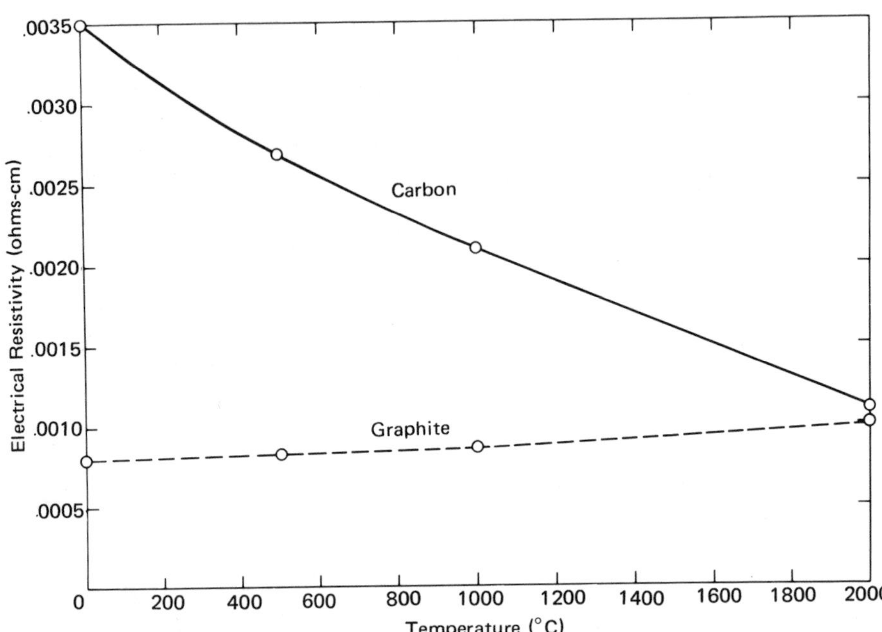

Fig. 3.4 Electrical resistivity of carbon and graphite as a function of temperature. (From Mellichamp and Finnegan.[5])

crystallites and at intercrystalline contacts. With a rise in temperature, scattering by thermal agitation of atoms is increased, but scattering at crystalline boundaries and contacts is decreased until at 2000°C carbon and graphite approach the same value for electrical resistivity. The resistivities of the spectrographic rods lie somewhere between that of amorphous carbon and ideal graphite.

Carbon rods are more difficult to shape into electrodes than graphite rods, because the smaller crystallites offer greater resistance to a cutting edge. There is more wear on cutting tools, and electrode dimensions are more difficult to maintain. This is a major determinant for the preference of graphite over carbon for electrodes.

3.6 DIRECT-CURRENT ARC EXCITATION

It is with dc arc excitation that the electrode material and design is of the greatest significance to the analytical results. With but few exceptions, graphite is the selected material, while the design is varied along with the arc current to control volatilization of the sample into the arc column. The selection of the electrode material and electrode design is based on the burning properties or consumption rate needed to obtain the optimum desired analytical results. For consumption rate, thermal conductivity is the most important physical property to be considered. Electrical resistivity is inversely proportional to thermal conductivity and can be used as an index to consumption rates of the anode.[22] With a decrease in thermal conduction there is a rise in electrical resistivity and consumption rate. Carbon has 10 times the electrical resistivity of graphite (0.0025 to 0.00025 Ω in.$^{-1}$), thus a faster consumption rate of the anode, as shown in Fig. 3.5. In a more recent study Szabo *et al.*[23] using a 20% O_2-80% Ar atmosphere with the dc arc concluded that the difference in electrode oxidation is due primarily to crystal structure and thermal conductivity.

The consumption rates for all grades of carbon anodes increases, as expected, with an increase in arc current. The grades with lower thermal conductivity increase at slightly higher rates. Figure 3.6 shows the changes in consumption with arc current for 0.121-in.-diameter anodes of two grades. Cathode consumption rates vary little with thermal conductivity as compared with the anode. Consumption rates of anodes increase with a decrease in diameter (cross-sectional area). Figure 3.7 shows how arc current can be varied to produce the same consumption rate for any given cross-sectional area. Curve (a) represents carbon and curve (b), graphite. Figure 3.8 shows the relationship between electrical resistivity, or thermal conductivity, arc current, and anode consumption.

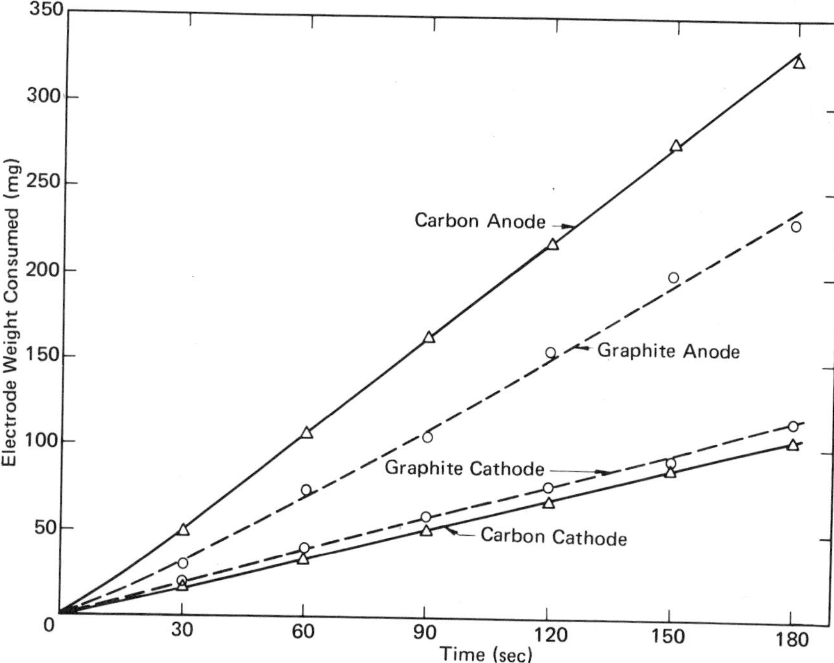

Fig. 3.5 Comparison between consumption rates of carbon and graphite electrodes (anodes and cathodes), 0.121-in. diameter, in a 10-A arc. (From Mellichamp and Finnegan.[51])

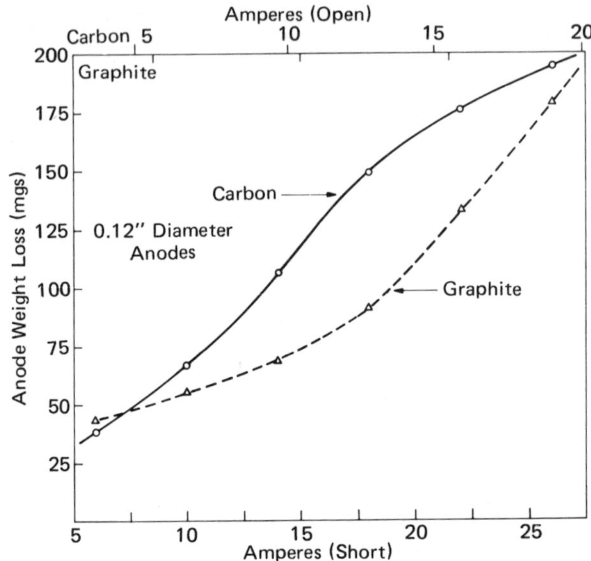

Fig. 3.6 Weight loss of carbon and graphite anodes as a function of arc current. (From Mellichamp and Buder.[52])

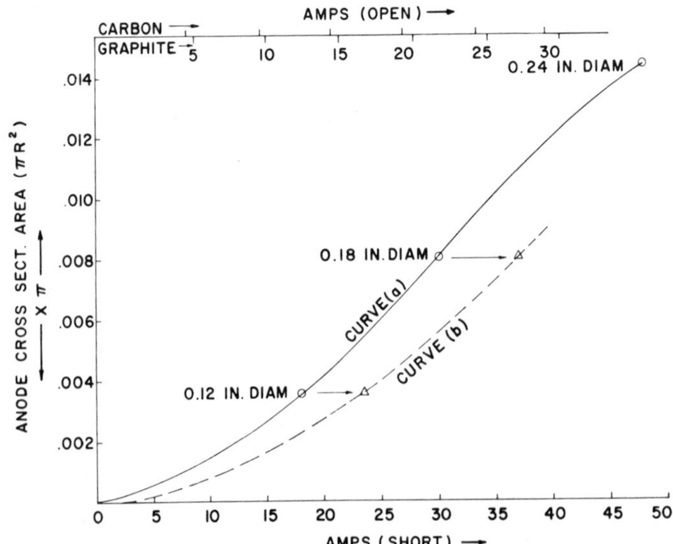

Fig. 3.7 Consumption rates of anodes as a function of cross-sectional area for (a) carbon and (b) graphite. (From Mellichamp and Buder.[52])

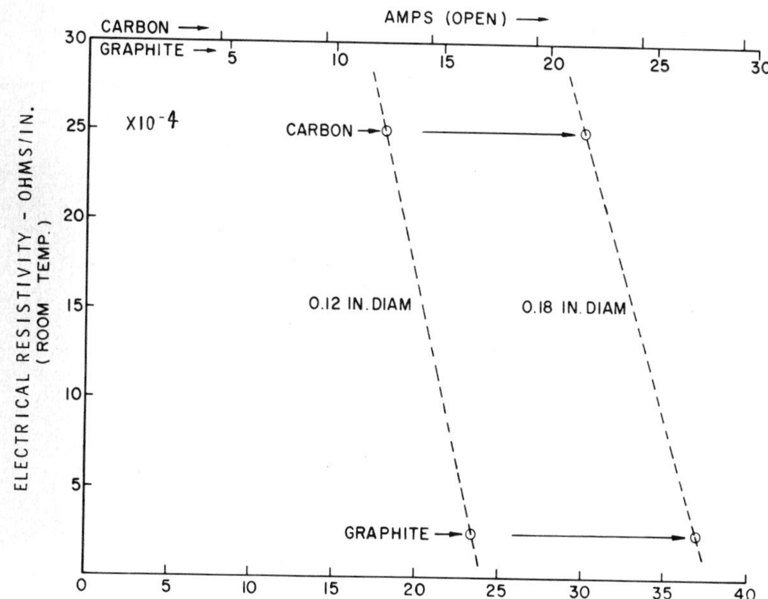

Fig. 3.8 Consumption rates of anodes as a function of electrical resistivity. (From Mellichamp and Buder.[52])

The three variables that determine the consumption rate of the anode (electrode grade, arc current, and cross-sectional area) are so interrelated that, to some extent, a change in one variable can be partially compensated for by an equivalent change in another. For example, a change of 0.01 in. in anode diameter corresponds to a current change of about 1 A. A change in electrical resistivity of the anode grade equivalent to the difference between carbon and graphite is equivalent to a current change of 5 A.

The maximum temperature reached at the anode is determined by the sublimation point of the anode material and is a fundamental constant for carbon (~3700°C)[24] as well as graphite electrodes. The anode reaches this maximum temperature only in the electrode's surface layer within the anode spot with a temperature gradient extending from this area. The anode is consumed mostly by reaction with the atmosphere and to a lesser amount by carbon evaporation into the arc stream.[25] This is the reason for the importance of a controlled atmosphere to consumption rate. About half of the energy loss is by thermal conduction through the electrode, and the remainder is by radiation to the atmosphere and by carbon evaporation.[26] Carbon electrodes with less thermal conductivity then graphite do not lose heat as readily by conduction. Thus, heat is concentrated more at the top of the anode and the temperature gradient down the electrode is increased. Figure 3.9 compares graphite

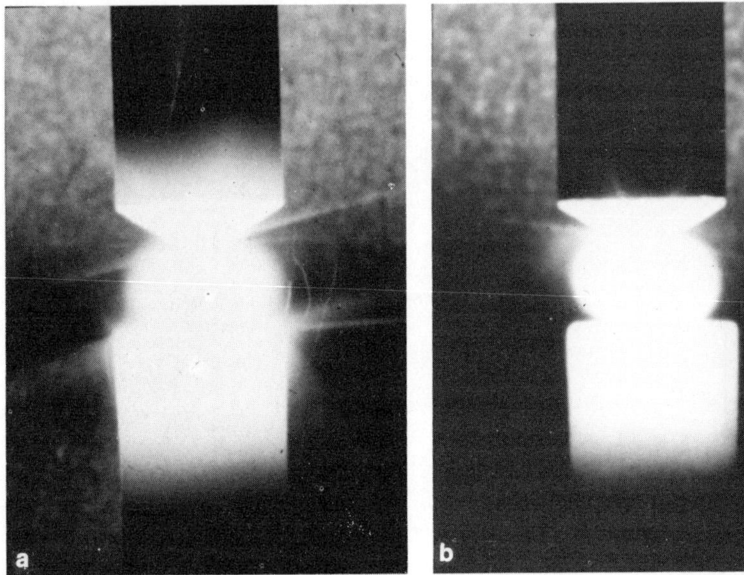

Fig. 3.9 Photographs of the arc between two grades of carbon showing difference in thermal gradient: (a) Union Carbide graphite grade AGKSP; (b) Union Carbide carbon grade L113SP. (Courtesy of Union Carbide Corps.)

electrodes with carbon electrodes under the same arc conditions to show the greater heat concentration with carbon electrodes. Chemical reaction with the atmosphere and carbon evaporation are increased, which accounts for the fact that a carbon anode is consumed faster than a graphite anode.

With a 5-mm arc gap, most of the carbon vapor originating from the anode material is destroyed in the arc stream by oxidation at the anode,[24] and very little vapor will reach the cathode. The reaction can be observed in the arc as a white zone at the anode, barely visible in graphite and more pronounced in carbon. The difference in carbon evaporation rates between carbon and graphite anodes, as indicated by the reaction-zone areas, can be a clue to why a carbon cathode opposite a carbon anode is consumed at a lower rate than a graphite cathode opposite a graphite anode. The increased amount of carbon vapor in the arc plasma suppresses cathode consumption to a greater extent.

The heat generated by the electrical resistance of the electrode materials is negligible when compared with the heat originating from the arc and does not materially contribute to the consumption of the electrodes. If the electrodes are shorted with the arc current on for several minutes, it can be seen that the heat obtained by this means is of little or no consequence to the burning of either carbon or graphite electrodes. However, the physical conditions that determine electrical resistivity also determine, to some degree, the thermal conductivity, which does influence consumption rates.

3.7 EFFECT ON DIRECT-CURRENT ARC ANALYSIS

Anode consumption rate is decreased with the addition of the sample material. The arc temperature is a function of the ionization potentials of the elements in the arc column and is lowered by the sample matrix elements. The sample material adds to the bulk and heat sink of the anode, and differences in consumption rates of electrode grades are not as marked as they are without the sample material.

Anode consumption and sample volatilization are directly related. Any factor that increases anode consumption will increase the rate at which the sample is introduced into the arc column. An example(27) is the difference in the volatilization of silicon by a change in electrode grade with no other changes in conditions. The sample was volatilized in 60 s instead of 90 s with a carbon anode in place of the graphite anode. There was also noted an enhancement in the ratios of the impurity lines to the internal standard line. Selective volatilization of impurity elements is also reduced as is shown in Fig. 3.10. In a second example,(28) volatilization of indium is reduced from 96 to 75 s with a change from graphite to carbon anodes. The consumption rate of indium was doubled with an increase of arc current from 8 to 14 A. Figure 3.11 shows how the volatilization rate of the sample material is directly related to the effective cross section of the anode cup (effective area is equal to the outside area minus the inside area). Consumption time in Fig. 3.11 is indicated by a drop in arc voltage with sample volatilization.(28)

The selective distillation of impurity elements from the sample material is dependent on how well the volatilization of the matrix elements can be controlled. The conditions selected to control sample volatilization vary with sample material and whether reproducibility or sensitivity is the more important. Another consideration is the matrix material that is to be analyzed. In general, a fast volatilization rate increases matrix effect because the atomic population in the arc column at a given instant is greater. The prevalent matrix elements will enhance elements with lower ionization potentials and suppress those with higher potentials. As noted before, in silicon the impurity lines are enhanced by the faster rate obtained with carbon in place of graphite. A uniform volatilization rate will reduce selective distillation and improve reproducibility, especially when a buffer element is added or an internal standard line is used. When the volatilization rate is too fast, the sample material will be sputtered into and away from the arc without being excited, with a resultant decrease in both reproducibility and sensitivity. On the other hand, a slow rate will emphasize fractional distillation of impurity. This slower rate can be used to advantage

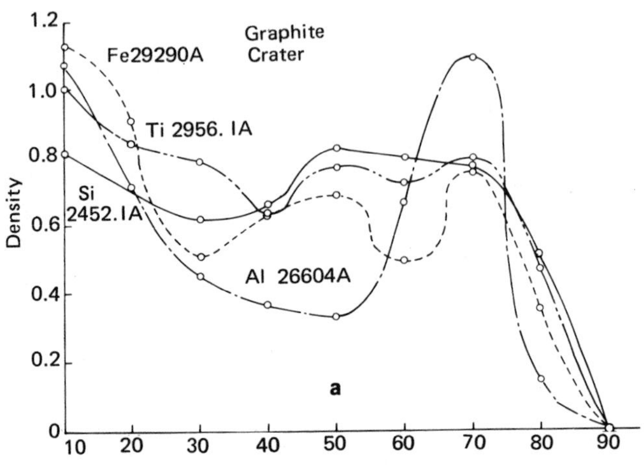

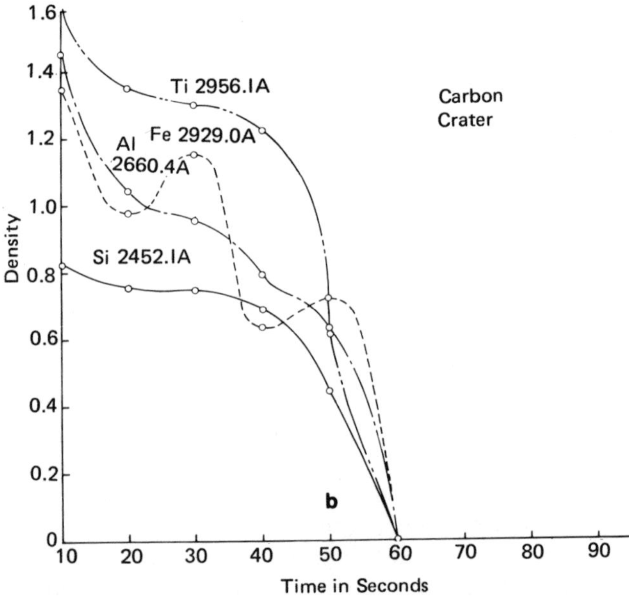

Fig. 3.10 Comparison of impurity volatilization from silicon using (a) graphite and (b) carbon anodes. Moving-plate study, camera racked at 10-s intervals. (From Mellichamp and Finnegan.[27])

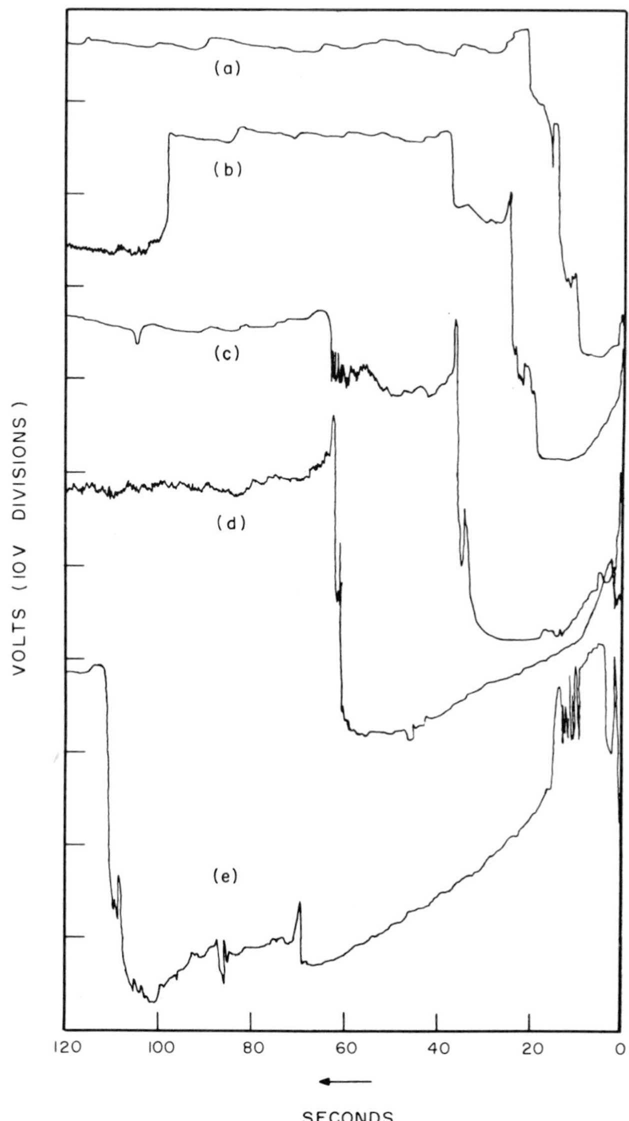

Fig. 3.11 Changes in volatility time fo 10-mg indium samples from anodes of varying cross-sectional areas in a 10-A arc. Effective cross-sectional areas (outside area minus inside area) in mm^2 arc: (a) 1.16, (b) 1.77, (c) 3.52, (d) 4.94, and (e) 7.05. (From Mellichamp.[28])

when maximum detection limits are needed. However, procedural reproducibility is often sacrificed for sensitivity. Differences in analytical results between carbon and graphite electrodes are described in the literature.[11,27,29–31]

The question may be raised whether the temperature in the arc depends on the electrode material. It is improbable that a large temperature difference between the graphite and the carbon arc will be found, since the energy balance of the column is virtually independent of the electrode characteristics.[32] When a sample is arced, the electrode material might have a secondary influence on the temperature, resulting from the different rates of evaporation, and consequently different plasma compositions obtained with a graphite arc and a carbon arc.

Because the cathode is affected by the anode and the sample material, little differences in analytical results are noted with different cathode materials. Cathodes constructed from carbon are preferred by some users[11,27] because carbon cathodes appear to reduce arc wandering more than graphite cathodes. Sample material volatilized from the anode tends to condense more along the sides than at the tip so that it remains relatively free from sample interference with electrical conductivity. Because of the ease of machining, graphite cathodes are more commonly used.

3.8 SPARK EXCITATION

The greatest single application of spectroscopy to chemical analysis is in the industrial production of metals.[33] In steelmaking and fabricating plants, spectroscopy and other instrumental methods have almost completely replaced conventional chemical methods for the routine determination of impurity elements and alloying constituents. Spectrochemical control is routine for such nonferrous metals as aluminum, magnesium, zinc, and tin and alloys of these metals. Spectral excitation is usually accomplished by means of a high-voltage condensed spark with the sample material as either one or both electrodes.

The use of metal samples as self-electrodes has several advantages. For one, the spectra of the matrix elements are normally the source of internal standard lines. Fabrication of electrodes is relatively simple because the electrodes can be cast at the foundry site with minimum contamination. Chemically analyzed standards in the desired concentration ranges and forms for most alloys are available from the National Bureau of Standards and commercial sources. Under favorable conditions, the excitation process may be precisely controlled with metallic self-electrodes to obtain the highest precision and accuracy of which spectrochemical analysis is capable.

Two basic types of electrodes are in common use: (1) pins or cylinders, which are either cast or formed from filings or shavings by compression in dies, and (2) flat discs, which are either cast or fabricated by powder-metallurgical methods. In the first type, a spark is produced between two similar pins, and in the latter, the disc is opposed by a pointed counter electrode of another conductive material and similarly excited. For optimum results all samples should be in as nearly the same physical state as possible. The spark reacts primarily with the surface; thus such factors as surface finish, physical form, metallurgical state, segregation, and porosity are important considerations. Standardized methods for sampling most metals for spark analysis are given by the ASTM.[13] In most "point-to-plane" procedures the recommended counter electrode is graphite. In general, a graphite grade of high density will give a satisfactory finish.[34] However, grades with medium density will increase the cross-sectional area of the spark scar and tend to average out inhomogeneities in the standard and sample.[7] In the "copper spark" procedure, the sample material is deposited on a copper electrode and sparked with a copper counter electrode.[35] Silver electrodes are used for spark excitation in the vacuum region for the detection of carbon and other nonmetallic impurities.[36] For all counter electrodes the sparking surface, density, dimensions, etc., must be consistent for optimum reproducibility.

3.9 MISCELLANEOUS APPLICATIONS

While the dc arc and the spark are the most frequent sources of atomic excitation, others, such as the uni-arc and ac arc are occasionally used to obtain certain results. In some cases the same electrode design is used with several different excitation sources. The flexibility of spectrochemical analysis that makes it such an effective analytical tool is due to the great variety of combinations between excitation conditions and the selection of electrode materials and designs.

The ability to analyze materials in solutions adds another dimension to spectroscopy. Solutions have a number of advantages: (1) previous sample history is eliminated, (2) maximum homogeneity is obtained, (3) the relative volatility of elements is reduced, and (4) preparation of standard solutions is facilitated. Three electrode designs—the porous cup, rotating disc, and flat top—are generally accepted for standard techniques, while two others—rotating platform and plastic cup—are less widely known but offer interesting possibilities. To broaden applications the five electrode designs are used in conjunction with different excitation conditions. A comparison of these designs and excitation conditions has been reported.[37] The plastic cup has since been modified into what is called a vacuum-cup electrode.[38] The rotating disc finds particular application in the analysis of engine oils for possible wear. The U.S. government has several programs for the standardization of the analysis of oils from ships and planes. Electrodes are a major consideration for the accuracy of results. Important to all solution techniques is the porosity and the closely related density of the electrode material. For graphite electrodes the porosity can be reduced by a coating of paraffin, Duco cement, or a similar substance, and increased by a preburn.

Some advantages occur when the sample material is in the cathode rather than the usual anode for dc arc excitation. The relatively slower sample volatilization at the cathode introduces only trace quantities, thus giving a maximum arc temperature and a high degree of ionization. Ion migration results in an enrichment layer at the cathode. Cathode layer excitation is most sensitive with small quantities of material, less than 10 mg. With larger amounts it has little advantage over anode excitation. Mitchell[(29)] has developed a procedure for the analysis of soils, plants, and related materials. His preference is carbon for the cathode material. Differences in various carbon grades are not obvious because of the slow burning rate. Apparently other factors, such as consistency of electrode dimensions and quality, are more critical. Scott[(39)] has made an extensive study of cathode-layer sensitivities as related to electrode dimensions and arc current. His conclusions are that dimensional variations not only alter spectral line intensities but also the ratios between lines. Consistency in electrode boring depth is of particular importance. Electrical characteristics of the arc are of less importance.

The high-voltage ac arc is not as extensively used as the dc arc or the spark. For some metallurgical analysis, however, the ac arc gives better accuracy than the dc arc and at the same time greater sensitivity than the spark. The ac arc is found to be more stable than the dc arc with less erratic volatilization and excitation.[(40)] It is also suited to the analysis of liquids. The ac arc between metallic electrodes is particularly good for trace determinations in metals. The spark is preferred for alloying constituents because of its greater accuracy. With a counter electrode of another material in the ac arc, a slight advantage of carbon over graphite has been detected.[(41)]

In the "carrier distillation" method, a relatively volatile compound is added to the sample so that impurity elements will be distilled along with the added compound. Thus, the matrix material is suppressed and a spectrum is recorded free from matrix interference lines. The method, originally developed for U_3O_8 analysis with Ga_2O_0 as the carrier[(42)] has been adapted to include a number of other refractory materials. In this method only a small amount of the anode material is consumed; however, thermal conductivity of the anode does influence distillation rates. Differences in sensitivity and precision between carbon grades have been observed in the analysis of nuclear-grade U_3O_8 with AgCl–LiF or Ga_2O_3 as the carriers.[(43)] Several grades produced good sensitivities as well as good precisions for most elements. For other elements, the analyst must weigh the relative importance of the two when choosing the electrode grade.

Carbon electrodes consumed in air emit cyanogen bands in the 3500- to 4200-Å spectral region that obscure analytical lines and contribute to the overall background. These bands can be suppressed or even eliminated by operating in an atmosphere free from nitrogen. A variety of methods for accomplishing

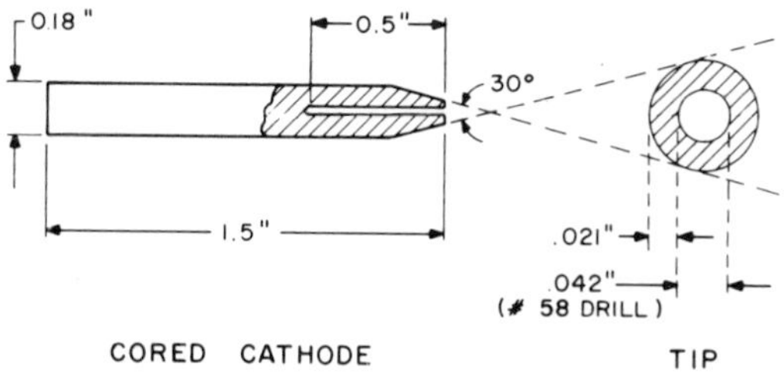

Fig. 3.12 Design of a cored cathode. (From Mellichamp.[47])

this have been reported. Among the simpler control chambers is that developed by Owen.[44] A study of the effects of operating in noble gas atmosphere is given by Thiers and Vallee.[45] In an attempt to improve the accuracy of quantitative analysis, Stallwood[46] introduced the air jet, whereby an upward flow of air removed the outer flamy fringe of the discharge. The method has since been modified to include burning in other atmospheres.

Designs of cathodes for dc arc analysis have been somewhat neglected when compared with anode designs. One reason is that arc wandering primarily determines the resultant shape of the cathode. During analysis a flat cathode is rounded while a tapered one is blunted. One method of suppressing arc wandering and at the same time improving stability is by the use of a cored cathode.[47] A narrow hole drilled along the axis of the cathode is packed with a 1:2 mixture of $BaCO_3$ and graphite powder. The design for a cored cathode is shown in Fig. 3.12. In the arc the added element slowly enters from the cathode and forms a stationary positive-ion cloud at the tip. This ion cloud acts as a stabilizer to electron flow, thus reducing current and voltage fluctuations. A silver chloride-impregnated anode has also been reported to improve sensitivity and precision.[48] The anode electrode is impregnated with silver diamine chloride, which yields a uniform distribution of silver chloride. This served as the carrier for the determination of tantalum and niobium from refractory oxides. The uniform release of the carrier was responsible for the greater sensitivity of the technique.

A study of metal cathodes—W, thoriated tungsten, Re, Ta, Mo, Nb, Pt, and graphite—in static argon atmospheres showed that arc wander is eliminated by a Ta-tipped graphite cathode for spectrochemical analysis.[49] Conditions similar to those of the study are used to determine effective work functions of various refractory metals by striking an arc to a metal wire cathode to form a melted ball having an emitting area defined by its diameter.[50]

3.10 DISCUSSION

Spectroscopy is the oldest instrumental technique in continuing practice for the identification and evaluation of the elements and chemical compounds. Over the years methods have developed to add sufficient versatility to include the analysis of most inorganic materials in their various states. Basic requirements for a good spectrochemical method are that it be accurate, reproducible, sensitive, simple, economic, and expeditiously accomplished. Accuracy derives largely from standardization and preparation, while reproducibility and sensitivity are dependent on the electrode material and the design into which it is shaped. Even the best methods are ineffective with poor electrodes; conversely, the best electrodes will not save a poor method.

Graphite in several of its many forms fulfills the needs for electrodes in most spectrochemical methods. Its unique properties are excelled by no other material. To assure good reproducibility, the physical properties of the electrodes must always remain the same, electrode to electrode, lot to lot, year in and year out. The key to this assurance is uniformity of electrode structure. The most important properties used in evaluating uniformity are particle size, pore volume, pore-size distribution, permeability, resistivity, hardness, and flexural strength. Available to the spectroscopist are commercial sources of high-purity material in either preformed shapes or as stock rods for laboratory preparation. The technology of production is in the hands of the manufacturers; however, they do not always know the exact requirements nor the effectiveness of their product until it is evaluated by the users.

The study of carbon electrodes profits from knowledge obtained from many sources because it finds application in many fields and new experiments are constantly being made. Still more data are necessary for a better interpretation of properties of carbon that relate to spectrographic problems. Not enough is known about the dependence of the final microstructure on the starting organic material and the heat-treatment procedures, and the connection between this microstructure with the different physical and chemical properties of the material. Properties are dependent on the formulation of the starting mix, such as the proportion of binder to filler, coke-particle sizes and shapes, as well as process procedures such as extrusion pressures, molding techniques, heat treatment, and other steps. When more of these data become available, commercially manufactured carbon and graphite electrodes will become even more effective for spectrochemical analysis than they are today.

3.11 REFERENCES

1. C. L. Mantell, *Carbon and Graphite Handbook*, John Wiley & Sons, Inc., New York (1968), pp. 247–288, 393–394.
2. A. J. Ihde, in: *Great Chemists* (E. Farber, ed.), John Wiley & Sons, Inc. (Interscience Division), New York (1961), p. 467.
3. E. L. Grove, *Analytical Emission Spectroscopy*, Marcel Dekker, Inc., New York (1971), Vol. 1, Part 1, p. 6.
4. H. I. Sharlin, *Sci. Amer. 204*, 107 (1961).
5. P. L. Walker, Jr., ed., *Chemistry and Physics of Carbon*, Marcel Dekker, Inc., New York, Vol. 1 (1965); Vol. 2 (1966); Vol. 3 (1968); and Vol. 4 (1968); *Proceedings of the*

Conferences on Carbon, First and Second at the University of Buffalo, Buffalo, New York, The Waverly Press, Inc., Baltimore, Md. (1956); Third and Fourth at the University of Buffalo, Buffalo, New York, Pergamon Press, New York (1959 and 1960); Fifth at Pennsylvania State University, University Park, Pa., Macmillan Publishing Co., Inc., New York (1962).

6. P. D. Coulter, *Amer. Lab. 3(3)*, 41 (1971).
7. A. J. Mitteldorf, *Spex Speaker 10*, 1 (1965).
8. J. Weinard, *Develop. Appl. Spectrosc. 1*, 137 (1962).
9. J. W. Robinson and G. D. Hindman, *Spectrosc. Letters 5*, 169 (1972).
10. R. Mannkoff and H. Fesser, *Spectrochim. Acta 28B*, 29 (1973).
11. L. H. Ahrens and S. R. Taylor, *Spectrochemical Analysis*, Addison-Wesley Publishing Company, Inc., Reading, Mass. (1961), pp. 57–61.
12. C. E. Harvey, Eastern Analytical Symposium, New York (Nov. 6, 1959); *Anal. Chem. 32*, 16A (1960).
13. *Methods for Emission Spectrochemical Analysis*, 6th ed., American Society for Testing and Materials, Philadelphia (1971).
14. D. V. Babarikin and A. E. Ilzin, *Uch. Zap. Latv. Gos. Univ. 176*, 80 (1972).
15. B. G. Voronov, *Zh. Prikl. Spectrosk. 18(6)*, 1085 (1973).
16. B. Tesarik, *Hutnicke Listy 27*, 660 (1972).
17. I. A. Grikit and E. G. Galushko, *Zh. Prikl. Spectrosk. 19*, 213 (1973).
18. H. Peter, *Jena Rev. 19*, 336 (1974).
19. Ya. D. Raikhbaum, K. F. Popov, and A. I. Kuznetsova, *Zh. Prikl. Spectrosk. 20*, 703 (1974).
20. L. Papp, D. Demeny, and M. Kadas, *Acta Phys. Chim. Debrecina 18*, 249 (1973).
21. J. G. Castle, Jr., *Proceedings of the First and Second Conferences on Carbon*, University of Buffalo, Buffalo, New York, The Waverly Press, Inc., Baltimore, Md. (1956), p. 13.
22. W. E. Allsopp and G. L. Shaw, Pittsburgh Conference on Analytical Chemistry and Applied Spectroscopy, 1951.
23. Z. L. Szabo, H. Fejerdy, and A. Buzasi, *Acta Chim. (Budapest) 80*, 365 (1974).
24. N. K. Chaney, V. C. Hamister, and S. W. Glass, *Trans. Electrochem. Soc. 67*, 107 (1935).
25. H. Steinle, *Z. Angew. Mineral. 2*, 28 (1949).
26. H. G. MacPherson, *Phys. Rev. 66*, 357 (1944).
27. J. W. Mellichamp and J. J. Finnegan, *Appl. Spectrosc. 11*, 158 (1957).
28. J. W. Mellichamp, *Anal. Chem. 37*, 1211 (1965).
29. R. L. Mitchell, *Commonwealth Bur. Soil Sci. Gr. Brit. Tech. Commun. 44*, (1948).
30. A. C. Oertel, *Spectrographic Analysis of Mineral Powders*, Commonwealth Scientific and Industrial Research Organization, Adelaide, Australia (1961), p. 44.
31. M. F. A. Hoens and J. A. Smit, *Proceedings of the Sixth International Conference on Spectroscopy*, Pergamon Press Ltd., London (1957), p. 192.
32. P. W. J. M. Boumans, *Theory of Spectrochemical Excitation*, Plenum Press, New York (1966), p. 314.
33. N. H. Nachtrieb, *Principles and Practice of Spectrochemical Analysis*, McGraw-Hill Book Company, New York (1950), p. 205.
34. W. A. Garee and J. Weinard, Pittsburgh Conference on Analytical Chemistry and Applied Spectroscopy, 1959.
35. M. Fred, N. H. Nachtrieb, and F. S. Tomkins, *J. Opt. Soc. Amer. 37*, 279 (1947).
36. G. Andermann, Applied Research Laboratories, Glendale, Calif. (1958).
37. W. K. Baer and E. S. Hodge, *Appl. Spectrosc. 14*, 141 (1960).
38. T. H. Zink, *Appl. Spectrosc. 13*, 94 (1959).
39. R. O. Scott, *Spectrochim. Acta 4*, 73 (1950).

40. F. Twyman, *Metal Spectroscopy*, Charles Griffin & Co. Ltd., London (1951), p. 193.
41. J. Connor and S. T. Bass, *Appl. Spectrosc. 14*, 55 (1960).
42. B. T. Scribner and H. R. Mullin, *J. Opt. Soc. Amer. 36*, 357 (1946).
43. C. E. Pepper, A. J. Pardi, and M. G. Atwell, *Appl. Spectrosc. 17*, 114 (1963).
44. T. E. Owen, *Appl. Spectrosc. 12*, 178 (1954).
45. R. E. Thiers and B. L. Vallee, *Proceedings of the Sixth International Conference on Spectroscopy*, Pergamon Press Ltd., London (1957), p. 179.
46. B. J. Stallwood, *J. Opt. Soc. Amer 44*, 171 (1954).
47. J. W. Mellichamp, *Appl. Spectrosc. 21*, 23 (1967).
48. N. I. Tarasevich, K. A. Semenenka, and E. K. Solodova, *Zh. Anal. Khim. 28*(*11*), 2247 (1973).
49. W. A. Gordon, *NASA Tech. Note NASA TN D-4236* (1967).
50. W. A. Gordon and G. B. Chapman II, *Surface Sci. 39*, 121 (1973).
51. J. W. Mellichamp and J. J. Finnegan, *Appl. Spectrosc. 13*, 126 (1959).
52. J. W. Mellichamp and R. K. Buder, *Appl. Spectrosc. 17*, 57 (1963).

Behavior of Refractory Materials in a Direct-Current Arc Plasma 4

NEW APPROACHES FOR SPECTROCHEMICAL ANALYSIS OF TRACE ELEMENTS IN REFRACTORY MATRICES

Reuven Avni

4.1 INTRODUCTION

4.1.1 Problem Description

4.1.1.1 Trace Elements in Refractory Materials

The Oxford dictionary defines "refractory" as "stubborn, unmanageable, not yielding to treatment, hard to fuse or work." This definition is suitable for the spectrochemical analysis of the trace-element concentrations (i.e., its impurities) in refractory materials. The usual or unusual problems encountered in a spectrochemical analysis of impurities in a refractory matrix are, for example: (1) their rates of evaporation from the electrode are extremely low and not constant during the arcing period; (2) refractory materials are very difficult to burn to completion; (3) the matrix element usually has a very dense spectrum which interferes with analytical spectral lines; and (4) they may form glassy material after a brief period of arcing, or react chemically with the electrode material.

Numerous methods have been developed to overcome some of these problems: (1) mixing the refractory matrix with graphite prior to the arcing, (2) adding various buffers or carriers, or (3) in extreme cases, separating the trace

Reuven Avni • Nuclear Research Center, Negev, Beer Sheva, Israel elements from their matrix by chemical or physical methods. In this introduction the factors affecting these problems will be elaborated systematically in order to understand the various effects and the variables governing them.

4.1.1.2 The Third Matrix

In the spectrochemical analysis of refractory materials by the dc arc technique, the electrodes usually differ from the substance to be analyzed; graphite or carbon electrodes are most commonly used. The arc burns between the electrodes in a gas atmosphere, usually in air. Both graphite and air, together with the reaction products between them, form the major plasma population. These two basic matrices of the plasma (i.e., carbon and air) were investigated by Smit,(1) Roes,(2) Huldt,(3) and others(4,5); they demonstrated spectroscopically the presence of the following species: N_2, N, N^+, N_2^+, O_2, O, O^+, O_2^+, NO^+, C_2, C, C^+, CN, and CO.

The plasma population establishes an energy balance between the electrical energy introduced and thermal losses per unit time. According to Elenbaas(6) and Heller,(7) this energy balance in a steady-state plasma is

$$j \cdot E = - \operatorname{div} (\rho \operatorname{grad} T) \tag{4.1}$$

where j is the current density, A cm^{-2}; E the electric field, V cm^{-1}; ρ the thermal conductivity, ergs cm^{-1} s^{-1} deg^{-1}; and T the absolute temperature. The solution of this energy equation, as will be shown further, indicates the nature of the parameters governing the plasma.

Another balance which can be established in the plasma is the material balance. It is described by four basic equations(2,8–10):

1. Saha's equilibrium constant for ionization and dissociation processes. For example, for a dissociation process A + B ⇌ AB:

$$K_d = \frac{n_A \cdot n_B}{n_{AB}} = 1.88 \times 10^{20} \left(\frac{M_A \cdot M_B}{M_{AB}} \right)^{3/2} T^{3/2} \frac{Z_A \cdot Z_B}{Z_{AB}} \times 10^{\frac{-5040 V_d}{T}} \tag{4.2}$$

in which n is the respective particle concentration, cm^{-3}; M the mass; Z the respective partition functions; T the absolute temperature; and V_d the dissociation potential of the molecule, eV.

2. The material balance for carbon is

$$n_C = n_C + 2n_{C_2} + n_{CN} + n_{CO} \tag{4.3}$$

3. The material balance of nitrogen and oxygen, as in air composition, is expressed by

$$\frac{n_N + 2n_{N_2} + n_{CN} + n_{NO}}{n_O + 2n_{O_2} + n_{CO} + n_{NO}} = 3.7 \tag{4.4}$$

4. The summation of partial pressures equals the total pressure in the system. This is expressed in terms of particle concentration by

$$\sum n = \frac{7.34 \times 10^{21}}{T} \tag{4.5}$$

where n represents the concentration of atoms, ions, and molecules in the plasma. For example, for a temperature of 6000°K, the overall particle concentration from the electrodes and the atmosphere is 1.2×10^{18} cm^{-3}.

The plasma between graphite and air is defined by the two balances, energy and material.

Addition of a new matrix, the refractory material to be analyzed, to the graphite and air plasma will result in various changes in these two balances. Generally, the addition of a third matrix, Al_2O_3 for example, in the graphite–air system gives rise to influences in two specific parts:

1. In the anode, owing to chemical and physical reactions between Al_2O_3 graphite and the air atmosphere.

2. In the plasma, where aluminum atoms, ions, and molecules in both the ground and excited states are added to the particles described above for graphite and air (N_2, N, CN, O_2, C, etc.).

The changes resulting from the presence of Al_2O_3 in the graphite–air system will be called the effects of the third matrix (E.T.M.). Qualitatively, the E.T.M. are easily understood, but quantitatively a connection between the trace elements and the energy and material balances in the plasma should be established. The spectrochemist is interested in "good" spectral line intensity of the trace element in the sample and a "good" correlation between the line intensity of the same element and its content in the standards.

The spectral line intensity of the trace element is directly related to the concentration of its particles in the excited state. In order to obtain a high line intensity, as it will be shown later, a high concentration of the particles of the element is needed in the plasma. In other words, the volatilization rate of the trace element from the electrode (in its solid state) should supply large amounts of free particles into the plasma, and the plasma parameters should be favorable for the excitation of those particles. The volatilization rate of the trace element, only a small fraction of that of the matrix, is governed by the volatilization rate of the matrix (Section 4.2.2). These criteria are also valid in the plasma, since the matrix will supply a higher population than the trace element,which influences the plasma parameters. This will be dealt with in Section 4.3.

For identical behavior between the standard and the sample, the description of E.T.M. above is irrelevant for the spectrochemical analysis of the trace element; however, in the majority of refractory materials, such an identity is never obtained. In this investigation the knowledge of the behavior of the E.T.M. allows one to diminish the contrast between standards and samples, thus paving the way to make reliable spectrochemical analyses of trace elements in refractory matrices. The matrices investigated and described here are the following: Al_2O_3,

TiO_2, WO_3, MoO_3, ThO_2, ZrO_2, U_3O_8, PuO_2, rare earth oxides, phosphates, and silicates.

4.1.1.3 E.T.M. in the Electrodes

Chemical reactions between refractory materials and graphite electrodes has been the subject of numerous investigations.[(11–16)] Based on the previous data, we shall dwell on the volatilization rate of trace elements and matrices. Being the only source which populated the plasma with the particles of interest, the volatilization rates represent an important parameter which has to be investigated.[17]

The volatilization rate, Q_j, of a substance (matrix) j is governed by the temperature gradients in the anode and the volume occupied by the matrix in the anode cavity. The nearer the substance is to the top of the crater of the anode, the higher its volatilization rate will be. The fixed height of the matrix, inside a cavity of given shape and dimensions, leads to a constant Q_j into the arc gap. The rate, however, is influenced by the following processes[(17)]:

1. Physical and chemical erosion of the anode, causing the matrix to gradually approach the plasma.
2. Changes in vapor pressure and heat of vaporization resulting from the new compounds formed by the matrix, the anode material, and atmosphere components.[(11,12)]
3. A change in the heat of conduction of the anode, due to the new compounds, and penetration of matrix particles into the crater walls.[(11)]

In view of processes 1–3, Q_j will depend markedly on the arcing time. The longer the arcing time, the more Q_j will be affected, as indicated in Fig. 4.1 (see further).

Average Q_j values have been estimated from the amount of material present in the anode crater and the total arcing time.[(18–19)] This method cannot be applied to refractory matrices, since they are not consumed even after a long arc period (5 min). Another evaluation of Q_j is by spectroscopic emission methods,[(18)] in which the particle concentration (atoms and ions) and flow rate were calculated. Such an evaluation of Q_j cannot be applied in the present problem because the matrix molecules have not been taken into account in the calculation of particle concentration.[(18)] The methods used for the evaluation of the volatilization rate will be described in Section 4.2. These are the chemical method,[(17)] in which the residue remaining in the anode after the arcing period is analyzed chemically, and the wire method,[(17)] in which the particles, collected by the wires passing through the plasma, are analyzed by neutron activation or spectrographic procedures.

In a refractory matrix the volatilization rates of the trace elements are strongly influenced by the volatilization rate of the matrix. That is shown in Tables 4.1a and b for some trace elements; they follow the matrix particles in their evaporation from the anode crater into the arc gap.

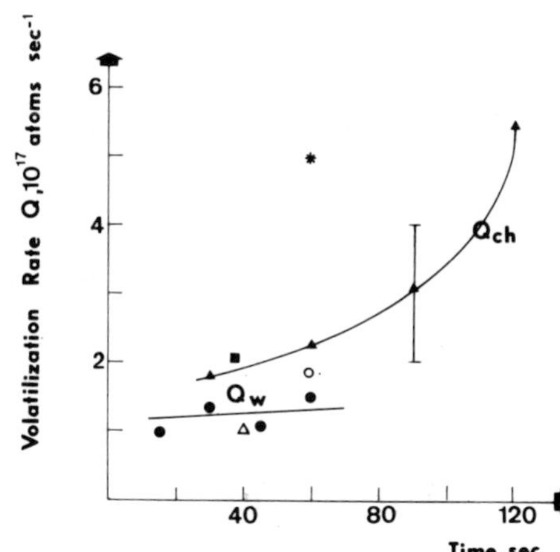

Fig. 4.1 Volatilization rate of some matrices as a function of arcing time: ▲, U_3O_8 by the chemical method; following are all by the wire method: •, U_3O_8; △, ThO_2; ○, ZrO_2; ■, La_2O_3; *, SiO_2.

4.1.1.4 E.T.M. in Plasmas

The energy balance, Eq. (4.1), shows the variables in the dc arc: electric and thermal. Presence of the third matrix will influence these variables, as will be shown in Section 4.3. A brief description of these variables is given in this section.

Smit,[1] Roes,[2] Finkelenburg and Maecker,[4] and Boumans[5] indicate the following correlations for the electric and the thermal variables.

4.1.1.4a Current Density. The current density is related to the electric conductivity, σ (Ω^{-1} cm^{-1}), and the electric field, E, in the plasma. This relation is expressed as

$$j = \sigma \cdot E \qquad \text{A cm}^{-2} \tag{4.6}$$

If we assume that more than 99% of the current in the plasma is carried by electrons, the electric conductivity can be defined as

$$\sigma = e\mu_e n_e \qquad \Omega^{-1}\ \text{cm}^{-1} \tag{4.7}$$

where e is the electron charge, coulombs; μ_e the electron mobility (its velocity in an electric field), cm^2 V^{-1} s^{-1}; and n_e the electron density, cm^{-3}.

The electrical conductivity, the electron density, and the electric field strength are gradients; in a cylindrical symmetry of the dc arc they have axial and radial values.

From Eqs. (4.6) and (4.7) with the knowledge of n_e and E values, the electrical conductivity σ and the current density j can be estimated. In Section 4.3 only n_e and E values in a plasma with and without the third matrix will be measured, while the other parameters (σ, μ_e, and $\bar{j}$) will be calculated.

4.1.1.4b Thermal Conductivity and Temperature Gradient. In Eq. (4.1) the thermal conductivity is defined as the rate at which heat crosses an area of 1 cm^2 perpendicular to the direction r of the temperature gradient dT/dr. The energy dissipated from the arc column is carried away by radial heat conduction and

Table 4.1a Volatilization Rates, Q_j, of Uranium and Trace Elements in U_3O_8 Matrix[a]

Arcing time (s):	10		20		40		60	
Element	Q_j[b]	Q_U/Q_T	Q_j[b]	Q_U/Q_T	Q_j[c]	Q_U/Q_T	Q_j[c]	Q_U/Q_T
U	200	–	150	–	200	–	300	–
Mn	1.3	150	1.0	150	1.4	143	1.9	157
Pb	1.5	133	1.3	115	1.4	143	2.6	115
Ni	0.3	666	0.2	750	0.5	400	0.05	–
Cr	1.3	150	1.0	150	1.2	167	2.0	150
Cu	1.5	133	1.2	125	1.3	150	2.0	150
Ga	1.5	133	1.0	150	1.3	150	2.0	150
In	1.2	167	1.0	150	1.5	133	2.0	150
Ge	1.6	125	1.3	115	1.5	133	3.5	85
K	1.5	133	1.3	115	1.5	133	3.0	100
Bi	1.3	150	1.0	150	1.4	143	2.0	150
Al	<0.05	–	<0.05	–	0.8	250	0.9	335
Mo	<0.05	–	<0.05	–	1.0	200	1.2	250
V	<0.05	–	0.7	215	1.0	200	1.3	230
Ti	<0.05	–	<0.05	–	<0.05	–	<0.05	–
Ba	<0.05	–	<0.05	–	1.0	200	1.3	230
Sr	<0.05	–	<0.05	–	0.8	250	1.3	230

[a] 6.0-mm arc gap at 12 A. Each trace-element concentration 500 ppm (25 μg in 50-mg substance in the anode cavity). Q_j, g s^{-1}; Q_T, Q_{Traces}.
[b] Relative standard deviation from four electrodes, 50%.
[c] Relative standard deviation from four electrodes, 25%.

convection; a small percentage is converted into radiation.[4,5] Maecker[20] showed that in the arc mantle, convection was more important for heat transport to the surroundings. Meixner,[21] Roes,[2] and others[4,5] described the conductivity (ρ) as being composed of (1) classical motions of atoms, molecules, ions, and electrons; (2) the thermal diffusion; and (3) conductivity due to the transport of energy of reaction (i.e., energy of dissociation and ionization). For a dc arc in air where the temperature range is between 5000 and 7000°K, the major factor governing thermal conductivity is the dissociation of nitrogen.[2,5]

The temperature gradient is obtained from the solution of Eq. (4.1) as described by Roes[2] and by Vukanović *et al.*[22–25] It can be determined axially and radially from line intensities of thermometric species such as Zn and Cu.[5]

The distribution of thermal conductivity and temperature in the plasma under the influence of the third matrix is described in Section 4.3.

4.1.2 Literature Survey

The spectrochemical literature shows a variety of methods used for the analysis of trace-element concentrations in refractory matrices. Reviewing only

Table 4.1b Volatilization Rates, Q_j, of Al or Si and Their Trace Elements in Al_2O_3 or SiO_2 Matrices[a]

Arcing time (s):	40				60			
	SiO_2 matrix		Al_2O_3 + 10% AlF_3 matrix		SiO_2 matrix		Al_2O_3 + 10% AlF_3 matrix	
Element	$Q_j{}^b$	$\frac{Q_{Si}}{Q_T}$	$Q_j{}^b$	$\frac{Q_{Al}}{Q_T}$	$Q_j{}^b$	$\frac{Q_{Si}}{Q_T}$	$Q_j{}^b$	$\frac{Q_{Al}}{Q_T}$
Si	100	–	–	–	80	–	–	–
Al	–	–	100	–	–	–	200	–
Mn	0.7	143	1.3	77	0.5	160	2.4	83
Pb	0.7	143	1.5	66	0.5	160	2.7	74
Ni	0.5	200	1.0	100	0.5	160	2.2	91
Cr	0.8	125	1.2	83	0.5	160	2.0	100
Cu	0.8	125	1.3	77	0.5	160	2.5	80
Ga	0.7	143	1.3	77	0.7	115	2.5	80
In	0.8	125	1.3	77	0.5	160	2.4	83
Ge	1.0	100	1.0	100	0.8	100	2.2	91
K	1.0	100	1.0	100	0.7	115	2.2	91
Bi	1.0	100	1.4	71	0.8	100	3.0	67
Mo	0.4	250	1.0	100	0.3	266	2.2	91
V	0.5	200	1.0	100	0.3	266	2.0	100
Ti	0.5	200	1.0	100	0.3	266	2.2	91
Ba	0.4	250	0.5	200	0.4	200	0.8	250
Sr	0.4	250	0.5	200	0.4	200	0.8	250

[a] 6.0-mm arc gap at 10 A. Each trace-element concentration 500 ppm (15 μg in 30-mg substance in the anode cavity. Q_j, g s^{-1}; Q_T, Q_{Traces}.
[b] Relative standard deviation from four electrodes: 30%.

trace analysis in uranium, thorium, zirconium, and natural silicates, the following classification of the various methods is obtained.

4.1.2.1 Chemical and Physical Separation

Chemical separations of trace elements from the matrices have been done, prior to the spectrochemical analysis (dc arcing). This includes solvent extraction by ether,(26–29) tributylphosphate,(30–31) or by other specific chemical reagents.(32–36) Anion or cation resins have been applied to the separation of trace elements from uranium, thorium, and zirconium matrices.(32–48) Physical separations have been used primarily in the USSR.(49)

Both chemical and physical separations prior to a spectrochemical procedure are inconvenient:

1. They are tedious and time consuming compared to the time required for spectrochemical analysis.

2. There is danger of contamination.

3. There can be a partial loss during concentration of the trace elements due to multiple manipulations in a separation method.

4. Separations are usually expensive, owing to the high cost of furnaces and high-vacuum equipment (physical separations) and of chemical reagents.

The major achievement of separation procedures is the elimination of matrix effects in electrodes and in the plasma. This simplifies the spectrochemical procedure.

If the separation procedures are to be avoided, the matrix effects have to be considered to obtain reliable spectrochemical results. This will be described in the following sections.

4.1.2.2 Buffers, Fluxes, and Internal Standards

When one states that the matrix effects have been overcome, what is usually implied is that an equality of parameters exist between the standards and samples so that line intensity is influenced to the same extent. In other words, chemical and physical composition of the standards used in a spectrochemical procedure should be identical to those of the samples. In that way each working curve obtained with the standards allows the interpolation of the line intensity of the unknown element in it. The ideal equality between standards and samples is rarely obtained. This can be exemplified by spectrochemical analysis of silicates. In nature the bulk composition of silicates varies widely (SiO_2 and other major elements, such as Ca, Al, Mg, Fe, and alkalies). To overcome this irksome problem, several international standards[50–52] have been prepared which are representative of the large compositional variations found in nature.[50, 53–57] Despite the availability of these reference materials, equality between standards and samples is obtained only by the addition of various chemical substances to both of them. These additives, called buffers or fluxes, greatly reduce the differences between the standards and samples that influence the line intensities of the analyte element.

This quasi-equalization is concerned with the volatilization rate from the anode crater and the behavior of particles in the dc arc plasma. Buffers such as lithium metaborate, persulfates, and others[58–77] are added to the natural silicates and standards in high concentration from 50% up to 500%.[59, 70, 71] This addition of such high concentration can be regarded as a change in matrices; the buffer becomes the new matrix. The use of various buffers as recorded in the literature for the same silicate samples suggests that equilization in all respects between standards and samples has not been achieved. New additions[53] in small concentrations, called the internal standards, help to achieve better equilization. Different internal standards[52] are added for various trace elements. The trace elements can be divided into various groups, according to the "best" internal standard and buffer. Extensive empirical and research work have been done to create such a classification.

However, instead of studying the behavior of buffers and internal standards, it seems more reasonable to investigate the behavior of the original samples with respect to their trace-element behavior, to avoid the use of additives.

4.1.2.3 Carrier Distillation

The use of carriers in spectrochemistry is attributed to Scribner and Mullin,(78) who were the first to develop an analytical method for the impurity elements in an uranium oxide matrix. The quantity 2% Ga_2O_3 was the carrier.

Carriers are widely employed for trace-element analysis of refractory matrices. In addition to Ga_2O_3, which was studied further by Pepper(79) and by Feldman,(80) other carriers have been introduced for different problems, to improve sensitivity or accuracy. Some examples are as follows: AgCl,(81–82) NaF,(83) In_2O_3,(84) NH_4F,(85) CsF,(86–87) and Ag_2O.(88) Such mixtures as AgCl + AgF,(89) AgCl + PbF_2,(90) Ga_2O_3 + SrF_2,(91) AgCl + GeO_2,(92) $LiCO_3$ + PbF_2 + NaCl,(93) and others(79,80) have also been used.

The addition of carriers to a matrix within a concentration range of 1–10% is estimated to play the following roles(49,94):

1. Depression of the plasma temperature(49,95–97) due to the lower ionization potential of the carrier element, thus lowering the abundance of the atom and ion spectral lines of the matrix. In the case of such matrices as Zr, Th, U, Pu, and the rare earths, which have a very dense atom and an ion spectrum that overlap the spectrum lines of the impurities, lowering the abundance is an important achievement. For matrices such as silicates of alumina, this effect is irrelevant.

2. Transportation of impurity elements from the matrix, located in the anode, into the plasma as a result of the higher vapor pressure of the carrier(98–100) and an increase of the mean time spent by particles in the discharge zone.(101–104) Vainshtein and his coworkers(98,99) employed radioactive tracers to study the spatial distribution of elements in the plasma. They showed qualitatively that the spatial distribution of most of the elements in the discharge zone were similar to that of the carrier element.

However, Samsonova(102) showed that the carriers did not influence the line intensity of the impurities via the evaporation mechanism of the constituent. Daniel(105) demonstrated that vaporization of Ga_2O_3-U_3O_8 mixtures was at about 1650C°, which is significantly lower than the melting points of the major constituents (Ga_2O_3 or U_3O_8). He classified the impurity elements into groups according to their ease of volatilization:

1. Trace elements that are readily affected by the volatilization of the carrier. These include the more volatile elements, such as boron, cadmium, and tin.

2. Those elements that show preferential compounds of eutectic formation with the carrier followed by volatilization. Such elements include aluminum, iron, and manganese.

3. Those elements whose compounds are quite nonvolatile at the arc temperature or which form more stable compounds with the matrix. Calcium and magnesium are examples.

4. Those elements that have volatilities in the same order as that of the mixture (carrier matrix) and therefore are not affected by the distillation of the carrier.[106] Examples include zirconium, thorium, and the rare earths.

The large variety of carriers used in spectrochemical analysis can be explained by their limited suitability[96,105] for various groups of the impurity elements. Common elements are determined with good precision and sensitivity[79] in refractory matrices such as oxides or uranium, thorium, zirconium, and the rare earth oxides. However, analysis of numerous refractory-type impurities must rely on supplemental methods.[106] For other refractory matrices, such as natural phosphates and silicates, the carrier distillation method has very limited application.[53,107]

4.2 NEW APPROACHES

Line intensity is the measured quantity in spectrochemical analysis. In order to choose the most suitable buffer or carrier, a study must be made of the factors affecting the line intensity by such additives. In other words, the volatilization rate of the substance from the electrode into the discharge zone and the plasma parameters must be investigated. Instead of studying the behavior of additives, it seems more practical initially to evaluate the effects of the matrix alone with regard to volatilization rates and plasma parameters. In this way it should be possible to establish a direct and reliable spectrochemical analysis for trace elements in refractory matrices.

4.2.1 Mandelshtam Scheme[108]

To proceed in a systematic way, the entire study will conveniently divide into several steps, as described by Mandelshtam[108] and by de Galan[109]:

$$C_j \overset{1}{—} Q_j \overset{2}{—} n_{t_j} \overset{3}{—} n_{\mathrm{ex}_j} \overset{4}{—} J \overset{5}{—} E$$

A substance j introduced into an electrode at concentration C_j is connected with the measured signal E by the following links:

1. The C_j or part of it is evaporated into the plasma. The number of particles leaving the electrode per unit time is the volatilization rate Q_j of the element.

2. The particles entering the plasma are subject to convection, diffusion, and electric fields which determine their time spent or residence time in the discharge

zone. The rate in which particles leave one region for another one is denoted by the transport parameter (ψ).[5]

$$\psi_j = \frac{Q_j}{n_{t_j}} \qquad \text{cm}^3\ \text{s}^{-1} \tag{4.8}$$

3. Particles in the plasma undergo various processes, such as excitation, dissociation, and ionization. The total concentration of these particles in the plasma is denoted by n_{t_j}.

4. The amount of energy emitted in unit time per unit volume along the optical axis (solid angle) resultant from a particular transition is called the emittance J: it is determined by the concentration $(n_{\text{ex}_j})_2$ of the upper-level excited particles, for example the concentration of excited particles in level 2.

5. The detector of emittance measures an integral of J for a period of exposure time; it is denoted by E.

Several of these links with reference to refractory matrices and trace elements will be treated quantitatively in the following sections.

4.2.2 Volatilization Rate

The relationship between the volatilization rates of trace elements and refractory matrices (Section 4.1.1.3) was quantitatively evaluated for U_3O_8, Al_2O_3, and SiO_2 in the following way:

1. In each matrix ("specpure") the trace elements Al, Ba, Bi, Cr, Cu, Ga, Ge, In, K, Li, Mn, Mo, Ni, Pb, Sr, Ti, and V as oxides were introduced to give a 500-ppm concentration for each element. Each matrix was arced in triplicates for 10, 20, 40, and 60 s.

2. After each period of arcing, the remaining substance in the electrode was analyzed chemically[17] for matrix losses and spectrochemically for the trace elements.[94]

3. Differences in contents prior and after arcing divided by the arcing period is the volatilization rate for the matrix and each trace element.

Results are given in Tables 4.1a and 4.1b. Electrode erosion results in the continuous approach of the substance in the anode crater to the plasma (i.e., continuous variation of Q_j); therefore, results given in Table 4.1 are integrated over the arcing time. The volatilization rates are also presented as a ratio of matrix to trace element. The results thus obtained (Table 4.1) show that volatilization rates of the trace elements follow that of the matrix from the constant ratios at different arcing time, with two exceptions: (1) for Al, Ba, Mo, Sr, and V, their Q_j follows that of the matrix after 20 s, and (2) the Q_j of Ti differs from the matrix even up to 60 s of arcing.

The reason for the two exceptions could be the formation of new compounds between those trace elements and the matrix. The new compounds for Al, Ba, Mo, Sr, and V volatilize only after 20 s of arcing (the temperature along the electrode increase with increasing arcing time) and after 60 s for Ti. It is assumed that the other refractory trace elements (i.e., rare earths) behave like Ti. Addition of 2–4% UF_4 to U_3O_8 matrix increases the volatilization rate[17,89] without affecting the temperature of the plasma or along the

Fig. 4.2 Setup for measuring side losses after the arcing period (Q_j^{ch}).

electrode.[17,94] Based on this result it can be assumed that the presence of fluoride in the matrix promotes the formation of the new compounds, and all trace elements and matrix volatilize together.[17,94,110] Few atomic or ionic spectra for U, Zr, and Th appear in the cathode region, so measurements are possible.[110] According to data given in Tables 4.1a and 4.1b, it can be concluded that the volatilization behavior of trace elements can be almost predicted by measuring only Q_j for the matrix. Two experimental methods of determining the matrix volatilization rate will be described: the chemical and the wire methods.[17]

4.2.2.1 The Chemical Method

In the chemical method (Q_j^{ch}) the electrode charge was weighed before arcing and chemically analyzed after arcing. The graphite anode was immersed in the adequate acid and the solution analyzed for the matrix content. Minimum arcing time for detection of losses was 30 s. The particles spattered (aggregates) from the crater (lateral losses) were collected by a silica dome surrounding the arc gap and by aluminum foil covering the floor, as shown in Fig. 4.2. The dome has two open areas, one around the cathode and another in the side wall, which ensures normal free convection to the arc. The dome and aluminum foil were immersed in the suitable acid and the solution analyzed for the matrix element content. A minimum of 60-s arcing is prerequisite to detect the lateral losses with good accuracy. The deposit on the dome was found to be negligible compared with that found on the soil. Results for U_3O_8, SiO_2, ThO_2, and La_2O_3 are given in Fig. 4.1.

4.2.2.2 The Wire Method[17]

The wire method (Q_j^w) consists in passing aluminum wires at uniform velocity through the plasma in the vicinity of the anode. In this way the volatilization rate of the matrix element can be evaluated by determining the amount of material collected by the wire. The accumulation of the wire was measured by neutron activation (for U and La matrices) or emission spectrochemical analysis (Th, Zr, and Si) of the wire.

The deposit on the wire can be related to the volatilization rate (Q_j^w), as follows. For a matrix j, the particle flux from the anode cavity into the plasma can be approximated[5] (if diffusion is ignored) by

$$n_{t_j} v_j = \frac{Q_j}{A_c} \tag{4.9}$$

where n_{t_j} (cm^{-3}) is the total particle concentration in the plasma, v_j (cm s^{-1}) the linear velocity of the particles in the anode vicinity, and A_c (cm^2) the cross section of the anode crater from which the particles enter the plasma.

Consider a homogeneous flux of particles $n_{t_j}v_j$ in the cylindrical symmetry of the plasma; then the number of particles N'_w colliding with the wire in a single pass is

$$N'_w = \int (n_{t_j}v_j)A(t)\,dt \tag{4.10}$$

where $A(t)$ (cm^2) is the area of the wire exposed to the plasma at time t (s).

The uniformity velocity of the wire in the x-axis direction is v_w (cm s^{-1}) and the length exposed to the plasma is l (cm). Substitution of $A(t)\,dt$ in Eq. (4.10) gives

$$N'_w = \int_{-R_c}^{R_c} (n_{t_j}v_j)\frac{\pi d_w}{2}\, l\, \frac{dx}{v_w} \tag{4.11}$$

where d_w (cm) is the wire diameter and R_c (cm) the cylinder radius.

Integration of Eq. (4.11) leads to

$$N'_w = n_{t_j} \cdot v_j \cdot \frac{\pi}{2} \cdot \frac{d_w}{v_w} \cdot A_c \tag{4.12}$$

Substituting n_{t_j} from Eq. (4.9) and assuming a proportionality factor (α) between N'_w and the absolute number of particles N_w on the wire, we obtain for the volatilization rate, from Eq. (4.12),

$$Q_j^w = \frac{2}{\alpha}\frac{N_w v_w}{\pi d_w} \tag{4.13}$$

For the derivation of Eq. (4.13) it has been assumed that: (1) the velocity of v_j in the anode vicinity is equal for atoms, ions, and molecules; and (2) the proportionality constant α, which is related to the sticking coefficient of the particles, is constant for different wire diameters and velocities.

Now, if the wire is cut into equal parts and the amount collected on its various parts are determined separately, the radial distribution of Q_j^w can be determined. An Abel inversion procedure similar to that used for measuring radial distributions of special lines intensities must be used.[5,111] Calculations are conveniently made in terms of particle flux $n_{t_j}(r)\,v_j$ (cm^{-2} s^{-1}), which is related to the volatilization rate $Q_j^w(r)$ by

$$Q_j^w(r)\,dr = 2\pi r n_{t_j}(r)\,v_j\,dr \tag{4.14}$$

Figures 4.3a and 4.3b show the wire arrangement in the carriage for the measurement of Q_w. In Fig. 4.3b a schematic representation of the wire displacement is given. Four wires located in the same wire carriage penetrate the plasma sequentially at the same height (0.2 mm) above the anode. The distances between the four wires are such that the second wire enters the plasma when the first one leaves it. Velocity of the carriage through the plasma was gained by the

Fig. 4.3a Carriage and wire disposal for the wire method.

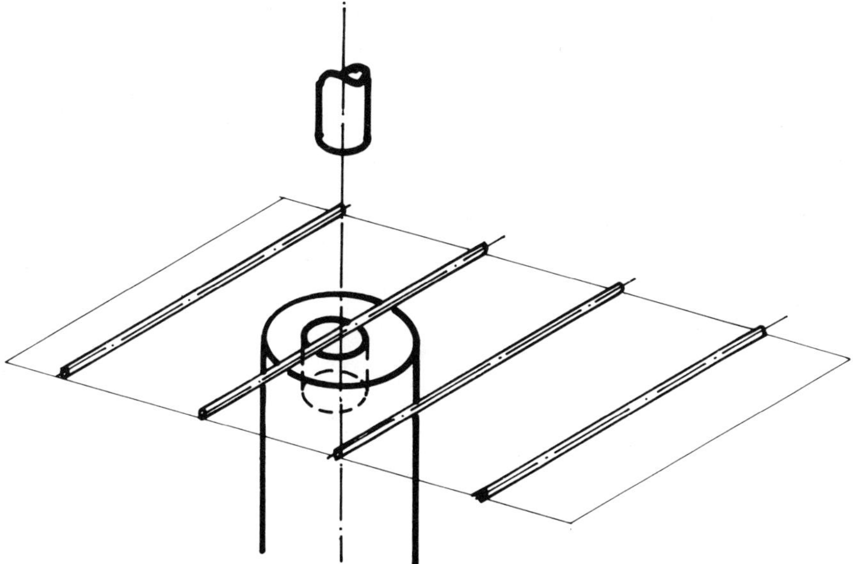

Fig. 4.3b Schematic drawing of the wires for Q_j^{ch} measurement.

impact of steel springs.[112] The carriage can be released through the arc gap in any desired time during the arcing period. The wires were subsequently treated in two ways: (1) complete wires were analyzed to compare the values of Q_j^w with those of Q_j^{ch}, and (2) wires cut into equal parts of 2.0 mm length were used to determine the radial distribution of the particle flux.

In previous publications,[17,112] it was demonstrated that the wire method is reliable and that Q_j^w obtained represents the absolute particle concentration of the matrix element in the anode region supplied from the anode crater per unit time.[16]

Figure 4.1 shows the values of Q_j^{ch} and Q_j^w for various refractory matrices, such as U_3O_8, La_2O_3, ThO_2, ZrO_2, and SiO_2, as a function of arcing time. Between 20 and 40 s, Q_j^w is almost equal to Q_j^{ch}. At a later stage this comparison is no longer valid, owing to electrode erosion. For arcing periods less than 20 s, only Q_j^w is valid because the error in Q_j^{ch} is too large.[17]

Figure 4.4 illustrates the radial distribution of the particle flux in the anode vicinity for U_3O_8 matrix. In the anode region (0.2 mm) of the plasma, the radial diffusion of the particles outside the arc axis is within two orders of magnitude. In other words, for U_3O_8 and similar matrices only 5% of the particles entering the plasma move vertically along the axis toward the cathode; 95% of the particles diffuse radically at the moment they enter the dc arc plasma. In spectrochemical analysis the arc core,[113] not the axis, is the most important part of the plasma, because it is the brightest. Thus it is necessary to measure the diffusion from the arc core in the anode vicinity. For $Q_j^w(r)$ the wires are cut into 2.0-mm portions; that is, radial diffusion can be measured at a 2.0-mm radius around the arc axis. Table 4.2 shows that about 30% of the particle flux diffuses

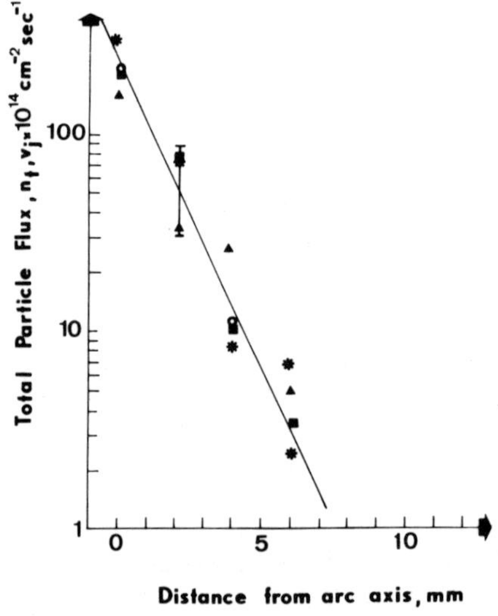

Fig. 4.4 Radial distribution of uranium particle flux in the anode region. Al wires: ∗, ●, ▲, ■, first to fourth, respectively.

Table 4.2 Radial Particle Flux of U_3O_8, ThO_2, and ZrO_2 on 2.0 mm Aluminum Wire Portions, in the Anode Region[a]

Radial distance from arc axis (mm)	Particle flux, $n_t(r)v_j \times 10^{-15}$ (cm^2 s^{-1})			Relative standard deviation[c] (%)
	U_3O_8[b]	ThO_2[b]	ZrO_2[b]	
0–2	12.0	12.5	11.5	25
2–4	3.5	4.0	3.0	25
4–6	0.7	0.8	0.5	25

[a] 8.0-mm arc gap at 12 A. Wire velocity 110 cm s^{-1}.
[b] Anode crater charge: 50 mg of U_3O_8; 60 mg of ThO_2; 35 mg of ZrO_2. Same level in the anode cavity for the three matrices.
[c] Calculated from four wires.

outward from the arc core (radius ~2 mm). These data pertain to the free particles of the matrix.

The next question: If the spatial distribution of the trace-element free particles is similar to that of the matrix free particles in the plasma, the radial diffusion for trace-element particles as for the matrix particles will reduce analytical detection limits in spectrochemical analysis. Measurement of the axial distribution of relative line intensity for both trace elements and matrix will provide the solution to our problem, as described in the next section.

4.2.3 Axial Distribution of Line Intensity

The axial distribution of spectral line intensities for trace and matrix elements between anode and cathode in a dc arc plasma was measured as described in a previous publication.[(94)] The dc arc gap was focused on an eight-hole diaphragm mounted on the entrance slit of the spectrograph (see Fig. 4.14). For a given spectral line, the relative intensity at a given height of the arc gap was normalized to the relative intensity of the same spectral line in the anode region of the plasma. Such a normalization reduces the systematic and random errors in line intensity measurements. The intensity ratio obtained represents the ratio of particle concentration and temperatures for two levels (z) in the plasma. This ratio can be explained as follows: Let 1 and 2 be two different heights in the plasma. These would be slices 1 and 2, as observed by openings 1 and 2 in the diaphragm, Fig. 4.14. The spectrograph samples equal volumes at these heights (z). At each z the absolute line intensity J for the same atom transition is described by

$$J_1 = \frac{h\nu}{4\pi}\frac{gA}{Z} n_1(1-\alpha)_1 \exp\left(-\frac{\Delta E}{kT_1}\right) \tag{4.15}$$

$$J_2 = \frac{h\nu}{4\pi}\frac{gA}{Z} n_2(1-\alpha)_2 \exp\left(-\frac{\Delta E}{kT_2}\right) \tag{4.16}$$

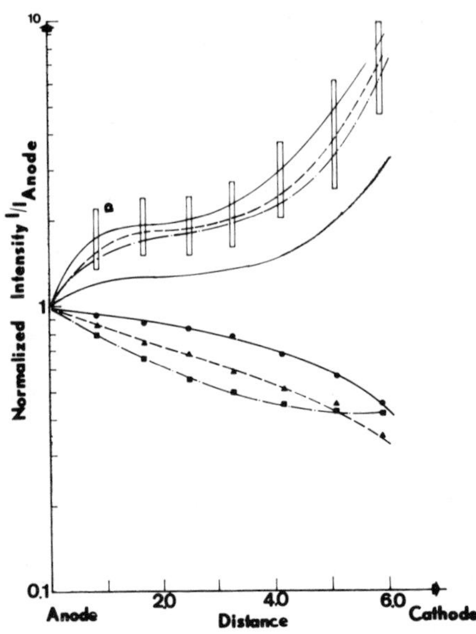

Fig. 4.5 Axial distribution of relative line intensity normalized to the anode region for U_3O_8, ThO_2, and ZrO_2 matrices; 6-mm arc gap; 13 A. ▯, Line intensity for common trace elements in U_3O_8 (—), ThO_2 (– · –), and ZrO_2 (---). —, rare earth trace element in the curve without marking points in the U_3O_8 matrix. Matrix-element line intensity: ●, U; ■, Th; ▲, Zr.

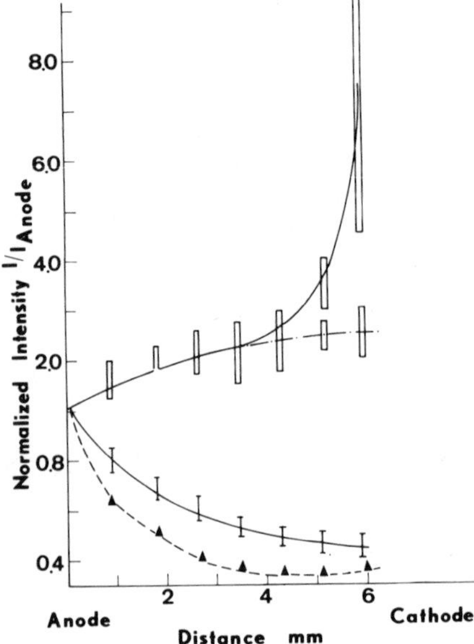

Fig. 4.6 Axial distribution of relative line intensity normalized to the anode region for rare earth oxide matrices; 6.0-mm arc gap; 13 A. –▯–, Normalized intensity for common trace elements. – · ▯ · –, Normalized intensity for As, Hg, P, An, and Cd. ▲, Normalized intensity of La, Y, and Sm. I, Normalized intensity of Ce, Dy, Eu, Gd, Go, Nd, and Yb.

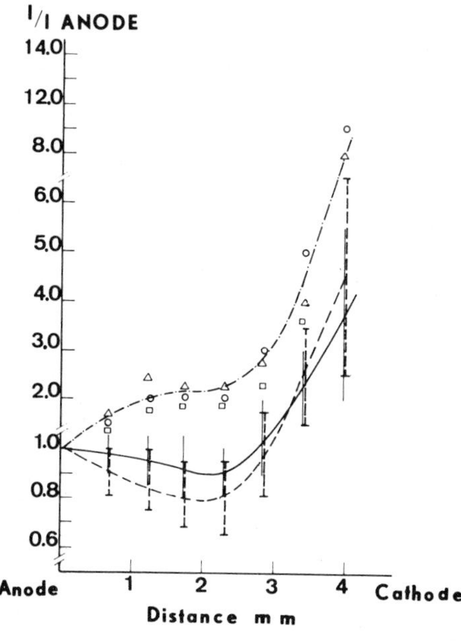

a

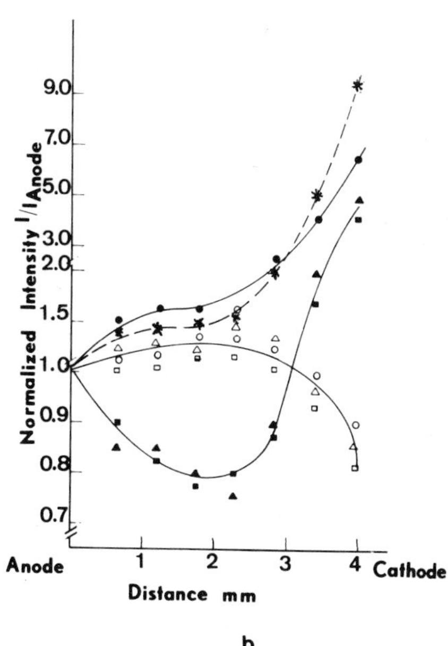

b

Fig. 4.7(a) Axial distribution of relative line intensity normalized to the anode region for rock phosphate; 4.0-mm arc gap; 10 A. |, Normalized intensities of atom lines of common traces in standard matrix, $Ca(PO_3)_2$. $\bar{\text{I}}$, Normalized intensities of common traces in rock phosphate: ○, Mg(II); △, Mn(II); □, Ba(II) in standard. (b) △, □, ○: Normalized intensity of P line in standard, rock phosphate, and standard +20% graphite, respectively. ▲, ■, Ca(I) in standard and rock phosphate. ●, Ca(II) in standard and rock phosphate. ∗, Ca(I) and Ca(II) in standard + 20% graphite.

where h is Planck's constant, ν the line frequency, g the statistical weight of the upper term, A the transition probability, Z the partition function, n the particle concentration (atoms + ions), α the ionization degree, k the Boltzmann's factor, and T the absolute temperature.

Therefore, the line intensity ratio

$$\frac{J_1}{J_2} = \frac{n_1(1 - \alpha)_1}{n_2(1 - \alpha)_2} \exp\left(\frac{T_2 - T_1}{T_1 - T_2}\right) \tag{4.17}$$

Equations (4.15) and (4.16) assume a local thermodynamic equilibrium at each level in the dc plasma and that only atoms and ions populate the plasma. The presence of molecules is conveniently neglected (see Section 4.3).

Figures 4.5–4.9 show the axial distribution of the normalized relative line intensities of several trace and matrix elements in Al_2O_3, U_3O_8, ThO_2, ZrO_2, PuO_2, La_2O_3, SiO_2, and $Ca_3(PO_4)_2$ matrices. From these figures the following conclusion can be derived for all the trace elements in the refractory matrices above:

1. According to normalized line intensities three distinct plasma regions are formed: anode, central, and cathode regions.

2. The axial distribution of spectral line intensity of the trace elements differs from that of the matrix element. The line intensities of the trace elements in the cathode region exceeds their intensities in the anode region. For the matrix element the maximum line intensity was obtained in the anode region and the minimum in the cathode region.

3. In the central region the normalized line intensity of the trace elements is always higher than that of their matrix element.

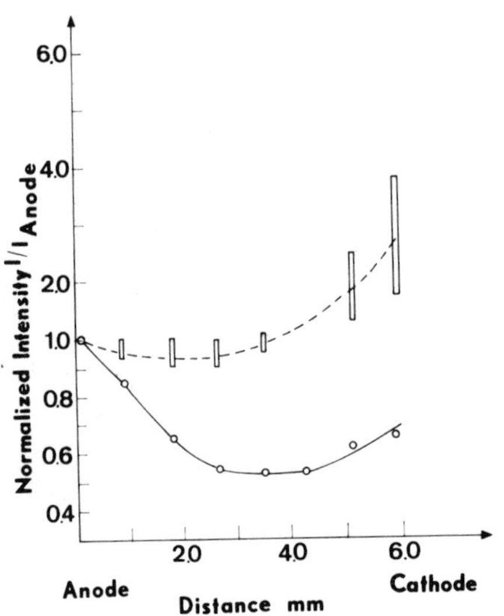

Fig. 4.8 Axial distribution of relative line intensity normalized to the anode region for a SiO_2 matrix; 6.0-mm arc gap; 13 A. ▯, Normalized intensity of common traces. ○, Normalized intensity of Si lines.

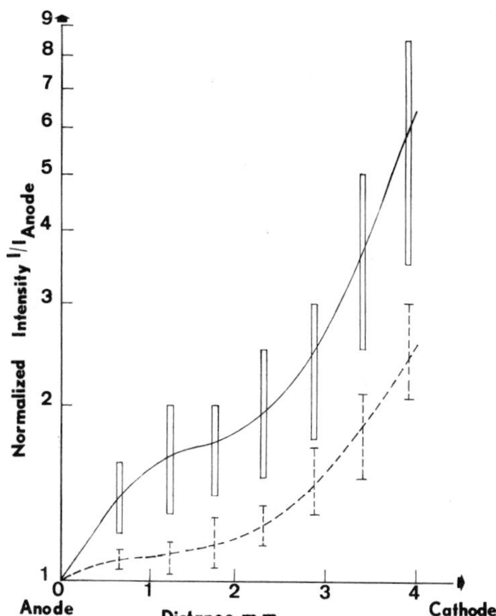

Fig. 4.9 Axial distribution of relative line intensity normalized to the anode region for a PuO_2 matrix; 6.0-mm arc gap; 13 A. ▯, Common trace elements. Ɪ, Rare earth trace elements.

4. The normalized line intensity of the trace element is not similar to the axial distribution of intensity of the matrix element in the plasma. The plasma shows a kind of separation process, Eq. (4.17), between the trace-element particles and their matrix particles.

A comparison of the data obtained from Figs. 4.5–4.9 with trace elements and matrix volatilization rates is given in Table 4.1. This suggests that the behavior of the trace elements differs from that of the matrix only in the plasma. Further investigations are needed to explain this separation process in the plasma. The investigation of the E.T.M. on plasma variables is essential for an understanding of trace-element behavior. Nevertheless, the results in Figs. 4.5–4.9 and Table 4.1 show a new approach to the spectrochemical analysis of trace elements in refractory matrices, that is, by exposing only the cathode region where the maximum separation between trace and matrix elements takes place. Such a method, called the cathode-region procedure, does not require buffers or carriers, chemical or physical separation prior to arcing. Before describing the cathode-region method, we shall describe plasma variables, which will throw light on trace-element behavior in a refractory matrix in the dc arc plasma.

4.3 PLASMA VARIABLES IN THE PRESENCE OF REFRACTORY MATERIALS

According to Mandelshtam's scheme (Section 4.2.1), the next link to be investigated after the volatilization rate is the transport phenomena of free particles in the plasma. The transport, which is controlled by the electric field, convection, and diffusion, can be evaluated using Eq. (4.8). Instead of describing the transport parameter at this point, we shall initially investigate the behavior of plasma variables (related to ψ_j) which are influenced by refractory materials, such as electric fields, temperature gradients, electron density, and particle concentrations.

4.3.1 Voltage and Electric Fields

In practice it was found that the voltage–current correlation in a dc arc does not obey Ohm's law.[1,4,5] For a given gap between the electrodes, a current increase is accompanied by a voltage decrease. Smith,[1] Finkelenburg and Maecker,[4] Boumans,[5] and Ecker[114] described the correlation voltage–current in a dc plasma as follows:

$$V = V_a + V_c + E_z \cdot l \tag{4.18}$$

$$V_a + V_c = a + \frac{c}{i} \tag{4.19}$$

$$E_z = b + \frac{d}{i} \tag{4.20}$$

where V is the voltage arc gap; V_a and V_c voltages on the anode and cathode, respectively; E_z the longitudinal electric field; l the distance between the electrodes; and a, b, c, and d are constants independent of current i and l but dependent on the gases and electrodes used.

The electric current, i, in the arc is determined by two factors: (1) the electric field created by difference in voltage between the electrodes; and (2) the conductivity caused by free electrons and ions present in the arc due to its high temperature. The difference in free electrons and ions concentration determines the space charge. With only a few doubly charged ions in the arc column, the concentration of ions is practically equal to that of the liberated electrons (i.e., the space charge in the arc column is low in respect with the total charge of free electrons).[1,114] The resulting space charge in the arc creates a field that promotes discharge.

The electric field in the arc consists of the following components (neglecting the microfield strength):

1. The longitudinal field, often called potential gradient, E_z. In the column of a carbon arc, this field is assumed to be homogeneous, since the column is almost cylindrical and the relation between voltage and arc length at constant current is practically linear [Eqs. (4.18)–(4.20)]. E_z is dependent on l, decreasing with increasing gap length,[1,115] and on the ionization potential of the gases.[1,5]

2. The radial field E_r, caused primarily by the small positive space charge inside the arc core. This field isolates electrons and ions in spite of their tendency to mix by diffusion. In ordinary arcs E_r never exceeds E_z.[1,4] According to the theory of gas discharge in tubes, E_r can be expressed[1] as

$$E_r = -\frac{kT}{e}\frac{1}{n_e}\frac{dn_e}{dr} \tag{4.21}$$

where n_e is the electron density (cm^{-3}) and r the distance from the core axis. E_r is supposed maximum at the edge of the arc core because dn_e/dr is largest there.[1]

Theoretically, the evidence of E_z and E_r is easily checked. For cylindrical symmetry the Laplace equation $\nabla^2 V = 0$, time independent, was solved for $V_a = 80$ V and $V_c = 0$ V (neglecting the electrodes phenomena.[114] Table 4.3 shows the results for a 10-mm arc gap. The Laplace solution clearly shows the existence of radial and axial electric fields within a fixed arc gap.

The experimental method for measuring the longitudinal electric field strength (E_z) is given by Eq. (4.18) and has been described elsewhere.[109,116] The handicap of the $V_a + V_c + E_z \cdot l$ method is the variability of the arc gap l during measurement; E_z is assumed constant in a given arc gap and changes only by changing l. Such an assumption is contradictory to the theoretical calculations given in Table 4.3; furthermore, no methods are available for measuring the E_r in a dc arc plasma.

Table 4.3 Spatial Distribution of Voltage in the Arc Gap, Using Laplace Equation $\nabla^2 V = 0$ in Cylindric Coordinates[a]

z (mm)		Arc voltage ± 0.01 V								
	r (mm): 0.0	0.6	1.0	2.0	3.0	4.0	5.0	6.0	8.0	10.0
Anode										
1.0	61.64	61.17	60.37	55.26	47.26	42.62	40.80	40.20	40.05	40.00
2.0	48.96	48.50	47.90	45.50	42.80	41.07	40.33	40.08	40.00	40.00
3.0	42.75	42.60	42.42	41.70	40.92	40.38	40.12	40.03	40.00	40.00
4.0	40.59	40.56	40.53	40.38	40.21	40.09	40.03	40.00	40.00	40.00
5.0	40.08	40.07	40.07	40.06	40.03	40.01	40.00	40.00	40.00	40.00
6.0	39.95	39.96	39.96	39.98	39.99	40.00	40.00	40.00	40.00	40.00
7.0	39.81	39.82	39.85	39.92	39.97	39.99	39.99	40.00	40.00	40.00
8.0	39.27	39.42	39.55	39.82	39.94	39.98	39.99	40.00	40.00	40.00
9.0	36.75	37.70	38.40	39.45	39.83	39.95	39.99	40.00	40.00	40.00
10.0	21.95	35.23	37.39	39.29	39.80	39.95	39.99	40.00	40.00	40.00
Cathode										

[a] Arc gap 10 mm; radius 10 mm; $V_a = 80$ V; $V_c = 0$ V; V center = 40 V.

We measured either E_z or E_r by passing a probe of 4 wires through the plasma. The method developed[115] is called the wire method. By passing wires through the plasma axial and radial voltages were measured, provided that:

1. Voltage values are not influenced by the material (element or alloy) and by the diameter (cross section) of the wire.
2. The voltages are not affected by the velocity of the wires passing through the plasma.
3. The values obtained from the wire methods in the vicinity of the anode and the cathode are identical to the voltage values which obtain when employing the $V_a + V_c + E_z \cdot l$ method.

In a previous publication[115] the author has affirmed conditions 1-3. Figure 4.10 shows voltage values measured by an oscilloscope employing the wire method. For details of the wire-method procedures, the reader is referred to Refs. 112–115.

Figures 4.11 and 4.12 show the variation of axial and radial electric field strengths when the third matrix (U_3O_8, ThO_2, and ZrO_2) is introduced in the graphite–air plasma for a 6.0-mm arc gap. Figure 4.13 shows the change in E_z when trace elements are added to refractory matrices. Addition of the refractory matrix to the graphite–air plasma results in an increase in both E_z and E_r values (Figs. 4.11 and 4.12), while the addition of trace elements and the third matrix results in decreases of E_z and E_r values of the plasma (Fig. 4.13).

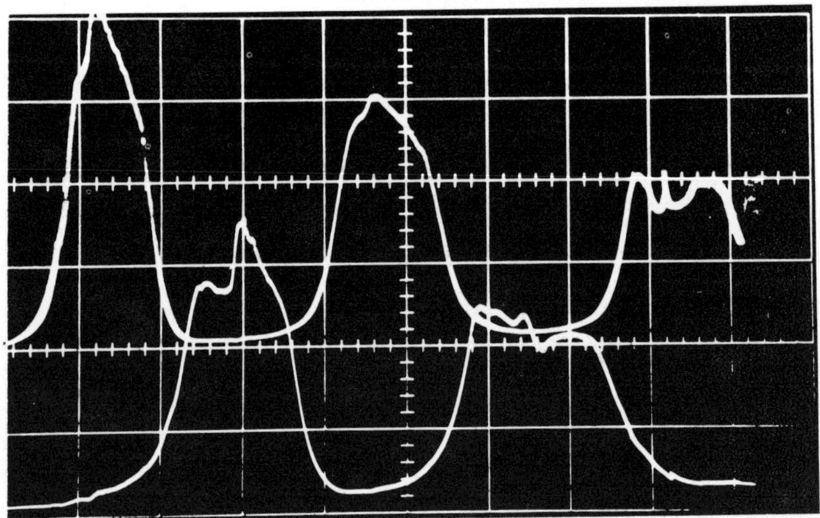

Fig. 4.10 Oscilloscopic picture for measuring voltages by the wire method. Left, anode region; right, cathode region. Abscissa, 5 m s/cm^{-1}; Ordinate, 10 V cm^{-1}.

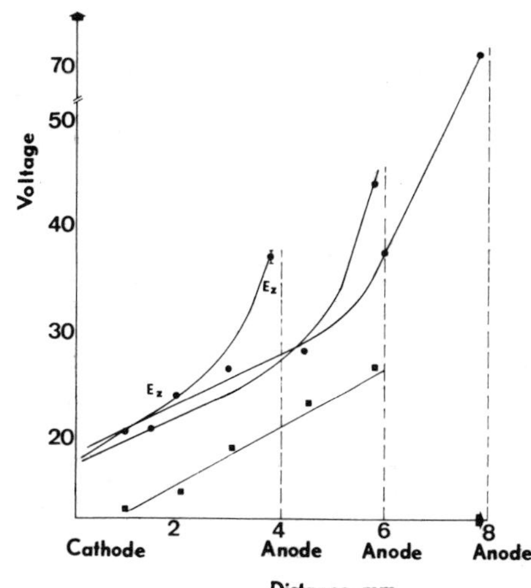

Fig. 4.11 Axial distribution of voltage by the wire method: ■, in graphite–air plasma; ●, with U_3O_8 at different arc gaps.

With the use of Eq. (4.6) (i.e., $E = j/\sigma$ for the interpretation of the results above), the following can be concluded:

1. Addition of the refractory matrix to the graphite–air plasma increases the electric field and, by that, increases the ratio of current density (j) to electric conductivity (σ). For the whole arc column, j became the current strength, i[5], which is kept constant in the dc arc. For constant i, the addition of the third matrix, increasing E, results in a decrease in the electric conductivity, σ, value. From Eq. (4.7) the product of electron mobility (μ_e) and the electron density (n_e) for the whole arc decreases by the addition of the refractory matrix to the graphite–air plasma.

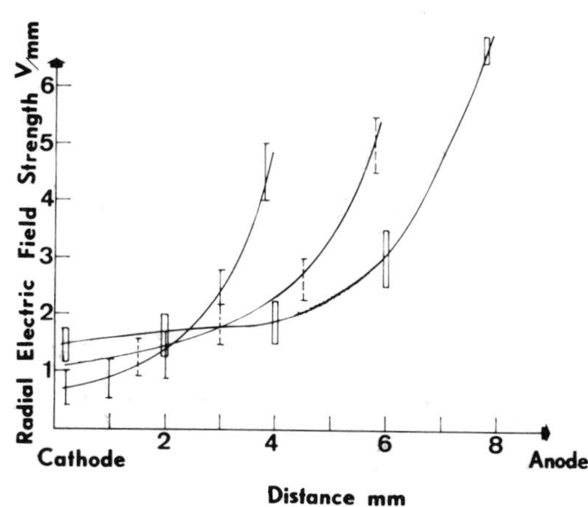

Fig. 4.12 Radial distribution of electric field strength in plasmas with U_3O_8, ThO_2, and ZrO_2 matrices; by the wire method. 1.0 mm; radius 7.0 mm. Arc gaps: I, 4.0 mm; I, 6.0 mm; ▯, 8 mm.

Table 4.4a Spatial Distribution in the Center Region of Temperature, dT/dr, Electron Density, Electric Conductivity ($\overline{\sigma}$), and Thermal Conductivity (ρ) as Calculated from the Energy Balance Equation, (4.26), for a Graphite–Air Plasma[a]

			Graphite–air plasma without third matrix							
			Measured			Calculated				
Plasma region	Radial distance R (cm)	Measured electric strength $\overline{E}$[b] (V cm^{-1})	T, Zn lines (°K)	$T/R \times 10^{-3}$ (°K cm^{-1})	$n_e \times 10^{-16}$, Saha's (cm^{-3})	ρ[c] (W cm^{-1}/deg^{-1})	Electron mobility $\mu_e \times 10^{-4}$ (cm^2 V^{-1} s^{-1})	$\overline{\sigma}$ (Ω^{-1} cm^{-1})	$n_e \times 10^{-16d}$ (cm^{-3})	$dT/dr \times 10^{-3}$ (°K cm^{-1})
Center	0–0.05		6200	124	0.20	0.022	2.85	34.0	0.7	97
	0.05–0.1	20.0	6000	120	0.08	0.017	2.75	11.0	0.15	121
	0.1–0.15		5700	114	0.03	0.011	2.50	4.1	0.07	116
Error		±20%	±200°K		±25%	±15%	±15%	±15%	±20%	±20%

[a] 6.0-mm arc gap at 10 A.

[b] $\overline{E}^2 = E_z^2 + E_r^2$. Measured by the wire method.[112] The values are for a cylinder of $R = 2$ mm.

[c] The values are calculated after Roes[2] for N_2.

[d] Calculated from the energy balance equation, (4.26), using the measured T/R values.

Table 4.4b Spatial Distribution of Temperature, dT/dr, Electric Density (n_e), Electron Conductivity ($\overline{\sigma}$), and Thermal Conductivity (ρ) as Calculated from the Energy Balance Equation, (4.26), for Plasma with Third Matrix and Trace Elements[a]

			Added U_3O_8							
			Measured			Calculated				
Plasma region	Radial distance R (cm)	Measured electric field strength $\overline{E}^b$ ($V\ cm^{-1}$)	T, Zn lines (°K)	$T/R \times 10^{-3}$ (°K cm^{-1})	$n_e \times 10^{-16}$, Saha's (cm^{-3})	ρ^c ($W\ cm^{-1}\ deg^{-1}$)	Electron mobility, $\mu_e \times 10^{-4}$ ($cm^2\ V^{-1}\ s^{-1}$)	$\overline{\sigma}$ ($\Omega^{-1}\ cm^{-1}$)	$n_e \times 10^{-16d}$ (cm^{-3})	$dT/dr \times 10^{-3}$ (°K cm^{-1})
Anode	0–0.05		6400	128	0.40	0.026	3.0	3.0	0.07	139
	0.05–0.1	90.0	6100	122	0.25	0.020	2.8	0.7	0.02	134
	0.1–0.15		5700	114	0.16	0.011	2.5	0.2	0.005	120
Center	0–0.05		6700	134	1.0	0.038	3.6	80.0	1.5	127
	0.05–0.1	20.0	6500	130	0.5	0.031	3.0	22.0	0.5	133
	0.1–0.15		6100	122	0.4	0.024	2.9	10.0	0.3	134
Cathode	0–0.05		7200	144	4.0	0.050	4.0	460.0	7.2	138
	0.05–0.1	10.0	6900	138	1.6	0.045	3.7	130.0	1.0	136
	0.1–0.15		6500	130	1.2	0.031	3.0	50.0	1.0	129
Error		±20%	±200°K		±25%	±15%	±15%	±15%	±20%	±20%

[a] 6.0-mm arc gap at 10 A.

[b] $\overline{E}^2 = E_z^2 + E_r^2$. Measured by the wire method.(112) The values are for a cylinder $R = 2$ mm.

[c] The values are calculated after Roes(2) for N_2.

[d] Calculated from the energy balance equation, (4.26), using the measure T/R values.

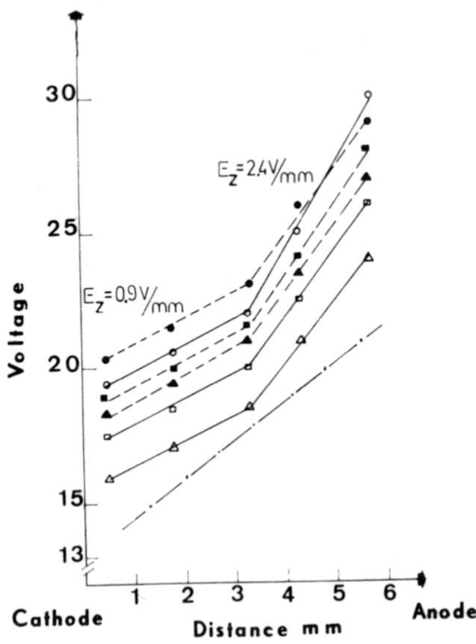

Fig. 4.13 Axial distribution of voltage as a function of trace-element concentration by the wire method. ○, □, △: 100, 500, and 1000 ppm of each common element, respectively, in U_3O_8. ●, ■, ▲: same concentrations, respectively, in ThO_2 matrix. —·—·—·: 500 ppm of each common element in graphite.

2. The addition of trace elements and the third matrix to the graphite–air plasma brings about a decrease of E_z so that the ratio i/σ also decreases (i.e., the product $\mu_e n_e$ increases for the whole arc at constant current; see Table 4.4). In other words, more ions form in the plasma from the trace element than from their refractory matrix.

4.3.2 Temperature

The method of measuring temperature exceeding 4000°K utilizes line intensity of emitted atomic or ionic spectral lines. For elaboration of the principles underlying the procedure, one is referred to Smit,[1] Boumans,[5] Orenstein and Brinkman,[117] Pearce,[118] Lochte-Holtgreven,[119] and others.[2,4,8–10] In this section we shall concentrate on temperature and its gradient in plasma with and without the third matrix.

The optical trains used for axial and radial temperature measurements, employing line intensities, are given in Fig. 4.14. For the radial measurement, an Abel inversion procedure was used.[5,111] The atom lines of zinc (3072, 3076, and 3282 Å) and copper (5105 and 5218 Å) were used as the thermometric species.[5,115] The temperature gradients were measured and compared with those calculated using the energy balance Equation (4.1).

Equation (4.15) was used for two different atomic spectral lines of the same element, Zn for example. Employing this two-lines method, temperature in

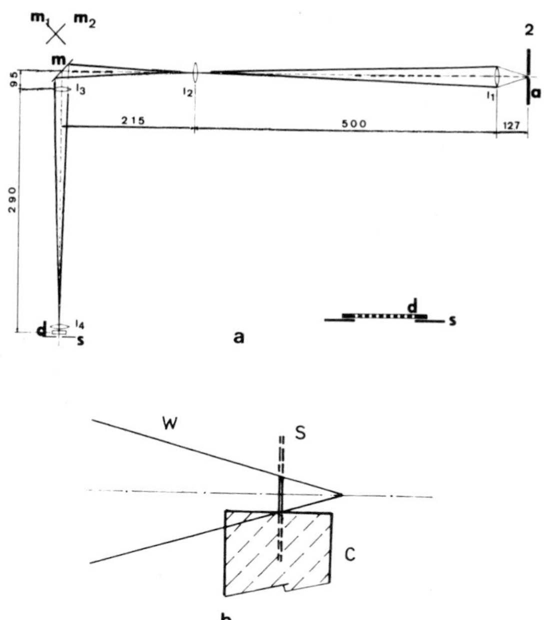

Fig. 4.14(a) Optical setup for Ebert 3.4-m spectrograph with arc gap focused on entrance slit. a, electrodes; l, lenses (fl_1 = 100 mm, fl_2 = 200 mm, fl_3 = 150 mm, fl_4 = 450 mm); m, mirror; d, diaphragm of eight holes; s, entrance slit; m_1 and m_2, mirrors on two different planes for radial measurement of line intensity. (b) Cathode region on the entrance slit. s, slit; c, cathode projection; w, wedge.

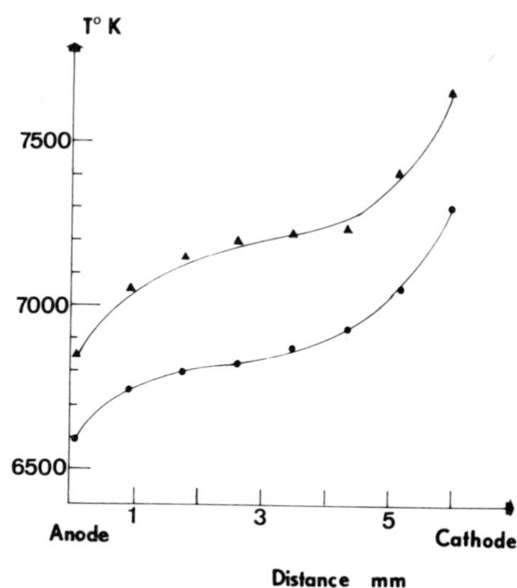

Fig. 4.15 Axial distribution of temperature in plasma at 6.0-mm arc gap. ▲, ThO_2 matrix; ●, ZrO_2 matrix.

Kelvin can be expressed as

$$T = \frac{5040(E_1 - E_2)}{\log [(Aghv)_1/(Aghv)_2] - \log (J_1/J_2)} \qquad (4.22)$$

where E_1 and E_2 are the excitation potentials in eV.

The third matrix was introduced together with trace elements. No measurable difference in temperature values were found up to a concentration of 500 ppm concentration for each trace element (B, Si, Ag, Mn, Pb, Lr, Ni, No, Ca, Sr, Ba, Bi, Co, Ge, Ga, Li, Na, Ti, and V) in the matrix.

Figures 4.15–4.20 show the measured axial distribution of temperature for different plasma matrices at a 6.0-mm gap. In the same arc gap, Figs. 4.21–4.23 show the radial distribution of temperature in plasmas with and without the refractory matrix for the same arc gap.

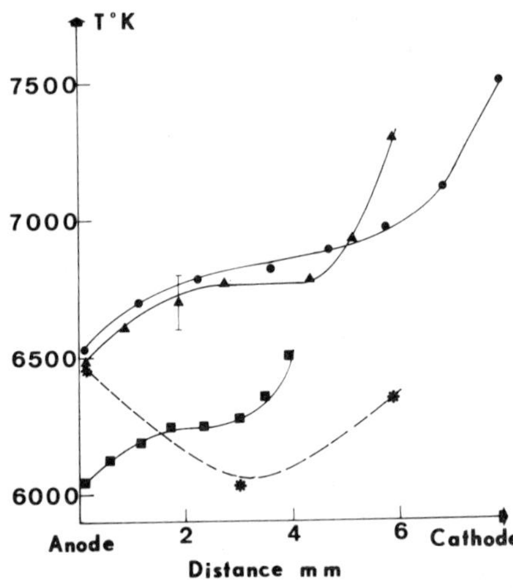

Fig. 4.16 Axial distribution of temperature in graphite–air plasma. *, with U_3O_8 matrix (▲, 6.0-mm gap; •, 8.0-mm gap); ■, with PuO_2 matrix.

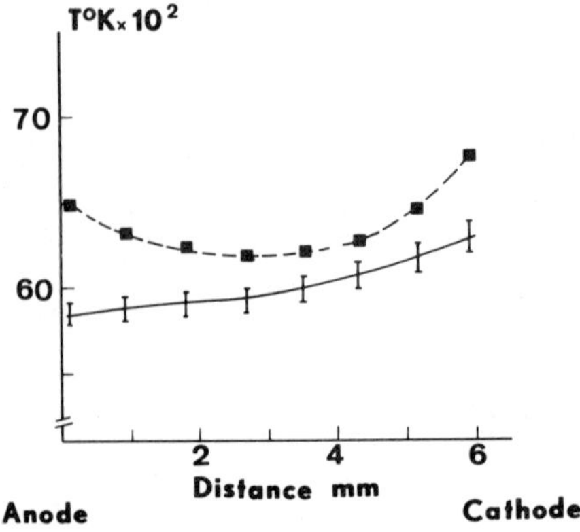

Fig. 4.17 Axial distribution of temperature in plasmas at 6.0-mm arc gap. ■, La_2O_3, Sm_2O_3 and Y_2O_3 matrix. I, Remaining rare earth oxides matrix.

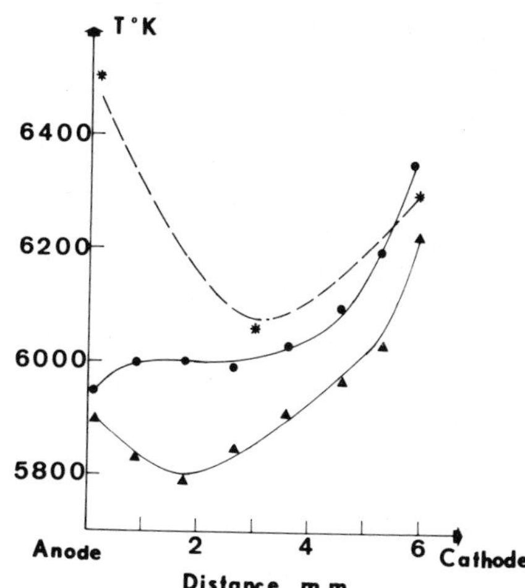

Fig. 4.18 Axial distribution of temperature in plasmas at 6.0-mm gap. *, graphite–air; •, with Al_2O_3, and ▲, with MgO matrices.

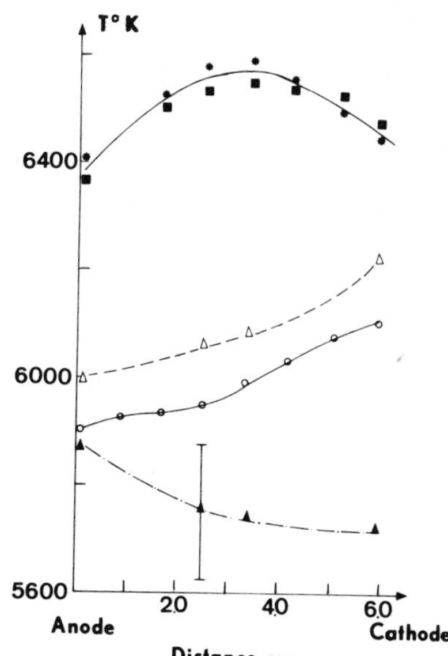

Fig. 4.19 Axial distribution of temperature in plasmas containing the following matrices at 6.0-mm arc gap. ○, SiO_2. △, SiO_2 + 5% each Ca, Mg, Al, and Fe. ▲, SiO_2 + 5% Na + 5% K. ■, SiO_2 + 12% (Na + K) + 500% graphite. *, SiO_2 + 12% (Na + K) + 300% graphite.

4.3.2.1 Axial Distribution

The axial distributions shown in Figs. 4.15–4.20 demonstrate the following:

1. Three distinct arc regions are formed: anode, central, and cathode regions. Their formation is independent of the presence of refractory matrix particles in the plasma.

2. The maximum temperature values of the graphite–air plasma occur in the anode region, while they occur in the cathode region on the addition of refractory matrices.

3. Disregarding the presence of the third matrix, temperature values in the anode region are almost the same. Outside the anode region, higher temperature values are obtained only in the presence of third matrix particles.

In a dc plasma with refractory matrix particles, the cathode region is characterized by maximum temperature values. This is in accordance with the maximum line intensity of trace elements (Figs. 4.5–4.9) in this region. However, the line intensities of the matrix elements (Figs. 4.5–4.9) in this high-temperature region are at their lowest value. This "paradox" is treated and explained in Sec-

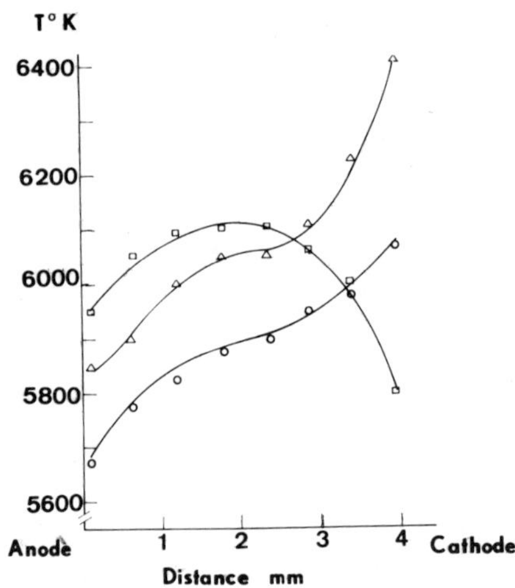

Fig. 4.20 Axial distribution of temperature in plasma with matrices at 4.0-mm arc gap. △, $Ca(PO_3)_2$; □, rock phosphate; ○, $Ca(PO_3)_2$ or rock phosphate + 20% graphite.

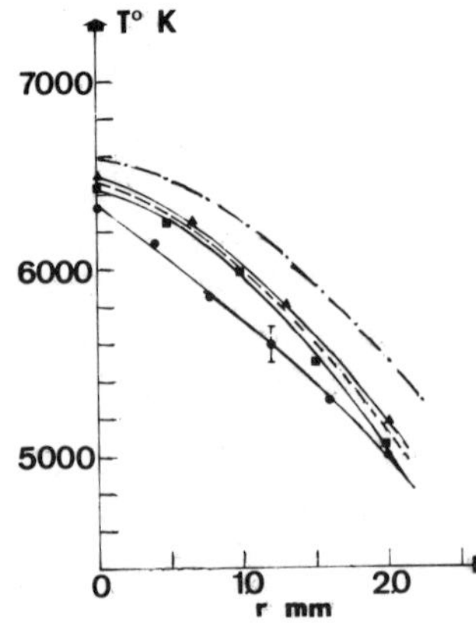

Fig. 4.21 Radial distribution of temperature in the anode region of plasmas with: U_3O_8 at ●, 4.0 mm; ■, 6.0 mm; and ▲ 8.0 mm; ZrO_2 at ---, 6 mm; and ThO_2 at —·—·—, 6-mm arc gaps.

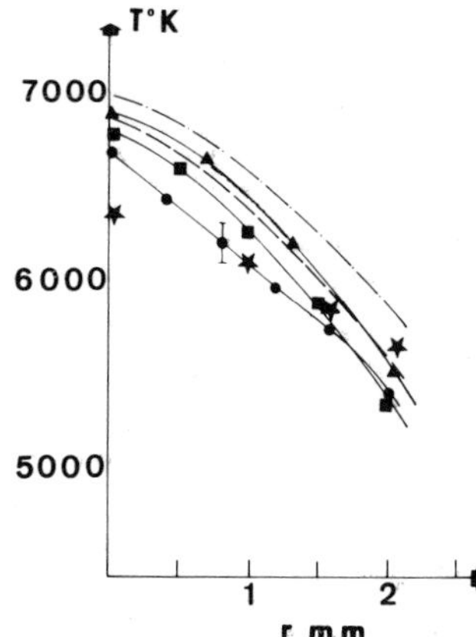

Fig. 4.22 Radial distribution of temperature in the central region of plasmas with: U_3O_8 at •, 4.0 mm; ■, 6.0 mm; and ▲, 8.0 mm; ZrO_2 at – –, 6.0 mm; ThO_2 at —·—·, 6 mm; and graphite-air at ★ 6.0-mm arc gap.

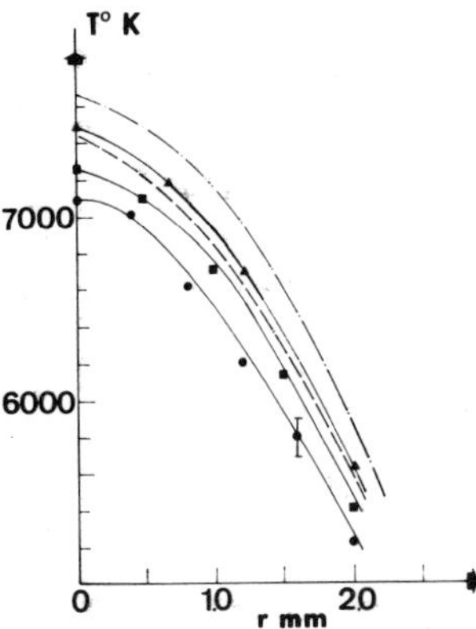

Fig. 4.23 Radial distribution of temperature in the cathode region of plasmas with: U_3O_8 at •, 4.0 mm; ■, 6.0 mm; and ▲, 8.0 mm; ZrO_2 at – – –, 6 mm; and ThO_2 at —·—·, 6.0-mm arc gap.

tion 4.3.4, in connection with total particle concentration in the plasma of both matrix and trace elements.

4.3.2.2 Radial Distribution

The radial distribution of temperature illustrated in Figs. 4.21–4.23 shows the following features:

1. For a graphite-air plasma the radial temperature gradient of the central region differs from those of the other two regions; in the center the gradient is smoother.

2. With the refractory matrix, the radial temperature gradient is almost the same in all regions of the plasma.

3. The rate of radial temperature decrease in the case of the refractory matrix is greater than for the plasma without it. In the graphite-air plasma, the

radial decrease is smooth up to a 2.0-mm radius, as opposed to the sharp decrease up to a 2.0-mm radius with the addition of the third matrix (Fig. 4.22).

In order to comprehend this behavior of the radial temperature gradient in such plasmas, we shall recall the energy balance equation, (4.1). In an ideal arc, where the symmetry axis of the electrodes coincides with that of the plasma, Eq. (4.1) for cylindrical (z, r) symmetry assumes the form

$$\frac{1}{r}\frac{\partial}{\partial r}(r\rho \text{ grad } T) = jE \tag{4.23}$$

Integrating and using $dT/dr = 0$ as a boundary condition along the z axis, Eq. (4.23) gives

$$-r\rho \frac{dT}{dr} = \int_0^r jEr\, dr \tag{4.24}$$

showing that the electric energy dissipation in a cylinder of unit height ($dz = 1$) is equal to the heat leaving this cylinder per second.

For a set of concentric (around the axis) cylinders of radius $r = 0$ to $r = R$ and unit height, it was shown[108] that using mean values for $j(\bar{j})$ and $E(\bar{E})$, Eq. (4.24), with $\bar{j} = 1/\pi R^2 \int_0^R$, then $j \cdot r\, dr$ takes the form

$$\bar{j} \cdot \bar{E} = -R\sigma \frac{dT}{dR} \cdot \frac{1}{\pi R^2} \tag{4.25}$$

Using Eq. (4.6), then Eq. (4.25) can be written for the radial temperature gradient as

$$s\frac{dT}{dr} = \bar{E}^2 \cdot \frac{\bar{\sigma}}{R} \cdot \rho \tag{4.26}$$

in which $s = \pi R^2$.

The calculation of the mean electric conductivity, accounting for mean electric field strength E and electron density (Section 4.3.3), was described in another publication.[112] Thermal conductivity for N_2 was evaluated as described by Roes[2] and Avni and Klein.[112] Therefore, dT/dr at a given height in the plasma can be evaluated according to Eq. (4.26). The radial temperature gradient and thermal and electric conductivity in the central region of the plasma with and without the third matrix are shown in Table 4.4. As shown in Table 4.4, the calculated dT/dr, Eq. (4.26), and the measured T/R values almost coincide. Moreover, the values of $\bar{\sigma}$, ρ, and dT/dr for plasma with and without third matrix show the following:

1. The mean electric conductivity $\bar{\sigma}$ for a graphite–air plasma has about one half the value of a plasma with the added U_3O_8 matrix and its trace elements.

2. In radial ranges from the arc axis, the excitation temperature T measured by the two-lines method will represent plasma temperature if and only if local thermodynamic equilibrium (T.E.) was achieved in such plasma regions.

3. The Elenbaas equation, Eq. (4.26), requires only steady-state conditions in the plasma. The dT/dr values equalize those of T/R in the same plasma region, so that the excitation temperature (two-lines) represents the plasma temperature in the involved region (i.e., L.T.E. were achieved in dc arcs with or without the refractory matrix).

4.3.3 Electron Density

The electron density (n_e) in a dc arc plasma is generally evaluated by the Saha method.[1,4,5,111,119] This method requires local thermodynamic equilibrium in the plasma for any intensity ratio of an ionic to an atomic spectral line of a given chemical element (trace or matrix) when used for the electron density evaluation. As for the temperature measurement, only few elements having well-established transition probabilities and partition function values are prerequisite. Another requirement is that the spectral lines show no self-absorption. Magnesium, manganese, and chrome spectral lines[5,109,112] are mainly used for the n_e evaluation.

The electron density obtained from line intensity measurements, using Saha's relationship, has the following form:

$$K_i = \frac{n_e \cdot n_i}{n_a} = \frac{(2\,KT\,m_e)^{3/2}}{h^3}\,\frac{2Z_i}{Z_a}\exp\left(-\frac{V_i}{kT}\right) \tag{4.27}$$

where K_i is Saha's constant for the ionization process; n_i/n_a the relative ion and atom densities; k the Boltzmann constant; h Planck's constant; m_e the electron mass; T the absolute temperature; Z_a and Z_i the partition functions of the atom and ion, respectively; and V_i the ionization potential of the element involved.

The n_i/n_a ratio is obtained from the spectral line intensity ratio J_i/J_a as given by the relation

$$\frac{n_i}{n_a} = \frac{J_i}{J_a}\,\frac{(\nu gA/Z)_a}{(\nu gA/Z)_i}\exp\left(E_i - \frac{E_a}{kT}\right) \tag{4.28}$$

where ν is the line frequency, g the statistical weight, A the transition probabilities, and E the excitation potential; indices i and a refer to ion and atom, respectively.

Equation (4.29) follows from Eqs. (4.27) and (4.28), from which the electron density is calculated:

$$n_e = K_i\left(\frac{n_a}{n_i}\right) \tag{4.29}$$

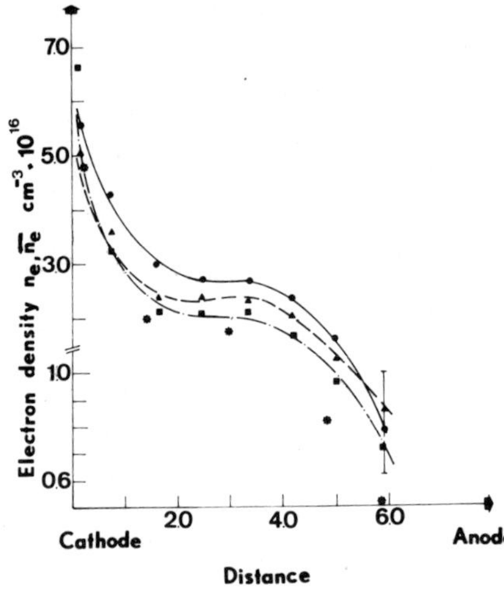

Fig. 4.24 Axial distribution of electron density (n_e) in plasmas at 6.0-mm arc gap. •, U_3O_8; ▲, ZrO_2; ■, ThO_2. *, Mean electron density (n_e) in plasma with U_3O_8.

The optical setups are the same for the axial and the radial measurements of line intensity as those used for temperature measurement (see Section 4.3.2). The same experimental data described for temperature measurement (Section 4.3.2) were used for n_e evaluations. The following spectral line ratios were used: Mn II 2576/Mn I 2794, Mg II 2795/Mg I 2852, Cr II 2835/Cr I 3021, and U II 4347/U I 4246. The third matrix in the anode crater contained 200 ppm of each trace element.

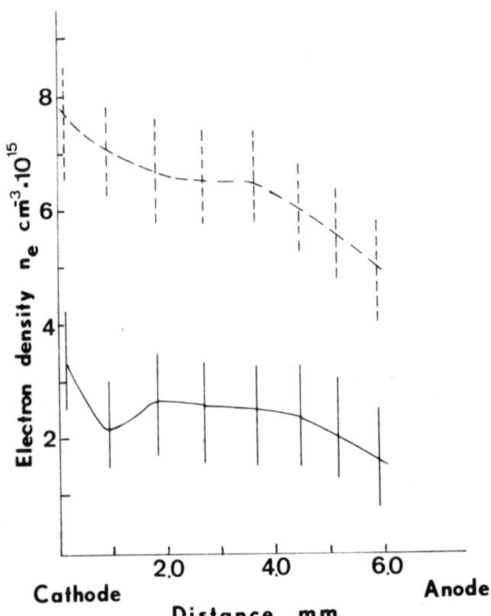

Fig. 4.25 Axial distribution of electron density (n_e) in plasmas with rare earth oxides at 6.0-mm arc gap. ¦ , La_2O_3, Y_2O_3, and Sm_2O_3; |, with each of remainder of rare earth oxides.

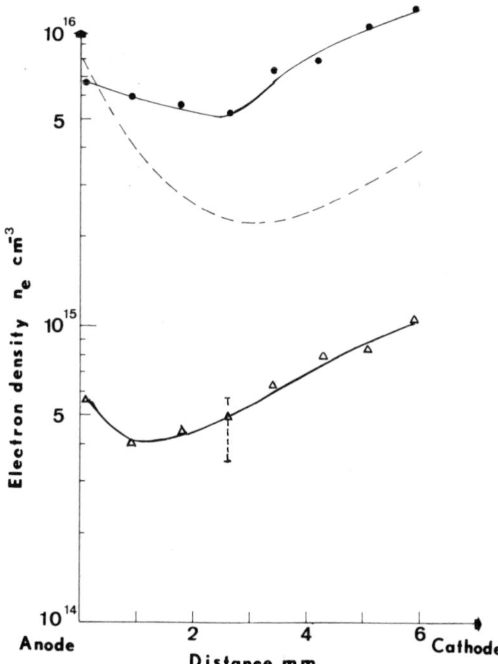

Fig. 4.26 Axial distribution of electron density (n_e) in the graphite–air plasma (----) and with •, Al_2O_3, or ▵, MgO at 6.0-mm arc gap.

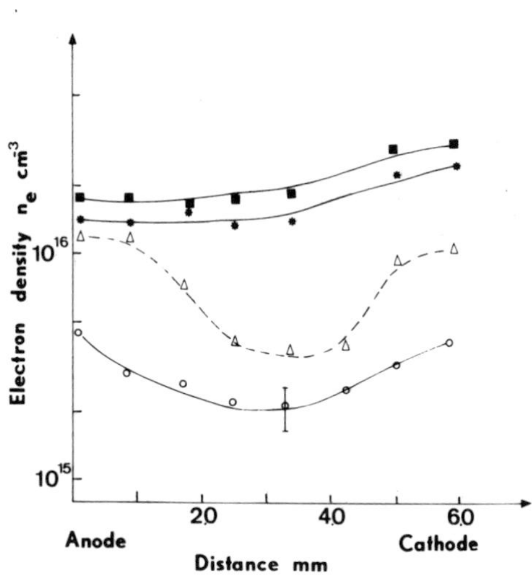

Fig. 4.27 Axial distribution of electron density (n_e) in plasmas with silicates at 6.0-mm arc gap. ○, SiO_2; ▵, SiO_2 + 5% each Ca, Mg, Al, and Fe; ▪, SiO_2 + 12% (Na + K) + 500% graphite; ∗, SiO_2 + 12% (Na + K) + 300% graphite.

Figures 4.24–4.27 show the axial distribution of electron density for 6-mm arc plasmas with and without various refractory matrices. For the same arc gap radial distributions of n_e are shown in Fig. 4.29.

4.3.3.1 Axial Distribution

The following features of the axial distribution of the free electron density are depicted in Figs. 4.24–4.27.

1. Distinct anode central and cathode regions were formed in plasmas with the refractory matrix particles. Those three regions are not well defined in the graphite–air plasma (Fig. 4.26).

2. In the graphite–air plasma, maximum n_e values occur in the anode region. Addition of the refractory material shifts the maximum values of n_e to the cathode region.

3. Electron density values in the plasmas containing a third matrix, except MgO, are greater by approximately one order of magnitude than those of the graphite–air plasma, i.e., $n_{e(T \cdot M)} > n_e$.

The fact that $n_{e(T \cdot M)} > n_e$ and that the highest value is located in the cathode region requires elaboration. According to Saha's relationship, higher

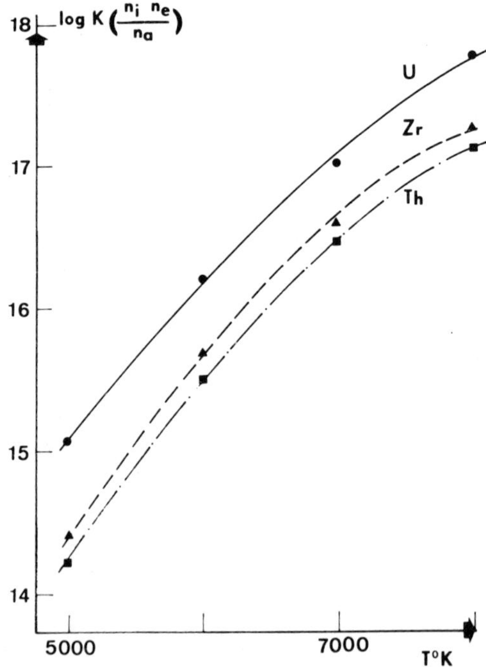

Fig. 4.28 Saha's constant for ionization (log K_i) as function of plasma temperature with ●, U_3O_8; ▲, ZrO_2; and ■, ThO_2.

values of electron density imply higher values of n_a/n_i ratios and/or high values for the ionization constant K_i, Eq. (1.20). Since K_i is a function of temperature, the higher temperature of the plasma of the third matrix (Section 4.3.2) also means an increase in the K_i values. Figure 4.28 shows the dependence of log K_i with temperature. For uranium an increase of 1000°K from 6000 to 7000°K results in an increase by one order of magnitude of K_i values. Then $n_{e(T \cdot M)} > n_e$ cannot be explained by an increase in the n_a/n_i ratio. An increase in this ratio means a decrease in the ion concentration (n_i) in the plasma. Such a decrease is paradoxical by definition because $n_e = n_i$ for the first ionizable atoms. The reason for the high n_e values is the high K_i values resulting from the increase in temperature of plasmas of the refractory materials. Moreover, from Eq. (4.7) and the data in Table 4.4 for the mean electric conductivity ($\bar{\sigma}$), high n_e values are accompanied by higher values for $\bar{\sigma}$.

The maximum electron density values found in the cathode region of a plasma with the third matrix as indicated by $n_e = n_i$ means that the maximum ion concentration should also occur in the region. Table 4.5 shows the normalized intensity ($J_{\text{cathode}}/J_{\text{anode}}$) for some trace-element atom and ion spectral lines. The ionic normalized intensity is almost twice as great as the atomic normalized intensity in the cathode region. This indicates that more ion particles are in the cathode region, Eq. (4.17).

Table 4.5 Normalized Spectral Line Intensity $J_{\text{cathode}}/J_{\text{anode}}$ for Some Ionic and Atomic Spectral Lines of Trace Elements in U_3O_8 and SiO_2 Matrices[a]

Spectral line (Å)	$J_{\text{cathode}}/J_{\text{anode}}$	
	U_3O_8	SiO_2
Fe (II) 2599	6.0	3.0
V (II) 3110	4.0	0.9
Cd (II) 2288	4.8	2.8
Cr (II) 2835	5.8	3.5
Mg (II) 2802	6.9	3.3
Mn (II) 2576	8.4	5.0
Fe (I) 2488	3.5	0.83
V (I) 3138	0.95	0.9
Cd (I) 3261	2.4	1.2
Cr (I) 3021	2.2	1.2
Mg (I) 2852	3.3	0.8
Mn (I) 2794	3.7	2.0

[a] 6.0-mm arc gap at 12 A; 100 ppm each trace element.

4.3.3.2 Radial Distribution

The energy balance equation, (4.25), with the measured value of T/R (temperature gradient) can now be used for evaluation of the mean electric conductivity, $\bar{\sigma}$, as described in a previous publication.[112] From $\bar{\sigma}$ values the mean electron density $\overline{n_e}$ was evaluated from Eq. (4.7). Tables 4.4a and 4.4b show the $\overline{n_e}$ radial values in the cathode, center, and anode regions with and without the addition of the third matrix. Figure 4.29 shows the radial distribution of both electron densities, one derived from atomic and ionic spectral line intensity (i.e., n_e) and second derived from the energy balance equation, (4.25), and Eq. (4.7) (i.e., $\overline{n_e}$).

The radial distribution of electron density in Fig. 4.29 and Table 4.4 shows the following features:

1. Good agreement was obtained between n_e and $\overline{n_e}$ values in the cathode

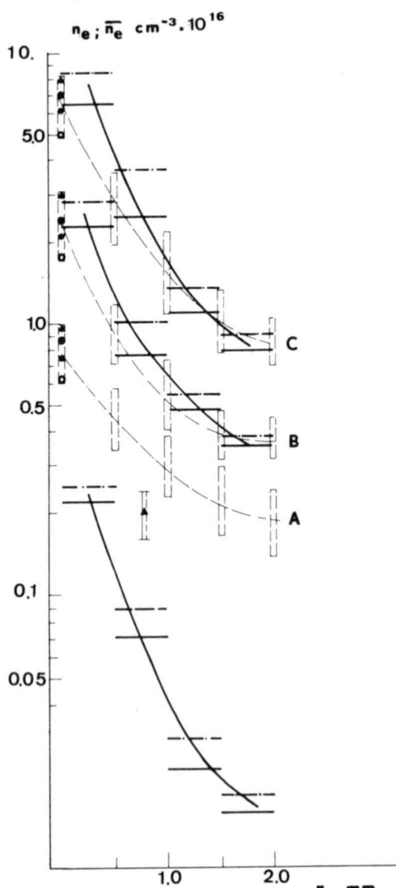

Fig. 4.29 Radial distribution of electron densities n_e and $\overline{n_e}$ for 6.0-mm gap in the cathode (C), central (B), and anode (A) regions. n_e measurements (--- curves) for plasmas with ▲, La_2O_3; •, U_3O_8; *, ZrO_2; and ○, ThO_2. $\overline{n_e}$ calculations (horizontal bars) for plasmas with —·—·— La_2O_3 or — ZrO_2.

and center regions (Fig. 4.29), while agreement is relatively poor in the anode region. The equality between n_e (Saha equation, which requires thermodynamic equilibrium) and $\overline{n_e}$ (energy balance, which requires steady-state conditions) indicate that L.T.E. conditions were reached in the cathode and central regions.

2. From the dc arc axis up to a radial distance of 2.0 mm, the electron density values decrease almost by one order of magnitude. The rate of electron density decrease is about the same in the anode, center, and cathode regions.

3. In the graphite–air plasma, the $\overline{n_e}(r)$ values are lower by a factor of 3 than its value in the presence of the refractory matrix (Tables 4.4a and 4.4b); that is, the mean electric conductivity ($\bar{\sigma}$) is greater in plasmas of the third matrix, in accordance with the conclusions reached in Section 4.3.1.

4.3.4 Free Particle Concentration

In this section we shall deal with two types of particle concentration, n_j and n_{t_j}. They can be defined as follows:

$$n_j = \sum n_{a_j} + \sum n_{i_j} \tag{4.30}$$

and

$$n_{t_j} = n_j + \sum n_{m_j} \tag{4.31}$$

The concentration n_j represents a summation over the concentration of atoms (n_{a_j}) and ions (n_{i_j}) of element j. The total particle concentration, n_{t_j}, also takes into account the summation of the molecular concentration (n_{m_j}).

The free particle concentration n_j in a plasma is evaluated from absolute line intensity measures[5, 109] [see Eq. (4.15)]. In this method one assumes only the presence of atoms and ions in the plasma (i.e., the element j does not form stable molecules). Presence of molecules of element j is neglected by convenience in the calculations, owing to two factors: (1) inadequate data on the molecular spectra of j[120]; and the fact that (2) calculation of n_{t_j} is complicated even if the molecular spectrum of j is known.[121] On dealing with trace elements, the n_j calculation is legitimate, since their concentration in the plasma is small. However, one should proceed with caution for elements which form stable molecules with oxygen, carbon, and nitrogen,[12] or with the refractory matrices. For a refractory matrix element, n_{t_j} has to be evaluated instead of n_j.

There is no suitable method available for the evaluation of n_{t_j}. For the present purpose, evaluation of n_{t_j} suffices without having to identify each type of matrix molecule present in the plasma. Whether the molecules are carbides, oxides, or nitrides is of no interest as long as they are computed as a whole.

A method was developed for measuring molecule, ion, and atom concentrations. The wire method, as described in Section 4.2.2, was used to measure the spatial distribution of particle flux $n_{t_j}(r, z)v_j$ of the matrix element by neutron activation analysis or by spectrographic analysis of the deposit on the wires.[17]

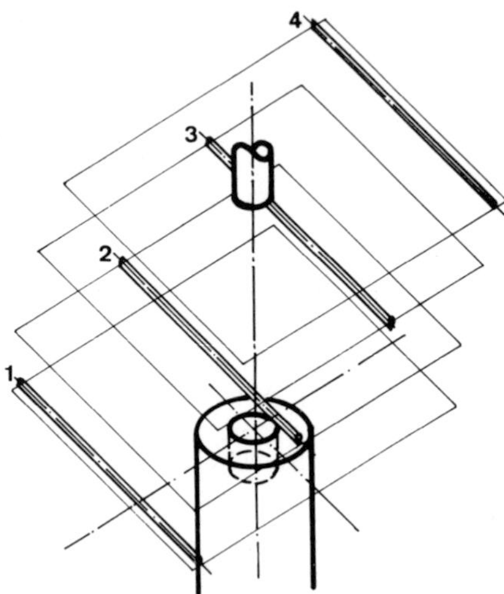

Fig. 4.30 Schematical wire disposal for measuring the total particle concentrations (n_t) in plasmas; one anode region through four cathode region.

Figure 4.30 shows the wire arrangement in the carriage released through the plasma. The particles on the wires were analyzed for U and La by neutron activation analysis[17] and for the other matrix elements by spectrochemical analysis (after immersing the wires in HNO_3). The result obtained from the wires corresponds to the particle flux $n_{t_j}(z, r)v_j$ of the matrix.[17] Figure 4.31 shows

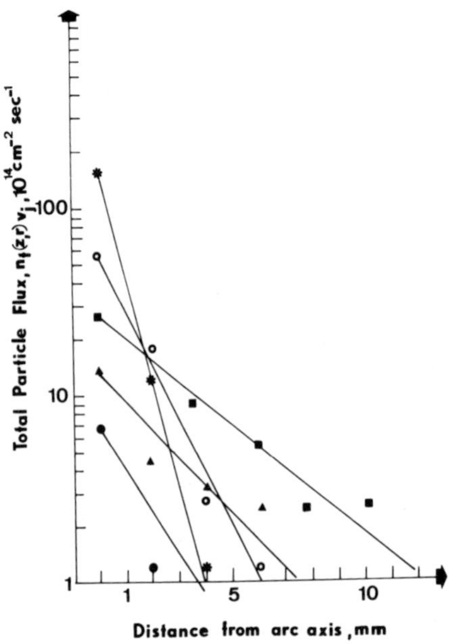

Fig. 4.31 Radial distribution of uranium (U_3O_8) total particle flux at different heights (z) in the plasma as derived from 2.0-mm wire portions. Gap, 8.0 mm; carriage velocity, 110 cm s^{-1}. Distance above anode: z = 0.2 mm, ∗; z = 2.2 mm, ○; z = 4.0, ■; z = 5.9 mm, ▲; and z = 7.8 mm, ●.

the radial distribution at various heights (z) of the particle flux. The total particle concentration was calculated from the flux, and the data from Fig. 4.31 were fit to exponential or Bessel functions. Thus, an experimental solution of the dn_t/dt model[5, 109] was obtained. This model requires the evaluation of the axial particle velocity v_j, as will be shown subsequently.

4.3.4.1 The dn_t/dt Model

The dn_t/dt model as described by Boumans[5, 122] considers flow of particles through an infinitesimal element of volume and can be expressed in the form of cylindrical coordinates (r, z) for a steady state:

$$\frac{dn_t}{dt} = D\left[\frac{1}{r}\frac{\partial}{\partial r}\left(r\frac{dn}{dt}\right) + \frac{\partial n^2}{\partial z^2}\right] - v_j\frac{\partial n}{\partial z} = 0 \tag{4.32}$$

where D and v_j are diffusion coefficients and the axial particle velocity, respectively. Solution of this partial differential equation for a 10-mm arc gap with a core radius (R) of approximately 5 mm was given by Ginsel[123] and Bavinck.[124] Ginsel[123] considered the supply of particles to the plasma from a point source; at appropriate boundary conditions[122] his solution has an exponential form:

$$n_t(z, r) = \frac{Q_j}{2\pi D\sqrt{z^2 + r^2}} \exp\left[-\frac{v_j}{2D}\left(\sqrt{z^2 + r^2} - z\right)\right] \tag{4.33}$$

Bavinck[124] considered a disc-shaped source for particle supply to the plasma: at appropriate boundary conditions (5) his solution is a first-order Bessel function:

$$n_t\left(\frac{z}{R}, \frac{r}{R}\right) = \frac{2Q_j}{\pi a D} \sum_{n-1}^{\infty} \frac{J\left(n\frac{a}{R}\right)}{B'\lambda n[J_0^2(\lambda_n) + J_1^2(\lambda_n)]} J_0\left(\lambda_n\frac{r}{R}\right)\exp\left(\frac{z}{R}B\right) \tag{4.34}$$

where $B' = \sqrt{W^2 + \lambda n^2} + W$, $B = \sqrt{W^2 + \lambda n^2} - W$, $W = v_jR/2D$, λn is an infinite series of positive numbers satisfying the boundary conditions, J_0 and J_1 and zero- and first-order Bessel functions, R is the radius of the arc core, and a is the radius of the anode crater. With R and Q_j known, the function $n_t(z, r)$ can be calculated from Eq. (4.34) using appropriate values of a, D, and v_j.

In a separate publication[125] it was demonstrated by the author that the results given in Figs. 4.31 and 4.32 can be fitted to Eq. (4.34). In this case the ratio $v_j/2D$ is obtained directly from the slope[125] of the exponential plots of particles flux $n_t(z, r)\,v_j$ in Figs. 4.31, 4.32, and 4.33 for uranium and lanthanum.

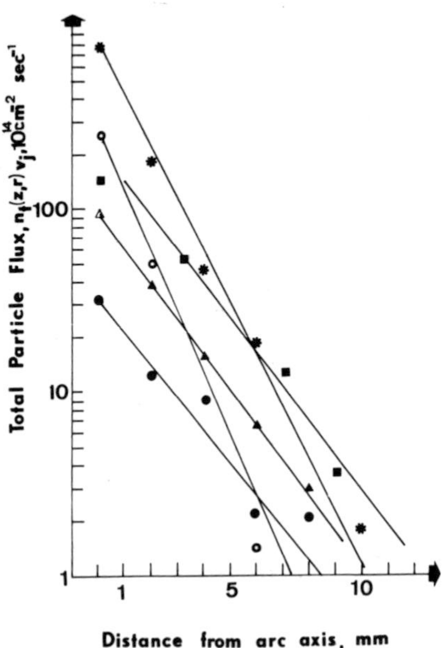

Fig. 4.32 Radial distribution of uranium (U_3O_8 + 4% UF_4) total particle flux at different heights (z) in the plasma, as derived from 2.0-mm wire portions. 8.0-mm gap; carriage velocity 110 cm s^{-1}. Distance above anode: z = 0.2 mm, ∗; z = 2.2 mm, ○; z = 4.0 mm, ■; z = 5.9 mm, ▲; and z = 7.8 mm, ●.

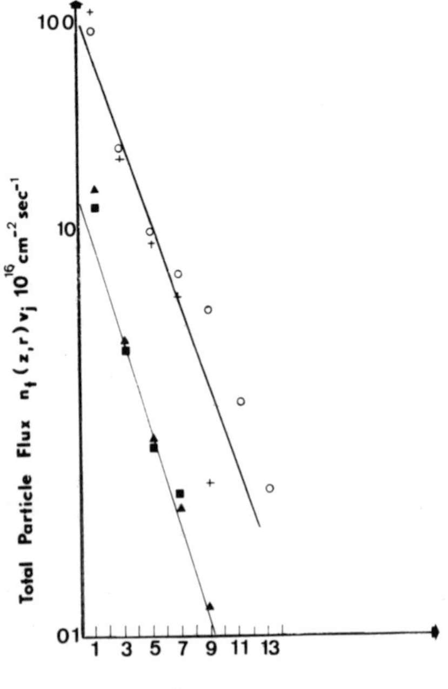

Fig. 4.33 Radial distribution of La total particle flux at different heights (z) in the plasma as derived from 2.0-mm wire portions. Gap, 8.0 mm; carriage velocity, 110 cm s^{-1}. ▲, ■; La_2O_3, cathode and anodes, respectively; +, ○; La_2O_3 + 4% LaF_3, cathode and anode regions, respectively.

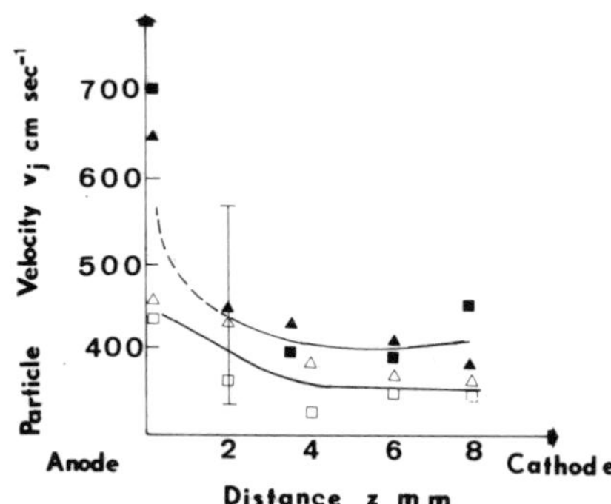

Fig. 4.34 Axial distribution of particle velocity (v_j). ▲, U_3O_8; ■, U_3O_8 + 10% UF_4; △, La_2O_3; □, La_2O_3 + 10% LaF_3.

4.3.4.2 Particle Velocity

Values of $v_j/2D$ were obtained from the slopes of the lines shown in Figs. 4.31 and 4.32 using Eq. (4.34). From this parameter, the axial particle velocity v_j is obtained after calculation of the diffusion coefficient D. The following equation[109, 125] was used for D:

$$D = \text{const} \cdot \frac{T^{1.75}}{\sqrt{M^* d_c}} \tag{4.35}$$

in which M^* is the reduced mass, d_c the diameter of the diffusing particle, and T the absolute temperature in the given region of the dc arc. The diffusion coefficient was calculated using Eq. (4.35) for different temperatures of the various regions in the plasma ($1 < r < 3$ mm and z up to 8.0 mm).

Figure 4.34 is a plot of the experimental values of the axial velocity of uranium and lanthanum particles. Within the experimental error, v_j is constant over the arc gap except for the anode region.

4.3.4.3 Total Particle Concentration of the Third Matrix

In a previous paper[16] the proportionality factor between the particle collected by the wire and the particle flux in the plasma was shown to be approximately unity within the experimental error [also see Eq. (4.13)]. In other words, absolute values of the particle flux in the plasma can be measured using the wire method. Using v_j values the experimental total particle concentrations were obtained from the particle flux. Table 4.6 shows the axial distribution of n_{t_j} together with the particle concentration n_j calculated from the absolute intensity of uranium (4289 and 4310 Å) and lanthanum (2791 and 2725 Å) lines. According to this table, the concentration of molecules of uranium or lanthanum in the arc plasma (R = 2.0 mm) is approximately 20 times that of their atom and ion concentrations.

Table 4.6 Axial Distribution of Matrix Element Particle Concentration (n_t and n_j) in U_3O_8, ThO_2, and ZrO_2 Matrices[a]

	Total particle concentration (cm^{-3})								
	$n_t \times 10^{-11}$ [b]			$n_j \times 10^{-11}$ [c]			$n_t - n_j/n_j$		
z (mm)	U	Th	Zr	U	Th	Zr	U	Th	Zr
Anode									
0.2	270.0	280.0	220.0	11.0	13.0	6.0	23.5	20.5	35.7
1.8	110.0	140.0	100.0	6.0	9.0	4.0	17.3	14.6	24.0
3.5	90.0	100.0	60.0	5.0	7.0	2.0	17.0	13.4	24.0
6.0	110.0	80.0	30.0	4.0	5.0	1.5	26.5	15.0	19.0
7.8	60.0	60.0	20.0	3.0	4.5	1.0	19.0	12.2	19.0
Cathode									

[a] 8.0-mm arc gap at 10 A; core radius, 2.0 mm; 35-s exposure.
Spectral line used: U(I) 4246 Å, U(II) 4310 Å; Th(I) 4012 Å, Th(II) 4025 Å; Zr(I) 4236 Å, Zr(II) 4231 Å.
[b] Wire method. Mean value over four carriages of wires. Relative standard deviation 25%.
[c] Mean values over four spectra (arc focused on slit). Relative standard deviation 20%.

The high concentration of molecules and the relatively low concentration of atoms and ions of the third matrix can explain the relatively high temperature of the plasma in spite of the relatively low first ionization potentials of these elements.

Table 4.7a shows the radial distribution of n_t of the third matrix element in the anode central and cathode regions of the plasma. High radial diffusion of U

Table 4.7a Radial Distribution of Total Particle Concentration (n_t) of Uranium (U_3O_8 Matrix) and Lanthanum (La_2O_3 Matrix)[a]

Plasma region	r (mm)	$n_t \times 10^{-11}$ (cm^{-3}) U_3O_8 matrix	La_2O_3 matrix	Relative standard deviation[b] (%)
Anode	0–2	270.0	30,000	25
	2–4	35.0	—	25
	4–6	5.0	—	25
Center	0–2	100.0	3,500	25
	2–4	45.0	—	35
	4–6	9.0	—	35
Cathode	0–2	60.0	27,000	25
	2–4	30.0	—	35
	4–6	7.0	—	35

[a] 8.0-mm arc gap at 10 A; 35-s exposure.
[b] Obtained from the passage of four carriages (i.e., 16 wires) in the anode region; 8 wires in the center region; 4 wires in the cathode region.

Table 4.7b Total Particle Concentration Ratio Normalized to the Value in the Anode Region[a]

Plasma region	Radial distance (mm)	n_t/n_t anode		
		U	Th	Zr
Anode	0–2	1.0	1.0	1.0
	2–4	1.0	1.0	1.0
	4–6	1.0	1.0	1.0
Center	0–2	0.37	0.32	0.32
	2–4	1.30	0.80	1.0
	4–6	1.80	0.90	1.25
Cathode	0–2	0.22	0.21	0.14
	2–4	0.86	0.80	0.67
	4–6	1.40	1.00	1.25

[a]Data calculated from Table 4.7a.

or La particles in the anode region decreases toward the cathode. Along the axis a lower particle concentration was found in the cathode region as compared to n_t values in the anode region. If we bear in mind that the width of the plasma in the cathode region exceeds that in the anode region,[(115)] the decrease toward the cathode can be explained by axial movement of particles outside the arc core. Therefore, for radii larger than 2 mm, n_t is almost constant at different heights in the arc gap,[(125)] as shown in Table 4.7b.

4.3.4.4 Particle Concentration of Trace Elements

The particle concentrations of several trace elements $n_{j_{\mathrm{Tr}}}$ were calculated only in the arc core ($R \simeq 2.0$ mm) as a function of height in the plasma. This axial distribution of $n_{j_{\mathrm{Tr}}}$ was calculated on the assumption that stable molecules do not form in the plasma. The axial distribution of $n_{j_{\mathrm{Tr}}}$ for trace elements in U_3O_8 and La_2O_3 matrices as well as a comparison with the n_j of uranium and lanthanum elements is given in Figs. 4.35 and 4.36. For a 100-ppm concentration for each trace element, $n_{j_{\mathrm{Tr}}}$ values are smaller by two orders of magnitudes than that of n_j of the matrix element in the central and cathode regions of the plasma. In the anode region $n_{j_{\mathrm{Tr}}}$ is by about three orders of magnitude smaller than n_j of the matrix indicating that radial diffusion of trace elements is smaller than that of the matrix element.

A comparison of $n_{j_{\mathrm{Tr}}}$ (Figs. 4.35 and 4.36) with n_t values for matrix elements in the anode region Table 4.6 shows that $n_t > n_{j_{\mathrm{Tr}}}$ by about 10^4. This is in good agreement with the 100-ppm trace-element content in the solid matrices.

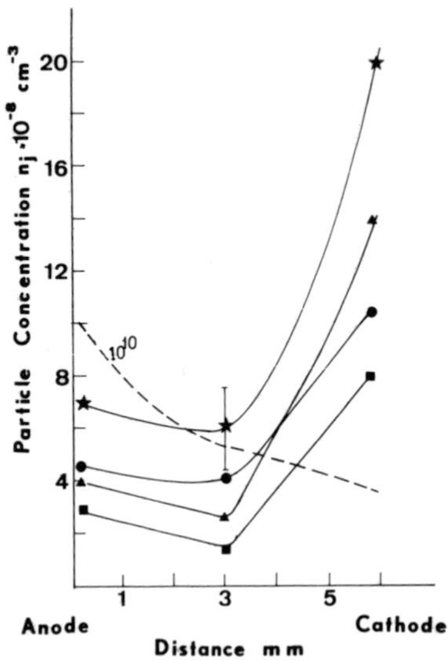

Fig. 4.35 Axial distribution of some trace-element particle concentrations (n_j) in U_3O_8 for ■, Cr (3021, 2838 Å); ●, Mn (2794, 2576 Å); ▲, Fe (2788, 2598 Å), and ★, Mg (2790, 2782 Å). ---, n_j of U atoms and ions.

In other words, the ratio of matrix particle to trace-element particle in the solid sample is similar to that in the anode-region plasma.

4.3.5 Transport Phenomena[125]

Equation (4.8) describes transport phenomena of free particles in the plasma. Theoretical models[5, 108, 122, 124] for calculating ψ_j are:

1. The velocity model, based on axial migration of particles.

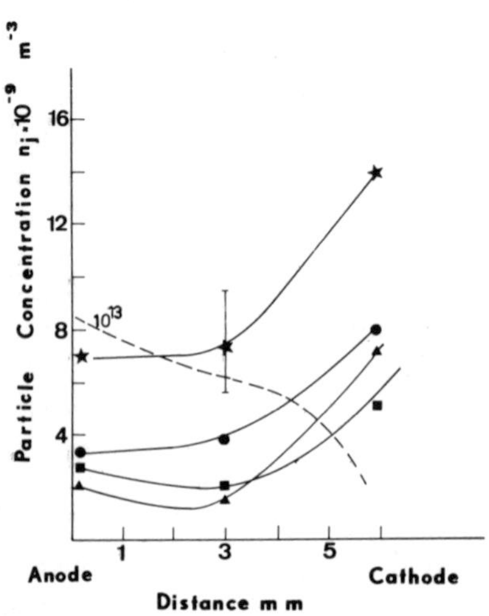

Fig. 4.36 Axial distribution of some trace-element particle concentrations (n_j) in La_2O_3 matrix for ■, Cr; ▲, Fe; ●, Mn; and ★, Mg. ---, n_j of La atoms and ions.

2. The dn/dt model, based on simultaneous effects of axial and radial migration of particles as described in Section 4.3.4.1.

Experimental methods for determining ψ_j or v_j are based on spectral-line-intensity measurements:

1. Measurement of the transit time[126] of particles in the plasma; the velocity v_j in the arc plasma is derived from the time of transit over a given distance.

2. Calculation of particle concentration from absolute values of line intensities.[109, 122] The concentrations thus obtained represent only atoms and ions (n_j). The value ψ_j was calculated from n_j using mean values of the volatilization rate Q_j, determined in the same experiment.

Both methods are applicable only when the test elements do not form stable molecules. This is illustrated by de Galan's[109] empirical correlation:

$$\log \psi_j = \log Q_j - \log \overline{n_j} = 2.5 + \alpha_j \tag{4.36}$$

where α_j is the ionization degree of the atom and $\overline{n_j}$ the average particle concentration of element j. Thus ψ_j depends on the degree of ionization only and not on the degree of dissociation of molecular species.

Axial and radial transports of the matrix particles were calculated using Eq. (4.8); Q_j and n_t were measured by the wire method (Sections 4.2.2 and 4.34). Table 4.8 shows the radial distribution of ψ_j in the anode, central, and cathode regions of the plasma. Increase of ψ_j outside the arc core ($R > 2$ mm) indicates radial transport in the arc mantle.

Figures 4.37 and 4.38 show the axial distribution of the transport parameter in the arc core for a radius of $R = 2.0$ mm for both matrix and trace elements. For the trace elements the volatilization rate was evaluated by different methods than the wire method, as described in Section 4.2.2 and Tables 4.1a and 4.1b.

Table 4.8 Radial Distribution of Transport Parameter (Ψ_j) for Uranium and Lanthanum Matrices[a]

Plasma region	r (mm)	$\psi_j = Q_j/n_t$ (cm^3 s^{-1})[b]	
		U_3O_8 matrix	La_2O_3 matrix
Anode	0–2	48.0	70.0
	2–4	370.0	—
Center	0–2	130.0	600.0
	2–4	290.0	—
Cathode	0–2	220.0	78.0
	2–4	420.0	—

[a] 8.0-mm arc gap at 10 A.
[b] Mean value over four carriages. Relative standard deviation 25% in the anode region and 30% in the center and cathode regions.

The value for ψ_j calculated in this investigation is compared with those given by de Galan,[109, 122] Table 4.9; de Galan's ψ_j is too high for both matrix and trace elements, for the following reasons:

1. In comparison to the matrix, the volatilization rate for trace elements used by de Galan[109] is high.

2. The $\overline{n_j}$ of the matrix element computed by de Galan is too low (i.e.,

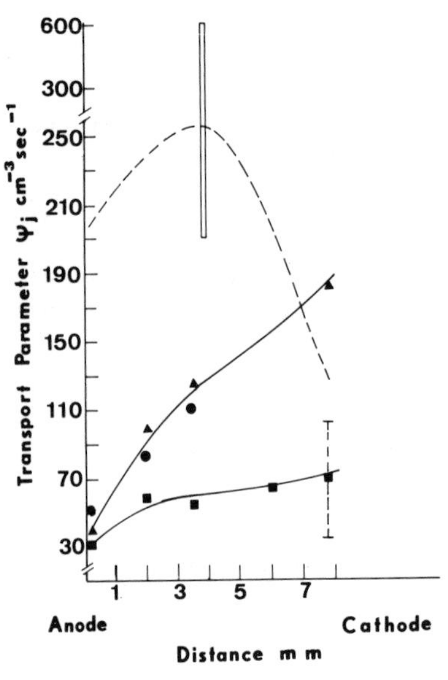

Fig. 4.37 Axial distribution of transport parameter (Ψ_j) at 8.0-mm arc gap. ▲, U_3O_8, 110 cm s^{-1} carriage velocity; ●, U_3O_8, 250 cm s^{-1} carriage velocity; ■, U_3O_8 + 4% UF_4, 110 cm s^{-1}; ---, for Cr, Fe, Mn, and Mg trace elements.

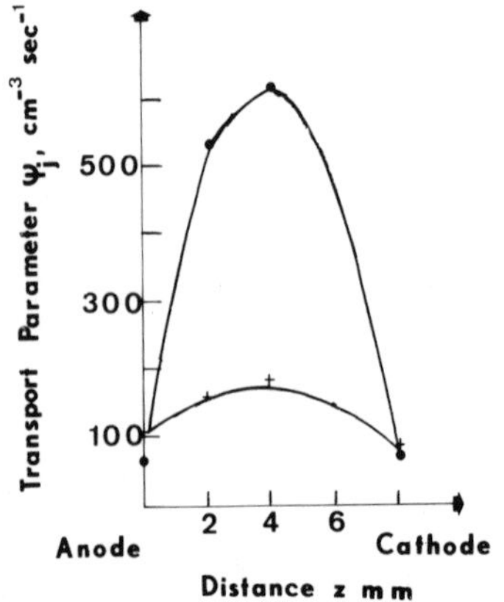

Fig. 4.38 Axial distribution of transport parameter (Ψ_j) at 8.0-mm arc gap. ●, La_2O_3; +, La_2O_3 + 4% LaF_3 at 110 cm s^{-1} carriage velocity.

Table 4.9 Comparison of Transport Parameter Values (log Ψ_j) Calculated by the Wire Method and the de Galan Method[109] for U_3O_8, ThO_2, ZrO_2, and Al_2O_3 Matrices in the Central Region of the Plasma

log ψ_j, wire method[a]				log ψ_j, de Galan method[b]			
U_3O_8	ThO_2	ZrO_2	$Al_2O_3/C = \frac{1}{3}$	U_3O_8	ThO_2	ZrO_2	$Al_2O_3/C = \frac{1}{3}$
2.5	2.3	2.7	2.9	3.8	3.6	3.9	4.1 6.41[c]

[a] Q_j and n_t values for the radius 0–2 mm.
[b] Q_j and n_j values for r 0–2 mm.
[c] de Galan value.[109]

Table 4.10 Neutral Particle Concentration Ratio for N_2, N, O_2, CN, C, and C_2 in Plasma Regions with and without U_3O_8[a]

Plasma region	Temperature (°K) With U_3O_8	Temperature (°K) Without U_3O_8	$n_{\text{with } U_3O_8}/n_{\text{without } U_3O_8}$ N_2	N	O_2	CN	C	C_2
Anode	6200	6500	1.7	0.6	4.5	0.5	0.4	0.7
Center	6700	6000	0.6	1.7	0.3	2.0	3	2.0
Cathode	7000	6200	0.5	2.2	0.2	3.0	12.0	5.0

[a] 6.0-mm arc gap at 12 A; radius, 0–2 mm.

molecules are not considered). Low values of n_j and high values of Q_j result in high values for ψ_j.

One cannot use Q_j for the matrix as the volatilization rate of the traces and calculate n_j of the traces and the particle concentration of the matrix.

4.3.6 Material Balance of Plasma Particles

Concentrations of N_2, N, NO, C_2, C, CN, CO, N^+, NO^+, C^+, and O^+ were also calculated as ratios in plasmas with and without addition of the refractory matrix, Tables 4.10 and 4.11. The role of the third matrix becomes more explicit when such calculations are made. Calculations of the particle concentrations above in a graphite-air plasma are based on the method and data given by Roes[2] for a carbon-air plasma. Roes' calculations are expressed as a function of plasma temperature. Our calculations for the graphite-air plasma with and without the third matrix are based on the different temperature values in the various regions of the plasma. These calculations were made for various axial regions in the arc cores (R = 2.0 mm; i.e., anode, central, and cathode regions). Trace elements were always present in the refractory matrices investigated.

Table 4.11 Ionized Particle Concentration Ratio for C^+, N^+, O^+, NO^+, and n_e in Plasma Regions[a]

Plasma region	Electron density $n_e \times 10^{-16}$ (cm^{-3})		$n^+_{\text{with } U_3O_8}/n^+_{\text{without } U_3O_8}$			
	With U_3O_8	Without U_3O_8	C^+	N^+	O^+	NO^+
Anode	0.6	0.5	0.2	0.3	0.6	0.5
Center	2.0	0.1	5.0	3.0	2.0	0.4
Cathode	7.0	0.1	40.0	12.0	4.0	0.3

[a]6.0-mm arc gap at 12 A; radius, 0–2 mm.

The data in the above tables are given as ratios of particle concentrations for plasma with and without the third matrix. Neutral particle concentration ratios are given in Table 4.10, while ionized particle concentration ratios, together with n_e values, are given in Table 4.11. From these tables the following is concluded:

1. Addition of the third matrix decreases the concentration of N_2, CO, O_2, and NO molecules, while the CN concentration remains almost constant. The concentration of N and C atoms and C_2 increases because of the presence of the third matrix.

2. The ion concentration of C^+, N^+, and O^+ increases on addition of the refractory matrix. The n_e ratio increases to a greater degree in the presence of the third matrix than that of the ion ratio values. The ions, as compared to the electron density in the presence of the refractory matrix, is explained by the Saha relationship, Eq. (4.27). This is shown in Table 4.12 using the specific reactions for ionization; the n_i/n_a ratio has higher values in the graphite–air plasma without the third matrix. Table 4.12 shows that the dominant reaction in a plasma containing the refractory matrix is the formation of atoms (i.e., $n^+ + en_a$).

Therefore, the major effect of the refractory matrix in the arc core ($R = 2$ mm) of a graphite–air plasma is an increase in the population of atoms + ions such as nitrogen, oxygen, and carbon instead of an increase in molecules such as N_2, O_2, and CO (Roes[2] and Boumans[5]).

The material balance in the arc core ($R = 2.0$ mm) of free particles of the refractory matrix can be estimated by the ratio of n_t of the central and cathode regions to the n_t in the anode region as measured by the wire method. Table 4.7b shows data of such normalized total particle concentrations for uranium, thorium, and zirconium.

Axial migration of n_t toward the cathode decreases by a factor of about 5. The radial diffusion of the refractory particles is about equal to their axial migration in the arc core.

Table 4.12 The E.T.M. on Ionization Reactions between Graphite and Air in the Central Region of the Plasma[a]

Reaction	V_i[b] (eV)	r (mm)	T (°K)	K_i[c]	Without U_3O_8: $\bar{n}_e \times 10^{-16}$ (cm^{-3})[d]	Without U_3O_8: $(n_i/n_a)_1 \times 10^{-3}$	With U_3O_8: $\bar{n}_e \times 10^{-16}$ (cm^{-3})[d]	With U_3O_8: $(n_i/n_a)_2 \times 10^{-3}$	$\frac{(n_i/n_a)_1}{(n_i/n_a)_2}$
$C \rightleftharpoons C^+ + e^-$	11.2	0–0.5	7000	2.3×10^{13}	0.20	11.5	5.5	0.4	29
		0.5–1.0	6500	3.9×10^{12}	0.27	14.5	2.0	0.2	72
		1.0–1.5	6000	6.9×10^{11}	0.10	6.9	0.5	0.14	49
$N \rightleftharpoons N^+ + e^-$	14.5	0–0.5	7000	4.0×10^{11}	0.20	0.2	5.5	0.073	26
		0.5–1.0	6500	3.3×10^{10}	0.27	0.12	2.0	0.016	75
		1.0–1.5	6000	3.3×10^{9}	0.10	0.03	0.5	0.007	47
$0 \rightleftharpoons 0^+ + e$	13.6	0–0.5	7000	3.2×10^{11}	0.20	0.16	5.5	0.06	27
		0.5–1.0	6500	3.2×10^{10}	0.27	0.12	2.0	0.016	75
		1.0–1.5	6000	3.9×10^{9}	0.10	0.04	0.5	0.008	49
$NO \rightleftharpoons NO^+ + e^-$	9.5	0–0.5	7000	1.6×10^{14}	0.20	80	5.5	3.0	27
		0.5–1.0	6500	2.8×10^{13}	0.27	100	2.0	1.4	75
		1.0–1.5	6000	5.7×10^{12}	0.10	57	0.5	1.1	52
$N_2 \rightleftharpoons N_2^+ + e^-$	15.5	0–0.5	7000	9.8×10^{9}	0.20	4.9×10^{-3}	5.5	1.8×10^{-4}	28
		0.5–1.0	6500	5.2×10^{8}	0.27	1.9×10^{-3}	2.0	2.6×10^{-5}	73
		1.0–1.5	6000	4.6×10^{7}	0.10	4.6×10^{-4}	0.5	0.9×10^{-5}	5

[a] 6.0-mm arc gap at 12 A.
[b] Ionization potential after Roes.[(2)]
[c] Saha's constant calculated.
[d] Electron density from the wire method. Relative standard deviation 30%.

4.3.7 Layout for Direct Spectrochemical Analysis

4.3.7.1 Summary of the E.T.M.

Based on the Mandelshtam scheme, Section 4.2.1, the volatilization rate, the transport parameter, and the total particle concentration were described and measured in preceding sections. Links such as excitation, ionization, dissociation, and intensities of lines or bands were described briefly. The last links of the Mandelshtam's scheme are discussed in detail by Boumans,[5, 127] de Galan,[109] Kaiser,[128, 129] and others.[120, 130, 131] Before proceeding to describe spectrochemical methods for analyzing trace elements in refractory matrices, a summary of the major effects of the third matrix on electrode and plasma parameters is necessary.

The volatilization rate of the solid sample from the anode crater is mainly effected by the refractory matrix in the electrode. The volatilization rates of the majority of trace elements is governed by that of the refractory matrix. High volatilization rates of the matrix means higher volatilization rates for the trace elements (Tables 4.1a and 4.1b). Moreover, the ratios between matrix and trace-element concentration in the electrode are almost equal to those of the anode region of the plasma (Figs. 4.35 and 4.36 and Table 4.6), suggesting that in these regions the behavior of trace elements follows that of matrix particles.

In the plasma itself, effects of the third matrix can be summarized as follows:

1. Temperature, mean electric conductivity, and electron density attain maximum values in the cathode region of the plasma within a radius of 2.0 mm from the axis (Figs. 4.16 and 4.24 and Table 4.4). These maxima are attributed only to the presence of the third matrix in the plasma.

2. The zone around the arc axis ($R = 2.0$ mm) is characterized by maximum line intensities for the trace elements (Figs. 4.5–4.9). The radial distribution of trace-element particles out of this zone is small comparing to radial diffusion of the matrix free particles (Tables 4.6 and 4.8).

3. In comparison to other regions of the plasma, the highest trace-element particle concentrations are in the cathode region, where matrix free-particle concentration is smallest (Figs. 4.5–4.9, 4.35, and 4.36 and Table 4.6).

4. The material balance of particle concentration (Section 4.3.6) within a radius of 2.0 mm from the arc axis demonstrates that in the presence of the third matrix, the cathode region has the highest concentration of ions and atoms of N, C, and C_2 (Tables 4.11 and 4.12) and the lowest for molecules, such as N_2, CO, O_2, and NO (Table 4.11).

5. From effects 1–4 it seems reasonable to assume that the inner zone, around the arc axis, of the cathode region is shielded by an outer zone with high concentrations of refractory matrix particles and molecules such as N_2, CO, O_2,

and NO. The high electron density in the cathode region is attributable to C^+, N^+, and O^+ and to the trace elements (Table 4.11).

6. An axial "separation" occurs in the plasma between the anode and cathode (Figs. 4.5–4.9). This "separation" between trace elements and matrix elements reaches a maximum in the cathode region. Using this separation and exposing only the cathode region on the photographic plate, a direct spectrochemical method for analyzing trace elements in refractory matrices is feasible.

4.3.7.2 General Method for Direct Spectrochemical Analysis

The refractory matrices treated in this investigation differ in various aspects: their voltailization rates are not the same; the chemical reactions of the third matrix with air and graphite differ from one another; additional refractory matrices can be formed during the arcing time. For example, natural materials such as phosphates or silicates consists of various major elements, such as Ca, Mg, Al, Fe, and Ti in addition to P or Si, and the behavior of the trace elements could be similar to any or all of these major components. Rock phosphate contains Ca and P as major constituents, both of which may influence the behavior of the trace elements; or one element may act as a matrix for several trace elements, while the other acts as a matrix for the remainder traces. Other individual refractory matrices, such as Al_2O_3, MoO_3, TiO_2, ThO_2, ZrO_2, U_3O_8, PuO_2, and rare earth oxides, can be classified together, since all have the following characteristics: (1) their volatilization rates are almost similar, and (2) their effects on plasma variables are similar. Thus if (1) and (2) are known, the use of internal standards are unnecessary. Furthermore, matrix factors can be used (i.e., trace-element standards in any matrix can be used for analyzing trace elements in another matrix).

These refractory matrices can be analyzed using a direct and general spectrochemical method. The sample is converted initially into its oxide form. After homogenization employing a "Wig-L-Bug," the sample is weighed in the anode crater (identical crater dimensions for all matrices) and pressed with a venting tool. During arcing (there is a different arcing period for each matrix) only the cathode region is projected through a diaphragm onto the entrance slit of a spectrograph. The electrodes gap is focused on the diaphragm and on the spectrograph collimator. Both current and arc gap are kept constant during arcing. The synthetic standards are treated in an identical manner. The spectra obtained from the cathode region on the photographic plates are measured densiometrically for trace-element content. Three different standards are exposed on each photographic plate for checking trace-element working crews.

As will be shown in subsequent sections, this cathode-region method of spectrochemical analysis of refractory matrices is rapid and reliable.

4.4 ANALYSIS OF TRACE ELEMENTS IN REFRACTORY MATERIALS

4.4.1 Apparatus, Operating Conditions, and Standards

4.4.1.1 Apparatus

Ebert 3.4-m grating spectrograph (Hilger and Watts)
Gratings: 600 grooves/mm blazed for 5.2 deg (Jarrell-Ash) 1800 grooves/mm blazed for 18.6 deg. (Jarrell-Ash)
dc arc source: 3-25A (Hilger and Watts)
Arc and spark stand (Spex)
Densitometers: GII (Zeiss Jena) and L-459 (Hilger and Watts)
Analog-to-digital converter with Kennedy tape recorder (N.R.C.N.); 3600 CD computer
Oscilloscope, type 551 (Textronix): for voltage measurement
Photographic plates: SA-1; SA-3; N-1 (Kodak); R-50 (Ilford)
Optical setup for measuring the cathode region, as shown in Fig. 4.4
Ultra-purity graphite (Ultra Carbon Co.)

4.4.1.2 Operating Conditions

The following conditions were constant during analysis:
Upper electrode (cathode): graphite, 3.17-mm diameter flat
Lower electrode (anode): graphite, 6.35 mm diameter crater–3.97 mm diameter depth up to 5.60 mm
Electrode gap: 4.0–8.0 mm
Diaphragm: 1.0 mm
Region exposed: 0.2–1.0 mm below the cathode (Fig. 4.14b)
Current: 10–13 A
Exposure: 35–40 s without preburn
Slit: 0.035 mm and up to 5.0 mm in height
Trace-element specpure (Johnson-Matthey) oxides.

4.4.1.3 Standard Sets

Sets of standards, including about 50 trace elements [Ag, Al, B, Ba, Bi, Ca, Cd, Co, Cr, Cu, Cs, Fe, Ge, Ga, In, K, Li, Mg, Mn, Mo, Na, Nb, Ni, Pb, Rb, Sb, Sc, Si, Sn, Sr, Ti, Tl, V, Zn (the "common" trace elements), and rare earths] were prepared in each refractory matrix as follows:

1. The common trace elements were mixed with each refractory matrix, giving a concentration range from 0.1 to 500 ppm each. Visual comparative analysis was used for standards in the concentration range 0.1-5 ppm each.

2. Li, Na, K, Rb, and Cs were mixed as above in the concentration range 0.1-100 ppm each.

3. As, Hg, and P were mixed separately, with each matrix in the concentration range as in standard 1.

4. Standards consisting of Al, Ba, Ca, Mo, Nr, Sr, Ti, and the rare earths were prepared in each matrix and 2–10% of the fluoride of the same matrix was added. The range of contents is the same as that of the common elements.

4.4.2 Uranium, Thorium, Zirconium, and Plutonium Oxides[94,110,115,132]

Samples of uranium, thorium, zirconium, or plutonium were transformed respectively in U_3O_8, ThO_2, ZrO_2, and PuO_2, by roasting in air at 900°C. Plutonium was treated in glove boxes.[132]

4.4.2.1 Similarities between the Matrices

The similarities between matrices were evaluated by measuring the following five parameters:

1. Volatilization rates. These data for U_3O_8, ThO_2, ZrO_2, and PuO_2 are given in Table 4.13. Addition of fluorides (2–4%) in the form of UF_4, ThF_4, ZrF_4, or Teflon spray increases the volatilization rates and detection limits for refractory trace elements such as rare earths.[110] The results in Table 4.13 show that U_3O_8, ThO_2, ArO_2, and PuO_2 behave in a similar manner.

2. The axial distribution of temperature given in Figs. 4.15 and 4.16 show that the matrices are similar within the limits of experimental error (200°K). Addition of fluorides to the refractory matrix does not markedly affect temperature values.

3. The axial distributions of total particle concentration are given in Table 4.6. The above matrices are similar. Particle concentration is insensitive to fluo-

Table 4.13 Volatilization Rate, Q_j, of U, Th, Zr, and Pu Matrices with and without Fluoride Additions[a]

Matrix	Anode load (mg)	$Q_j \times 10^{-17}$ (atoms s^{-1})			
		Wire method[b]		Chemical method[c]	
		Without F	With F	Without F	With F
U_3O_8	50	1.4	3.5	2.3	4.0
ThO_2	35	1.0	2.5	—	—
ZrO_2	60	1.5	3.7	—	—
PuO_2	50	—	—	2.0	3.5

[a] 6.0-mm arc gap at 12 A; arcing time, 60 s. Ten electrodes each matrix.
[b] Relative standard deviation 30%.
[c] Relative standard deviation 35%.

ride addition. This conclusion was also obtained from the transport parameter (ψ_j) of the free particles of the matrix (Section 4.3.5).

4. The axial distribution of electron density in Fig. 4.24 shows again the similarity between the matrices. As for the temperature values, the addition of fluorides leaves the electron density value unaffected.

5. In Figs. 4.5–4.9 similarities between trace elements in each refractory matrix are shown by the axial distribution of normalized line intensities (I/I_{anode}).

In an earlier publication[110] it was shown that rare earth trace elements cannot be analyzed using the method applied to common trace elements. Addition of 20% graphite and 4% fluorides is essential for better sensitivities of the rare earth traces.[110] Graphite addition increases volatilization rates of matrix and trace elements and the concentration of their free particle in the plasma,[110] leaving the axial distribution of normalized line intensities almost unchanged, as shown in Figs. 4.5–4.9.

On the basis of parameters 1–5, the cathode region is the most suitable region of the plasma for trace-element analysis. The similarities of the matrices allow application of matrix factors (i.e., the use of the standards in U_3O_8 for analyzing trace elements in ThO_2, ZrO_2, or PuO_2 matrices). This will be pursued in the subsequent section.

4.4.2.2 Matrix Factors in the Cathode Region

Because of local thermodynamic equilibrium in the cathode region, the theoretical and experimental matrix factors can be calculated in the following way. Using Eqs. (4.15) and (4.16) for the same spectral line in two different matrices, say M_1 and M_2, line-intensity ratios are

$$\log\left[\frac{J(M_1)}{J(M_2)}\right] = \log\left[\frac{n_j(M_1)}{n_j(M_2)}\right] + \log\left[\frac{Z(M_2)}{Z_1(M_1)}\right] + 5040\,V\left[\frac{1}{T(M_2)} - \frac{1}{T(M_1)}\right] \tag{4.37}$$

where V is the excitation potential in eV of the spectral line and n_j is the particle concentration of the trace element under consideration. The ratio $n_j(M_1)/n_j(M_2)$ is calculated from relative line intensities of the trace element in the two matrices. Partition function ratios $Z(M_2)/Z(M_1)$ were corrected for temperature values in the two matrices [i.e., $T(M_2)$ and $T(M_1)$] in the cathode region. In this way the theoretical m.f. is expressed by the calculated ratio $J(M_1)/J(M_2)$, Eq. (4.37).

The experimental m.f. were obtained by measuring $J(M_1)$ and $J(M_2)$ densitometrically and dividing one by the other. Table 4.14 shows the matrix factors for U_3O_8/Z_rO_2, U_3O_8/ThO_2, and U_3O_8/PuO_2 for several trace elements. An average matrix factor was used in the trace-element analysis of each matrix.

Table 4.14 Calculated and Experimental Matrix Factors for Some Impurities in U_3O_8, ZrO_2, ThO_2, and PuO_2 in the Cathode Region

Impurity element[a]	Calculated factor			Experimental factor[b]		
	J_U/J_{Zr}	J_U/J_{Th}	J_U/J_{Pu}	J_U/J_{Zr}	J_U/J_{Th}	J_U/J_{Pu}
Pb	1.0	1.52	0.86	1.10	1.70	0.77
Cr	1.0	1.45	1.49	1.05	1.55	1.85
Sn	1.0	1.42	1.04	1.08	1.35	0.94
Ge	1.0	1.37	1.33	0.88	1.53	1.19
Fe	1.0	1.37	1.23	0.92	1.50	1.11
Ni	1.0	1.35	1.40	1.15	1.50	1.55
Bi	1.0	1.34	1.10	1.08	1.55	0.99
Si	1.0	1.45	6.38	1.05	1.37	5.96
Ca	1.0	1.67	—	0.85	1.60	—
Zn	1.0	1.97	2.40	1.25	1.42	3.15
Mn	1.0	1.42	1.26	0.95	1.39	1.19
Ba	1.0	1.21	1.0	1.10	1.30	0.94
Sr	1.0	1.21	1.0	0.91	1.35	0.79

[a] For wavelength used, see Table 4.15.
[b] Relative deviation from the calculated factor 2% for Mn minimum and 25% for Zn as maximum.

Trace elements such as As, Hg, and P were analyzed separately in the central region of the dc arc plasma. No matrix factors were applied.

4.4.2.3 Analytical Results—Cathode Region

4.4.2.3a Detection Limits. Spectral lines used together with their detection limits are given in Table 4.15a, b. Detection limits were established by visual comparison using a master plate of standards varying from 0.1 to 5 ppm. In Table 4.15a detection limits obtained by the cathode-region method are compared with those obtained by carrier distillation. Comparison shows that the detection limits of the common trace elements are similar. However, for the rare earths the detection limits (Table 4.15b) obtained by the cathode-region method are almost one order of magnitude better than for carrier distillation.

4.4.2.3b Standard Working Curves. Exposures were made for 35 s without preburn through a seven-step (1:2) rotating sector. A 1800 grooves/mm grating was used for the rare earth elements and a 600 grooves/mm grating for the common trace elements. A computerized self-calibration method using Seidel densities and the Kaiser's method[128–129] was utilized.

Slopes of the working curves $[d(\log J)/d(\log C)]$ and coefficients of correlations are given in Table 4.16 for the U_3O_8 matrix. The coefficients of correlations show that reliable working curves can be obtained without the use of internal standards.

Table 4.15a Detection Limits of Common Trace Elements in U_3O_8, ThO_2, ZrO_2, and PuO_2 Matrices in the Cathode Region

Element	Wavelength (Å)	Detection limits (ppm)								
		U_3O_8	ZrO_2	ThO_2	PuO_2	U_3O_8 + F	ZrO_2 + F	ThO_2 + F	PuO_2 + F	U_3O_8 C.D.[73]
Ag	3280	0.5	0.1	0.2	1.0					0.1
	3382	0.5	0.1	0.2	—					
Al	3082	—	—	—	—	0.2	0.5	0.5	1.0	1.0
	3092	—	—	—	—	0.2	0.5	0.5		
B	2496	0.1	0.1	0.1	0.1					0.1
Ba	4554	—	—	—	—	0.5	1.0	1.0	5.0	1.0
Bi	3067	0.1	0.5	0.5	0.5					0.3
Ca	4226	—	—	—	—	0.5	1.0	1.0	2.0	1.0
Cd	2288	0.1	0.2	0.2	—					0.1
	3261	0.5	1.0	1.0	2.0					
Co	2432	1.0	1.0	1.0	2.0					1.0
Cr	3021	1.0	0.5	0.2	1.0					1.0
	4254	1.0	—	—	—					
Cu	3247	0.1	0.1	0.1	0.5					0.1
Fe	2843	1.0	1.0	1.0	—					1.0
	3021	0.5	0.1	0.1	0.5					
Ga	2943	0.5	0.3	0.1	1.0					1.0
	2944	2.0	0.8	0.5	—					
Ge	2651	0.2	0.2	0.2	—					0.3
	3039	0.2	0.2	0.2	1.0					
In	3039	0.5	0.2	0.2	0.5					
	3256	0.3	0.1	0.1	—					0.3
Li	3232	0.5	0.2	0.2	0.5					1.0
Mg	2795	0.2	0.1	0.1	1.0					0.1
Mn	2798	0.1	0.1	0.1	0.5					1.0
Mo	3132	—	—	—	—	0.5	2.0	2.0	2.0	0.1

Na	5889	0.1	0.1	0.1	0.1					1.0
Nb	4058	—	—	—	—	5.0	10.0	10.0	20.0	4.0
Ni	3050	0.5	0.5	0.5	1.0					1.0
	3002	0.5	0.5	0.5	—					
Pb	2801	1.0	0.5	0.5	1.0					0.5
	2833	0.5	0.3	0.3	—					
Sb	2598	2.0	1.0	0.5	2.0					1.0
Si	2516	0.5	0.5	0.5	2.0					1.0
	2881	0.5	0.5	0.5						
Sr	4607	—	—	—	—	0.5	1.0	1.0	5.0	1.0
Ti	3349	—	—	—	—	0.5	1.0	1.0	1.0	1.0
	3234	—	—	—	—	0.5	1.0	1.0	—	
Tl	2767	1.0	1.0	1.0	1.0					1.0
V	3182	—	—	—	—	0.5	0.5	0.5	<1.0	0.5
Zn	3282	5.0	3.0	5.0	2.0					10.0
	3345	5.0	3.0	5.0	—					
K	7664	0.05	0.1	0.1	2.0					1.0
Rb	7800	0.05	0.5	0.5	5.0					10.0
Cs	8251	0.5	0.5	0.5						1.0

Table 4.15b Detection Limits of Rare Earths in Uranium, Thorium, Zirconium, and Plutonium Matrices

		Detection limits (ppm)			
Element	Wavelength (Å)	U_3O_8 + UF_4 + C	ZrO_2 + ZrF_4 + C	ThO_2 + ThF_4 + C	PuO_2 + Teflon spray + C
Ce	3716.37	10	10	20	—
	4222.60	20	—	—	—
	4289.91	—	15	—	—
Dy	3385.03	5	10	—	—
	4000.45	5	10	10	5
Er	3372.75	5	—		
	3692.64	10	10		
	4007.94	5	10	10	50
Eu	3907.11	5	5	—	
	3971.96	5	5	5	
	3930.50	—	5	5	5
Gd	3362.23	5	—	—	—
	3422.47	20	10	20	10
Ho	3398.98	15	10	—	50
	3456.00	15	20	20	
La	3337.49	5	10	5	5
	3995.75	—	10	10	
	3949.11	10	—	10	
Lu	2615.42	5	5	10	5
Nd	4012.25	15		20	50
	4247.38	20	10	—	—
	4303.58	20	20	—	—
Pr	4222.98	100	100	100	
	4408.84	100	100	100	
Sm	4256.39	10		10	
	4280.79	10	10	10	50
	4424.34	—	10	—	—
Tb	3509.17	15	—	—	100
	3650.40	10	—	—	—
	3702.85	—	20	20	—
Tm	3291.00	10	5	—	—
	3362.61	5	5	—	5
	3462.20	10	5	5	—
Yb	3289.35	1	1	1	1
	3694.20	5	1	—	—
	3987.99	—	1	—	—
Y	3633.12	—	2	5	
	4374.94	2	2	5	2

(Continued)

Table 4.15b (*Continued*)

Element	Wavelength (Å)	Detection limits (ppm) U_3O_8 + UF_4 + C	ZrO_2 + ZrF_4 + C	ThO_2 + ThF_4 + C	PuO_2 + Teflon spray + C
Sc	3353.73	1	1	2	
	3630.75	1	1	2	10
Zr	3438.23	1	—	5	—
Th	4019.13	10	10	—	—

Table 4.16 Statistics of Working Curves and Coefficient of Variation for Some Trace Elements in U_3O_8, ThO_2, and ZrO_2 Using Matrix Factors in the Cathode Region

Analytical line	Analytic working curves: Slope value $d(\log J)/d(\log C)$	Coefficient of correlation, R^a	Samples S1–S10: Number of determinations	Coefficient of variation[b]
Al (+ F)	0.75	0.95	15	15–17
Ba (+ F)	0.65	0.92	16	17–20
B	0.80	0.98	17	12–15
Ca (+ F)	0.65	0.93	15	17–20
Cd	0.4	0.98	14	12–13
Cr	0.80	0.99	17	10–14
Cu	0.70	0.94	17	15–17
Fe	0.78	0.96	17	9–12
In	0.82	0.99	15	9–12
Mg	0.85	0.94	17	15–20
Mn	0.85	0.96	17	9–15
Mo (+ F)	0.70	0.98	15	9–15
Ni	0.65	0.98	17	8–10
Si	0.80	0.95	12	15–20
Sn	0.85	0.98	12	10–12
Ti (+ F)	0.72	0.98	17	8–10
V (+ F)	0.75	0.98	17	10–13
Zn	0.55	0.99	15	20–25
La (+ C + F)	0.72	0.95	20	20–25
Ce (+ C + F)	0.65	0.95	20	25–30
Yb (+ C + F)	0.80	0.98	20	17–20
Nd (+ C + F)	0.70	0.97	20	18–20

[a] $R = \dfrac{N \sum \log X \log Y - \left(\sum \log Y\right)\left(\sum \log X\right)}{\left[N \sum (\log X)^2 - \log \sum (\log X)^2\right]\left[N \sum (\log Y)^2 - \sum (\log Y)^2\right]}$; slope of $Y = a + bX$.

[b] Minimum and maximum values.

4.4.2.3c Evaluation Using Matrix Factors. For U_3O_8, ThO_2, and ZrO_2 matrices, 10 samples each, marked S1–S10, were analyzed for their trace-element content. The respective oxides were prepared from nitrate solutions (S1–S2) metals or alloys (S3–S7). The remainder (S8–S10) were initially oxides. Owing to the lack of Pu samples, only one plutonium specimen was investigated.[132] Each trace element was determined about 15 times in each matrix. In the case of the Pu specimen, only 5 determinations were made, because of its inconvenient recovery. Determinations were made only with matrix factors. Average factors from Table 4.14 are as follows:

1. U_3O_8 to ZrO_2 : $(m.f.)_1 = 1.0$ for common traces and 1.7 for rare earths.
2. U_3O_8 to ThO_2: $(m.f.)_2 = 1.5$ for common traces and 2.2 for rare earth traces.
3. U_3O_8 to PuO_2: individual $(m.f.)_3$ values for each common trace elements are indicated in Table 4.14.

The analytical results of the samples in their respective coefficients of variation (C.O.V.) are given in Table 4.16. Using mean values for various m.f.'s without background corrections, the C.O.V. for almost all trace elements in the three matrices—except PuO_2—was found to be approximately 15% for the concentration range from a few ppm up to 500 ppm.

4.4.3 Rare Earth Oxides[133]

The rare earth (R.E.) compounds or elements were transformed into their respective oxides by roasting at 900°C for 2 h. After homogenization, 30–50 mg of oxide was introduced into the anode crater. The level of the substance in the anode crater was constant for all R.E. matrices. The substance was pressed into the crater by a venting tool. The following R.E. oxides were investigated: La_2O_3, Y_2O_3, Sm_2O_3, Nd_2O_3, Ce_2O_3, Dy_2O_3, Eu_2O_3, Gd_2O_3, and So_2O_3. For this group trace-element spectrochemical analysis can be divided in two parts: (1) common trace analysis in the R.E. matrices, and (2) rare earth trace elements in the R.E. matrices. For analysis of the former, the R.E. oxides are used without additives while for the latter, 20% graphite and 4% fluorides were added.

4.4.3.1 Similarities between the Rare Earth Matrices

The scheme in Section 4.4.2 is adhered to:

4.4.3.1a Volatilization Rate. Q_{RE} values for each matrix are given in Table 4.17. They were measured by wire and chemical methods, as indicated in this table. Results in Table 4.17 shows that the additions of 20% graphite and 4% fluoride increases the volatilization rate of the R.E. elements by a factor of 2.

4.4.3.1b Axial Distribution of Temperature. Temperature values are given by Zn, Cu thermometric species. Behavior of the axial temperature gradients for

Table 4.17 Volatilization Rates of Rare Earth Matrices with and without Fluorides (4%) and Graphite (20%) Additions[a]

Matrix	$Q^{ch} \times 10^{-17}$ (atoms s^{-1})[b]		$Q^{w} \times 10^{-17}$ (atoms s^{-1}),[c] rare earth alone
	Mixture	Rare earth alone	
La_2O_3	—	1.9	2.0[d]
La_2O_3 + C + F	8.0	4.5	3.7[d]
Sm_2O_3	—	1.5	—
Sm_2O_3 + C + F	7.2	4.0	—
CeO_2	—	2.3	—
CeO_2 + C + F	7.0	4.3	—
Nd_2O_3	—	2.2	2.5[e]
Nd_2O_3 + C + F	9.0	5.0	4.0[e]

[a]6.0 mm arc gap at 12 A; arcing time, 40 s.
[b]Relative standard deviation 30%.
[c]Relative standard deviation 25%.
[d]Neutron activation method.
[e]Spectrochemical method of wire deposit.

the various R.E. matrices confirm the existence of two distinct groups: Y_2O_3, Sm_2O_3, and La_2O_3 as one group and the rest of R.E. matrices as another group. In both groups the temperature attained a maximum value in the cathode region.

Addition of 20% graphite and 4% fluorides to the matrices eradicates the differences between the two R.E. groups within the experimental error. Figure 4.17 shows that the additives result in a different axial distribution of temperature, in which its maximum value is located between the central and cathode regions of the plasma.

4.4.3.1c Axial Distribution of Voltage. Figure 4.39 illustrates the axial distribution of voltage as measured by the wire method. Two different axial electric field strengths, E_z, were found; 120 V cm^{-1} in the anode region and 13 V cm^{-1} in the central and cathode regions of the plasma. Within the experimental error E_z values are independent of the nature of the R.E. matrix within the experimental error. Furthermore, E_z values for all R.E. matrices, with and without 20% graphite and 4% fluorides, are similar.

4.4.3.1d Axial Distribution of Electron Density. Axial values of n_e are given in Fig. 4.25. The electron densities of pure Y, Sm, and La oxides (without graphite and fluoride) are equal. The internal equality is also true for other pure R.E. oxide matrices. In both groups, electron density maximum occurs in the cathode region.

Addition of 20% graphite and 4% fluorides results in equalization of the electron densities for all R.E. matrices investigated.

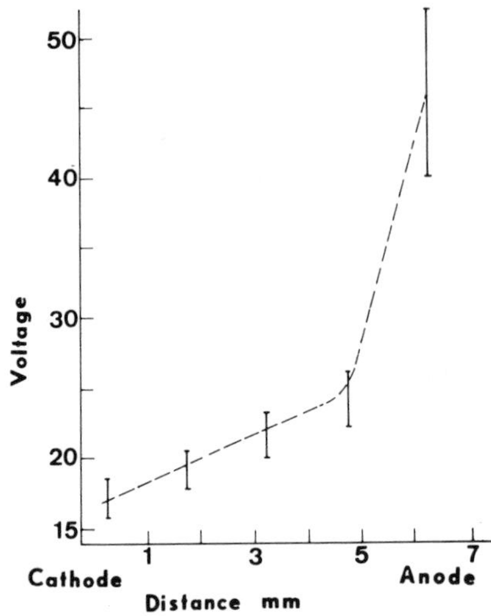

Fig. 4.39 Axial distribution of voltage in plasmas with Yb_2O_3, Pm_2O_3, Gd_2O_3, Ho_2O_3, Dy_2O_3, La_2O_3, or Nd_2O_3 matrices at 6.0-mm gap and 13 A.

4.4.3.1e Axial Distribution of Total Particle Concentration. n_t according to the wire method and n_j values obtained for the matrix particles are given in Table 4.18 for the cathode region. Each R.E. group is internally equivalent without the additives.

Addition of 20% graphite and 4% fluoride results in a two- to tenfold increase of the particle concentrations. These additives result in an equalization of all the R.E. matrices.

Table 4.18 Matrix Particle Concentration in the Cathode Region of the Plasma with Rare Earth Matrices[a]

Matrix	$n_t \times 10^{-13}$ [b] (cm^{-3})	$n_j \times 10^{-13}$ [c] (cm^{-3})	$\frac{n_t - n_j}{n_j}$
La_2O_3	270	4.0	66.0
La_2O_3 + 20%C + 4% LaF_3	420	10.0	41.0
Sm_2O_3	–	7.0	–
Sm_2O_3 + 20%C + 4% LaF_3	–	12.0	–
Nd_2O_3	100	1.4	70.0
Nd_2O_3 + 20%C + 4% NdF_3	320	9.0	35.0
CeO_2	–	0.7	–
CeO_2 + 20%C + 4% NdF_3	–	8.0	–
Eu_2O_3	–	0.9	–
Eu_2O_3 + 20%C + 4% NdF_3	–	8.5	–

[a] 8.0-mm arc gap at 10 A; 35-s exposure.
[b] Wire method. Mean value over four carriages of wires. Relative standard deviation 30%.
[c] Mean value over four spectra (arc focused on slit). Relative standard deviation 20%.

4.4.3.1f Axial Distribution of Normalized Line Intensities. The similarities of the common and rare earth trace elements in the various R.E. matrices are shown in Fig. 4.6 of normalized line intensities. Without or with the introduction of graphite and fluorides, the result is maximum line intensity for each trace element in the cathode region of the plasma.

As in the cases of U_3O_8, ThO_2, and ZrO_2 (Section 4.4.2.1), common and rare earth trace elements in R.E. matrices are analyzed separately. An admixture of 20% graphite and 4% fluorides is necessary for better detection limits of rare earth traces in R.E. matrices.

The results (Sections 4.4.3.1.1-4.4.3.1.6) indicate that the cathode region is the most suitable one for the spectrochemical analysis of common and rare earth traces. The parities above allow the use of matrix factors as follows: the use of standards in La_2O_3 and Nd_2O_3 for analysis of the common trace elements in each respective group of R.E. oxides, and the use of standards in La_2O_3 for analysis of rare earth traces in all R.E. matrices.

4.4.3.2 Matrix Factors in the Cathode Region

Equation (4.37) was used to calculate the matrix factors. The measured line-intensity ratio $(J_{(M_1)}/J_{(M_2)})$ for the same spectral line of the same element in two different matrices was used as the experimental matrix factor.

Table 4.19a shows the matrix factors for the common trace elements in each matrix with respect to La_2O_3 and Nd_2O_3. Table 4.19b shows the m.f. for several rare earth trace elements in R.E. matrices with respect to La_2O_3. The average values given in Table 4.19 were used for each trace element in each R.E. matrix.

Table 4.19a Experimental Matrix Factors for Some Trace Elements in Rare Earth Oxides in the Cathode Region

Trace element	Group 1		Group 2			
	J_{La}/J_Y	J_{La}/J_{Sm}	J_{Nd}/J_{Ce}	J_{Nd}/J_{Pr}	J_{Nd}/J_{Eu}	J_{Nd}/J_{Yb}
Ba	1.10	1.30	1.15	0.70	1.12	1.30
Cr	0.90	1.20	1.30	0.85	1.20	1.50
Ga	1.05	1.25	1.20	0.80	1.15	1.45
Fe	0.90	1.20	1.20	0.85	1.25	1.50
Ni	0.95	0.10	1.15	0.90	1.20	1.60
Bi	1.0	1.20	1.15	0.80	1.25	1.35
Si	1.05	1.25	1.20	0.95	1.20	1.55
Mn	0.90	1.20	1.20	0.8	1.15	1.50
Averages from all trace elements	0.9 ±20%	1.20 ±15%	1.20 ±25%	0.8 ±15%	1.20 ±20%	1.50 ±17%

Table 4.19b Experimental Matrix Factors for Some Rare Earth Trace Elements in Rare Earth Matrices in the Cathode Region[a]

Rare earth trace element	Wavelength (Å)	Matrix factor				
		J_{La}/J_{Y}	J_{La}/J_{Nd}	J_{La}/J_{Sm}	J_{La}/J_{Eu}	J_{La}/J_{Ce}
Tm	3425	1.4	1.3	1.4	1.0	1.0
	3362	1.6	1.3	1.5	–	–
Er	3312	0.90	–	0.82	–	0.92
	3372	–	0.65	–	0.88	–
Gd	3362	–	–	–	1.1	0.85
	3350	0.73	–	0.82	–	–
Ho	3456	1.08	1.05	1.2	1.10	1.0
Dy	3407	–	1.09	–	0.92	0.76
	3385	1.02	–	0.96	–	–

[a]6.0-mm arc gap at 13 A; trace concentration 500 ppm. Grating used: 1800 grooves/mm; two orders. 20% graphite and 4% fluorides were added to rare earth matrices. Relative standard deviation from four spectra, 20%.

4.4.3.3 Analytical Results—Cathode Region

4.4.3.3a Detection Limits. Table 4.20 shows the spectral lines and the detection limits of each element in each matrix. Detection limits were determined as described in Section 4.4.2.3. In general, detection limits are several ppm.

4.4.3.3b Standard Working Curves. For common trace elements, the cathode region as shown in Fig. 4.14b (0.2 mm beneath the cathode) was exposed for 35 s after 5 s of preburn, through the seven steps of the rotating sector. A 600 grooves/mm grating was used. Rare earth trace elements were analyzed, exposing the cathode region (0.8 mm beneath the cathode) for 30 s preceded by 25 s of preburn; 1800 grooves/mm was used.

The slopes $d(\log J)/d(\log C)$ of the working curves within the concentration range 5–500 ppm are given in Table 4.21 for La_2O_3 and Nd_2O_3 matrices. From the coefficient of correlation, a good reliability was obtained despite ejection of internal standards.

4.4.3.3c Statistics Using Matrix Factors. Rare earth samples were derived from two sources: those marked with the letter B were obtained from the Institut of Energia Nuclear, São Paulo, Brazil; those marked with the letter S were R.E. oxides, prepared in the laboratory.

Table 4.21 shows the coefficient of variation C.O.V. for each common trace element analyzed, based on 10 determinations of each element in each sample. Only respective mean matrix factors were employed. The C.O.V. varies from 10 to 20% for contents up to 500 ppm.

Table 4.20 Detection Limits (ppm) of Common and Rare Earth Trace Elements in Some Rare Earth Oxides in the Cathode Region[a]

Element and wavelength[b] (Å)	Rare earth matrices							
	La_2O_3	CeO_2	Yb_2O_3	Nd_2O_3	Sm_2O_3	Gd_2O_3	Sc_2O_3	Y_2O_3
Yb 3289	1	1	–	1	1	2	2	2
Tm 3425	2	2	2	2	–	–	5	–
Tm 3362	–	–	–	–	2	1	–	2
Er 3312	–	10	–	–	10	–	10	10
Er 3372	5	–	5	5	–	5	–	–
Gd 3422	–	–	–	–	10	10	2	10
Gd 3350	5	20	5	10	–	–	–	–
La 3337	–	10	5	5	10	10	2	10
Dy 3407	–	25	–	–	10	10	10	10
Dy 3393	10	–	5	10	–	–	–	–
Ho 3456	5	10	5	5	5	5	10	10
Nd 4012	5	5	5	–	5	5	10	10
Eu 3971	2	–	2	2	2	–	2	10
Eu 3930	–	10	–	–	–	5	–	–
Sm 3634	3	5	3	5	–	5	2	5
Ag 3280	0.2	0.2	0.2	0.2	0.2	0.2	0.2	0.2
Al 3082	2	2	2	2	2	2	2	2
B 2496	0.5	0.5	0.5	0.5	0.5	0.5	0.5	0.5
Ba 4554	20	20	20	20	20	20	20	20
Bi 3067	1	1	1	1	1	1	1	1
Cd 2288	1	1	1	1	1	1	1	1
Co 2432	1	1	1	1	1	1	1	1
Cr 2835	1	1	1	1	1	1	1	1
Cu 3247	2	2	2	2	2	2	2	2
Fe 2843	3	3	3	3	3	3	3	3
Go 2943	1	1	1	1	1	1	1	1
Ge 2651	1	1	1	1	1	1	1	1
In 3256	0.5	0.5	0.5	0.5	0.5	0.5	0.5	0.5
Mg 2779	1	1	1	1	1	1	1	1
Mn 2798	0.5	0.5	0.5	0.5	0.5	0.5	0.5	0.5
Mo 3132	5	5	5	5	5	5	5	5
Ni 3050	5	5	5	5	5	5	5	5
Pb 2833	1	1	1	1	1	1	1	1
Si 2576	2	2	2	2	2	2	2	2
Sr 4607	20	20	20	20	20	20	20	20
Ti 3234	10	10	10	10	10	10	10	10
V 3184	10	10	10	10	10	10	10	10

[a] 6.0-mm arc gap at 13 A.
[b] For rare earth trace elements, grating of 1800 grooves/mm, two orders, was used. 20% graphite and 4% fluorides were added to the matrices for rare earth trace elements.

Table 4.21 Working-Curve Slopes, $d(\log J)/d(\log C)$, and Coefficient of Variation for Common Trace Elements in La_2O_3, CeO_2, Nd_2O_3, Gd_2O_3, Sc_2O_3, and Y_2O_3 Using Matrix Factors in the Cathode Region[a]

Element	Working curves[b]		Number of determinations		Coefficient of variation[d] (%)
	Slope value	Coefficient of correlation, R^c	B sample	S sample	
Al	0.81	0.88	–	10	15
B	0.83	0.96	6	6	10
Ba	0.70	0.92	10	10	20
Co	0.74	0.92	10	–	12
Cu	0.79	0.90	10	4	10
Cr	0.69	0.95	8	8	15
Fe	0.76	0.96	8	8	10
Mg	0.82	0.99	5	5	10
Mn	0.80	0.99	10	–	8
Mo	0.68	0.92	10	4	15
Ni	0.72	0.98	12	–	15
Pb	0.83	0.98	5	10	10
Si	0.70	0.94	–	15	16
Sr	0.68	0.95	10	–	20
Ti	0.73	0.90	5	5	20
V	0.69	0.96	5	5	20

[a] 6.0-mm arc gap at 12 A; preburn 5 s; exposure 35 s.
[b] Working concentration range, 5–500 ppm.
[c] See Table 4.16 for formula.
[d] Both B and S samples.

4.4.4 Rock Phosphate[(107)]

In a previous publication[(107)] the author and his coworkers developed the cathode-region method for the spectrochemical analysis (dc arc) of trace elements in natural rock phosphate. This previous article, together with new results, will be reviewed in this section.

Rock phosphate samples differ from refractory matrices, previously mentioned, as follows:

1. The samples are composed of several major components, as given in Table 4.22.

2. The contents of these major components are variable (Table 4.22) as a function of geological processes. In trace elements, spectrochemical analysis of such materials requires numerous buffers,[(134–138)] diluents,[(139–140)] and internal standards.[(141)]

In this section the fundamentals described in previous sections were used with the aim of ascertaining which of the major components affect the volatilization rate of the sample and the plasma parameters. If such a major component

Table 4.22 Changes in the Major Constituents of the Israel Rock Phosphate[a]

Constituent	Oron site (%)	Arad site (%)
P_2O_5	14–35	22–33
CaO	35–55	45–53
F	2–4	2.5–3.5
SO_4	2–4	3–6.5
NaCl	1–2	1–2

[a]Analyzed at the Laboratories of the Nuclear Research Center, Negev.

exists in the rock phosphate, it might be feasible to adapt a single matrix for the spectrochemical analysis of trace elements in natural phosphates.

4.4.4.1 The Third Matrix Elements

In natural phosphate the matrix elements can be conveniently defined as that major constituent which affects the plasma and directly influences trace-element behavior. Figure 4.7 shows the axial distribution of the normalized line intensities of Ca, P, Na, and trace elements. According to the definition above, calcium is found to behave as the third matrix element (Fig. 4.7). The axial distribution of all the trace elements is similar to that of calcium (Fig. 4.7). It was assumed that the calcium molecules supplying the measured atoms and ions, in the plasma, consist of either or both of CaO and $Ca_2P_2O_7$, since they are stable compounds of calcium. $Ca(OH)_2$, $CaCO_3$, and $Ca(PO_3)_2$ (metaphosphate) were chosen as possible standard matrices, and their volatilization rates compared with that of the natural rock phosphate. Q_j^{ch} was measured by weight loss after an arcing period of 30 s (also see Section 4.2.2.1). Table 4.23 shows that

Table 4.23 Volatilization Rate (Q_j^{ch}) in Phosphates[a]

Matrix	Anode charge (mg)	Leaving the anode (mg)	Q_j (mg/s^{-1})
Standard	~50.0	20.0	0.68 ± 10%[b]
Rock phosphate	~50.0	22.0	0.73 ± 10%[b]
Standard +20% C	~50.0	40.0	1.33 ± 20%[b]
Rock phosphate +20% C	~50.0	46.0	1.55 ± 20%[b]

[a]4.0-mm arc gap; arcing time, 30 s.
[b]Standard deviation calculated from 10 experiments.

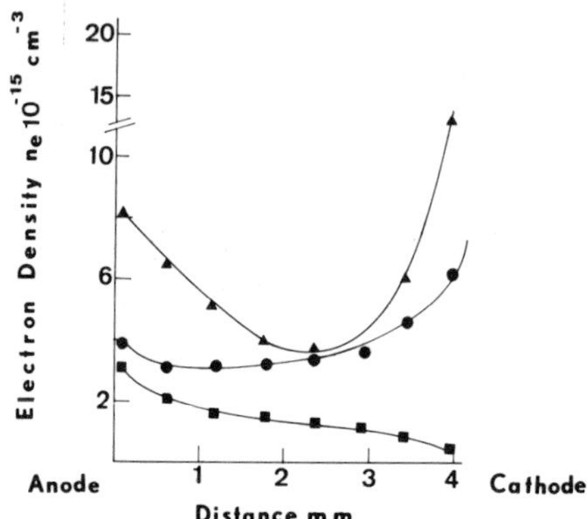

Fig. 4.40 Axial distribution of electron density (n_e) in plasmas with phosphate at 4.0-mm arc gap and 10 A. ■, Rock phosphate; ▲, $Ca(PO_3)_2$; and ●, standard matrix or rock phosphate + 20% graphite.

the volatilization rate of calcium metaphosphate is similar to those of the natural samples [i.e., $Ca(PO_3)_2$ can be used as the standard matrix].

On the basis of data given in Table 4.23, the plasma variables of calcium metaphosphate and natural phosphate matrices were measured. The axial distributions of temperature and electron density are given in Figs. 4.20 and 4.40, respectively. Equalization of the plasma variables for the metaphosphate and natural phosphate matrices was obtained by adding 20% graphite to both. The role played by the 20% graphite is twofold:

1. Eliminate the factors that cause different volatilization rates (Table 4.23) and plasma variables (Figs. 4.7, 4.20, and 4.40) between the metaphosphate and natural samples.

2. Increase the trace-element particle concentration in the plasma by increasing the volatilization rate of the matrix (Table 4.23).

The cathode region of the plasma is most suitable for quantitative trace-element analysis of rock phosphate using $Ca(PO_3)_2$ as the standard matrix. In the cathode region (0.5 mm-0.8 mm below the cathode) the maximum values of temperature, electron density, and relative line intensities for all trace elements (except As, Hg, and B) were found (Fig. 4.7).

4.4.4.2 Standard and Sample Preparation

1. To 30 g of $CaCO_3$ (Johnson-Matthey specpure) was added 4.0 ml of pure H_3PO_4.[107] After drying at 110°C for 2 h, calcium dihydrogen phosphate formed which was converted to $Ca(PO_3)_2$ by heating at 400°C for 2 h. A mixture of 20% graphite and 80% calcium metaphosphate was homogenized and two sets of standards prepared: (a) Rare earths Sc, Y, Th, U, and Zr in concentrations from 1 to 500 ppm each element; and (b) all common trace elements (Table 4.24) from 0.5 to 500 ppm for each element.

2. 0.5 g (< 150 mesh) of natural rock phosphate was treated with 0.5 ml of H_3PO_4. The sample was dried at 110°C and at 400°C and mixed with 20% graphite.

The anode crater charge was 50 mg either standard or natural phosphate. The operating conditions were described in Section 4.4.1 or in a previous publication.[107]

4.4.4.3 Analytical Results—Cathode Region

4.4.4.3a Detection Limits. The spectral lines used for the trace elements, together with their detection limits, are given in Table 4.24. Detection limits were obtained by densitometry of the lowest concentration using a seven-step rotating sector.

4.4.4.3b Standard Working Curves. Working curves were prepared as described in Section 4.4.3.3b, with the sole exception of a 30-s exposure without a preburn. The slopes of the working curve within the concentration range are classified according to the behavior of the trace elements. Slope values and coefficients of correlation are given in Table 4.25. A comparison between results obtained with Pd as internal standard and without it shows that good results can be obtained without the use of this internal standard (Table 4.25).

4.4.4.3c Results and Statistics for Samples. Approximately 300 natural samples were analyzed using the cathode-region procedure. The samples were obtained from various phosphate sites in Israel. Several samples with varying P_2O_5 content were analyzed 5 to 10 times to assess the analytical precision, given as C.O.V. in Table 4.25. For concentrations that exceed 500 ppm, samples were diluted with the mixture of $Ca(PO_3)_2$ graphite and analyzed as described previously.

The results in Table 4.25 show that the cathode region is an accurate and reliable method for quantitative trace-element analysis of rock phosphates.

The cathode-region analysis of several standard reference materials is now in process and analytical data will be published elsewhere.

4.4.5 Silicate Rocks and Minerals[54–51]

Table 4.26 shows the large compositional variation of natural silicates. Comparison of the variations with those of phosphate (Table 4.26) shows that silicate rocks are more complex than rock phosphates. In order to establish a general spectrochemical method of silicates, the analysis steps adapted are similar to those for the rock phosphate. In this case, by measuring the volatilization rate and the axial distribution of plasma variables, the aim is to localize a region in the plasma where the influence of the major components on their trace elements is favorable for a general quantitative spectrochemical method.

Table 4.24 Detection Limits of Trace Elements in Rock Phosphate in the Cathode Region

	Element	Wavelength (Å)	Detection limits (ppm)
1	Ag	3280.68	0.1
2	Al	3082.16 3092.71	0.5
3	Ba	4554.03	2
4	Bi	3067.72	0.5
5	Cd	2288.02	0.5
6	Co	3453.50	2
7	Cr	3021.56	1
8	Cs	8251.10	2
9	Cu	3247.54 3273.96	0.1
10	Fe	3020.49 3020.64 3021.07	1
11	Ga	2943.64	0.1
12	Ge	3039.06	0.2
13	In	3039.36	0.1
14	K	7664.91	0.5
15	Li	6707.84	0.1
16	Mg	2779.83	1
17	Mn	2794.82	1
18	Mo	3132.59	0.5
19	Na	3302.32	2
20	Ni	3050.82	1
21	Pb	2802.00	0.2
22	Rb	7800.23	0.5
23	Sb	2598.05	1
24	Sc	3353.73 3372.15	0.5
25	Sn	2839.99	1
26	Sr	4607.33	2
27	Ti	3234.52	2
28	Tl	2767.87	0.2
29	V	3181.41	0.5
30	Y	3216.69 3242.28	1

(Continued)

Table 4.24 (*Continued*)

	Element	Wavelength (Å)	Detection limits (ppm)
31	Zn	3302.59	10
32	Be	2348.61	0.5
33	Ce	3716.37	5
34	Dy	4045.99	5
35	Eu	3907.10	5
36	Er	3007.97	1
37	Gd	3422.47	1
38	Ho	3456.00	5
39	La	3337.49	2
40	Lu	2615.42	2
41	Nd	4303.58	10
42	Pr	4408.84	100
43	Sm	3621.23	5
44	Tb	3703.92	10
45	Th	4019.13	50
46	Tm	3462.20	1
47	U	3890.36	25
48	Zr	3391.98	0.5
		3438.23	12
49	As[a]	2780.22 2860.44	10
50	B[a]	2496.78 2497.73	1
51	Hg[a]	2536.52	10

[a]In the center region.

4.4.5.1 Volatilization Rate

The most important factor which has to be examined when considering the volatilization of silicate material into an arc gap is the vitrification of the sample (i.e., mineralogical changes). The formation of glass, a refractory material, hinders volatilization of the sample. Glass formation can be eliminated by adding, for example, graphite to the sample prior to arcing.

Figure 4.41 shows the Q_j^{ch} in mg s^{-1} of SiO_2 from the anode crater into the arc gap. For pure SiO_2, glass formation in the crater hinders volatilization (0.06 mg s^{-1}). On addition of 12% (Na + K) to SiO_2, Q_j^{ch} increases to 0.35 mg s^{-1}, but glass formation persists. The addition of graphite increases the values of Q_j^{ch}. Samples containing 33% graphite and 12% alkalies were totally consumed after a

Table 4.25 Working-Curve Slopes, $d(\log J)/d(\log C)$, and Coefficient of Variation for Trace Elements in Rock Phosphate in the Cathode Region[a]

	Working curves			100 samples	
	Slope value				
Element[b]	With I.S.[c]	Without I.S.[c]	R^d	Number of determinations	Coefficient of variation (%)
Al	0.74	0.75	0.95	5	15
Ba	0.53	0.52	0.97	10	20
Bi	0.90	0.89	0.98	5	10
Cd	0.87	0.88	0.99	5	10
Co	0.42	0.42	0.98	5	15
Cr	0.75	0.73	0.96	15	10
Cu	0.84	0.84	0.98	15	9
Fe	0.43	0.42	0.98	15	12
Ga	0.85	0.86	0.95	5	10
In	0.75	0.75	0.88	5	13
Mg	0.89	0.88	0.96	10	18
Mn	0.80	0.82	0.95	10	8
Mo	0.70	0.70	0.95	10	12
Ni	0.69	0.68	0.94	10	10
Pb	0.74	0.75	0.98	20	8
Sc	0.75	0.75	0.92	10	10
Sr	0.53	0.53	0.95	10	20
Ti	0.73	0.72	0.97	10	15
V	0.73	0.71	0.92	15	15
Y	0.80	0.82	0.90	20	15
Ce	0.50	0.51	0.96	5	12
La	0.55	0.54	0.98	15	13
Yb	0.49	0.48	9.95	5	18
Nd	0.43	0.45	0.95	5	17
Sm	0.41	0.40	0.92	5	17
Eu	0.59	0.57	0.93	5	12
Gd	0.59	0.59	0.95	5	15
Ho	0.76	0.75	0.88	5	20
Lu	0.60	0.59	0.95	5	18
U	0.50	0.51	0.96	20	20

[a] 4.0-mm arc gap at 10 A.
[b] Wavelength indicated in Table 4.24.
[c] Pd as internal standard (I.S.).
[d] Without internal standard (see Table 4.16 for formula).

100-s arcing period. The volatilization rate of the synthetic SiO_2 standards was identical to that of natural silicate only after 3 parts graphite was added to both matrices. Thus, by equalizing the volatilization rates, the silicate refractory matrix has been converted into a less refractory one.

Table 4.26 Compositional Variation (%) of Major Components in Several Petrological Types of Silicate Rocks[a]

Constituents	G.2 Granite	AGV-1 Andesite	BCR-1 Basalt	PCC-1 Peridotite
SiO_2	69.2	59	54.5	41.9
Al_2O_3	15.3	17	13.7	0.85
Fe as Fe_2O_3	2.8	6.8	13.5	8.53
MgO	0.8	1.5	3.3	43.6
CaO	2.0	4.9	7.0	0.53
Na_2O	4.2	4.3	3.3	0.05
K_2O	4.5	2.9	1.7	0.01
TiO_2	0.5	1.1	2.2	0.02

[a]From Flanagan.(142)

4.4.5.2 Plasma Variables

The axial distribution of the normalized line intensity is shown in Fig. 4.42 for the synthetic silicate matrix. As in the case of refractory matrices investigated, the cathode region shows the maximum line intensity for the trace elements and the minimum for the matrix element, Fig. 4.8. Addition of major components, such as Ca, Mg, Al, and Fe to SiO_2 does not significantly alter the normalized line distribution, as shown in Fig. 4.42. The addition of 5% (Na + K) to SiO_2 suppresses the normalized line intensity in the cathode region for all elements. Only the central region is not influenced by the presence of alkali, as depicted in Fig. 4.42. In general, the alkalies suppress the relative line

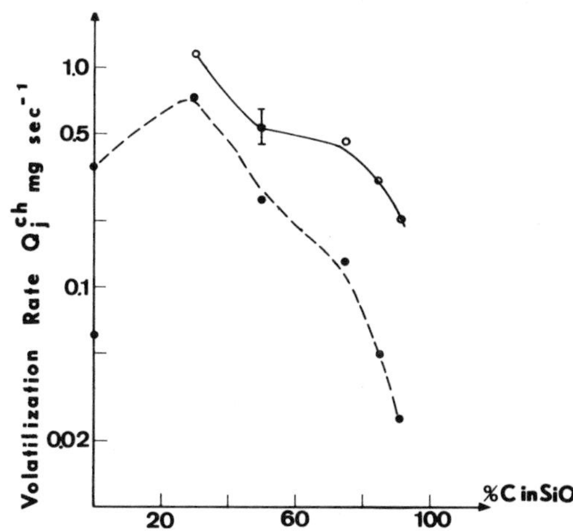

Fig. 4.41 Volatilization rate of (Q_j^{ch}) of silicates: •, SiO_2; ---•---, SiO_2 + 12% (Na ± K) calculated, without graphite; ○, SiO_2 + 12% (Na + K) + graphite.

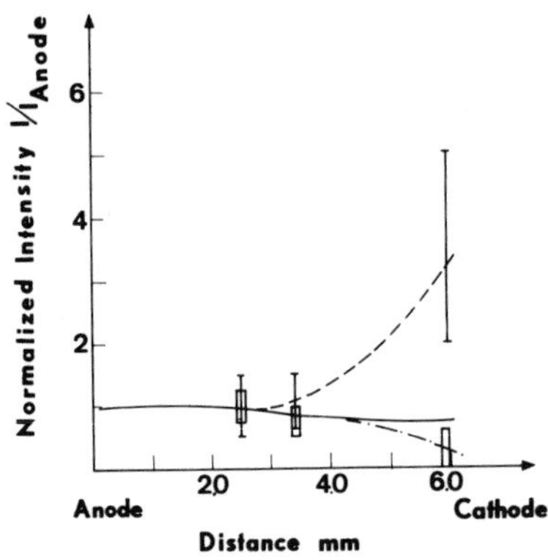

Fig. 4.42 Axial distribution of relative line intensity normalized to the anode region in silicates. --I--, Major and common trace elements in SiO_2 + 5% each Ce, Mg, Al, and Fe matrix; —·⫿·—, Na and common trace elements in SiO_2 + 5% (Na + K) matrix; —, Si in SiO_2 + 5% each Ca, Mg, Al, and Fe matrix.

intensities of all components (i.e., lower sensitivities are obtained for the trace elements).

Axial distribution of temperature (Fig. 4.19) and electron density (Fig. 4.27) confirm the conversion of the refractory SiO_2 matrix into a graphite one as a result of the addition of 3 parts graphite to SiO_2.[54–57]

The results from Figs. 4.19, 4.27, 4.41, and 4.42 can be summarized as follows:

1. The central region (2 mm out of a 6.0-mm gap) of the plasma[53] is the most suitable region for the spectrochemical determination of trace elements in silicate rocks and minerals.

2. Dilution of the synthetic standards (SiO_2 containing 8% Na and K) and the natural silicates with 3 to 5 parts graphite[54] results in a parity between the parameters of these two matrices.

4.4.5.3 Analytical Results—Central Region

4.4.5.3a Detection Limits. The spectral lines used for the trace and minor elements, together with their detection limits, are given in Table 4.27. These limits were measured as described in Section 4.4.4.3a for the rock phosphate.

4.4.5.3b Standard Working Curves. An exposure of 50 s without preburn was used. Emulsions were calibrated with a computerized self-calibration method using Seidel densities. Slope values of the analytical working curves and their coefficients of correlation are given in Table 4.27. Only minor increases in analytical accuracy can be obtained by the use of Pd as the internal standard, as shown in Table 4.27. Analysis without internal standards was possible only by matching the parameter values of the synthetic SiO_2 and silicate samples (Figs. 4.19, 4.27, 4.41, and 4.42 and Table 4.27).

4.4.5.3c Results and Statistics for Samples. Table 4.28 shows the analytical data for a large variety of international silicate standards. The results obtained by our method are compared with values obtained by a variety of laboratories

Table 4.27 Analytical Lines, Working Ranges, and Statistics of Working Curves in Silicates

Analytical line	Working range (ppm)	Working curves: Slope value[a]		Coefficient of correlation, R^c	
		With I.S.[b]	Without I.S.[b]	With I.S.[b]	Without I.S.[b]
Cr 2843	5–5000	0.4645	0.4709	0.95	0.95
Cr 4254	5–5000	0.5971	0.5952	0.97	0.97
Mn 2933	10–5000	0.6681	0.6816	0.98	0.98
Mn 4034	50–5000	0.8307	0.8442	0.98	0.99
V 3184	5–1000	0.6209	0.6170	0.98	0.98
V 4379	100–1000	0.7733	0.7486	0.96	0.96
Ti 3242	0.01–3.5% TiO_2	0.5775	0.5815	0.99	0.98
Ti 3990	0.05–3.5% TiO_2	0.7793	0.5168	0.99	0.88
Cu 3274	5–1000	0.7344	0.6929	0.98	0.96
Ni 3414	2–3000	0.5869	0.5855	0.98	0.98
Co 3453	2–1000	0.8336	0.8115	0.99	0.98
Pb 2833	5–1000	0.6160	0.6012	0.94	0.94
Sr 4077	5–2000	0.6120	0.6078	0.98	0.98
Sr 4607	20–1500	0.5840	0.7204	0.83	0.95
Ba 4554	5–3000	0.7771	0.7711	0.99	0.99

[a] $\log Y = b \log x + \log a$.
[b] Pd as internal standard (I.S.).
[c] See Table 4.16 for formula.

and methods. The number of determinations per sample usually exceeds 10. The C.O.V. for each element is given in Table 4.28. It is apparent that the overall accuracy is good—about 15% for all common trace elements analyzed.

4.4.6 Aluminum and Titanium Oxides

Trace elements in Al metal or alloys can be analyzed by various spectrochemical methods using spark excitation,[(149)] x-ray fluorescence,[(150)] and atomic absorption.[(151)] The refractory oxides were analyzed mainly for trace elements by dc arc techniques employing chemical separation prior to arcing,[(152)] or the addition of buffers[(153)] and carriers.[(154,155)] For TiO_2, spectrochemical methods with buffers[(156)] and carriers[(157)] have been used.

The direct cathode region was developed for the following reasons: (1) to check the versatility of the cathode-region method and extend it to other refractory matrices; and (2) to provide a general direct analysis of trace elements in Al_2O_3 and TiO_2 matrices.

The metal or alloy samples of Al were dissolved in purified HCl and transformed to Al_2O_3 by roasting at 900°C for 2 h. The alumina samples were directly

Table 4.28a Contents of Several Trace Elements in International Silicate Standards (ppm)

Sample	Source[a]	Cr	Mn	V	% TiO_2	Cu	Ni	Co	Pb	Sr	Ba	Ga	Zr	B	Cd
AGV-1	a	15	785	115	1.14	68	14	17	40	675	1300	20	250		
	b	13	728	121	1.08	64	18	13	35	657	1410	18	227		
BCR-1	a	12	1400	385	2.22	19	14	37	20	340	770	21	215		
	b	16	1350	384	2.23	22	15	36	18	345	790	22	185		
DTS-1	a	4300	1060	12	0.0094	9	2400	140	8	1	6				
	b	4200	963	19	0.02	8	2330	132	8		6				
G-2	a	26	280	34	0.51	12	7.5	4	32	519	1900	22	260		
	b	37	270	37	0.53	11	6	5	29	463	1950	20	316		
GSP-1	a	15	330	53	0.71	38	13	7	60	255	1300	24	528		
	b	13	326	52	0.69	35	11	8	52	247	1360	18	544		
PCC-1	a	3000	860	35	0.0085	11	2370	126	8	1	5				
	b	3090	889	31	0.02	10	2430	112	6	0.3	7				
G-1	a	18	195	12	0.27	13	4	2	50	240	950	18	220		
	h	22	270	16	0.26	13	2	2.4	49	250	1200	18	210		
W-1	a	125	1250	248	1.0	125	70	43		190	188	18	115		
	h	120	1300	240	1.07	110	78	50	8	180	180	16	100		
SCO-1	a	65	420	117	0.75	30	25	11	28	224	750	14	133	65	
SGR-1	a	34	297	125	0.27	70	28	13	41	316	322	12	52	30	
BR	a	410	1400	250	2.67	69	235	53	8	1200	1070	20	275		
	c	420	1600	240	2.62	72	270	50	8	1300	1050	26	240		
VSN	a	900	800	825	1.12	850	890	700	1100	1260	1500	384	710	250	1000
	f	820	700	650	1.0	800	786	734	930	820	960	372			

Fe-mica	a	80	2660	140	2.65	4.3	22	18		7	150	120	885
	g	90	2710	135	2.55	4	35	20		6	140	95	
Mg-mica	a	96	2000	100	2.15	3.5	85	24		26	4200	34	
	g	80	1900	82	1.67	4	105	20		25	4700	30	
SY-1	a	50	3200	88	0.46	25	37	25	450	250	330	24	3080
	d	52	3100	88	0.48	22	37	18	495	286	282	20	2900
SY-2	a	9	2500	52					75	290	470	29	
	e	20	2250		0.14				64	270	430	33	
SY-3	a	14	2700	55	0.13	13	14	4.5	105	250	370	19	
	e	n.d.[b]	2500	30	0.13	18	n.d.	n.d.	120	300	410	43	
BCS-267	a	130	1200	18	0.20	57	26	7					137
	f	50	890	1	0.10	64	14.5	7					140
BCS-269	a	150	230	176	1.31	68	79	23		109	520		206
	f	170	230	180	1.47	64	88	23		120	660		230

[a] The following letters refer to the References at the end of this chapter: a, this chapter; b, (140); c, (141, 142); d, (51); e, (143); f, (142); g, (145); h, (144).
[b] n.d., not detected.

Table 4.28b Coefficients of Variation of Trace-Element Data in International Silicate Standards

Sample		Cr	Mn	V	TiO_2	Cu	Ni	Co	Pb	Sr	Ba	Ga	Zr	B
AGV-1	C[a]	15	9	12	10	12	20	16	15	13	10	14	13	
	N	40	40	40	40	35	30	30	20	35	35	20	20	
BCR-1	C	30	12	5	7	12	20	9	18	10	10	12	14	
	N	24	28	20	28	26	18	22	14	48	48	26	25	
DTS-1	C	9	9		20	16	13	6						
	N	20	26	3	10	15	25	26	4					
G-2	C	20	12	13	10	15	20	30	13	9	11	10	18	
	N	18	28	30	40	18	18	12	16	28	20	13	20	
GSP-1	C	18	8	10	9	9	20	30	15	14	8	13	12	
	N	18	40	40	26	26	22	10	25	40	30	21	13	
PCC-1	C	13	11	15	10	15	10	6						
	N	22	28	10	5	19	14	18	3					
G-1	C	20	12	7	12	15	30	30	14	10	10	12	10	
	N	9	15	15	12	10	5	6	15	16	12	15	5	
W-1	C	14	10	8	6	10	11	10		12	15	12	11	
	N	24	28	16	25	30	37	37		28	20	17	27	
SCO-1	C	14	10	7	6	12	12	24	7	9	8	15.5	10.5	11
	N	10	34	26	18	26	26	12	12	12	11	12	11	9
SGR-1	C	16	4	10	10	8	8	12	8	6.5	4.6	20	16	10
	N	12	12	22	21	20	22	12	12	12	12	12	8	11
BR-1	C	10.5	8	8	8	12	15	15	15	10	12	10	6	
	N	50	50	40	50	50	50	50	50	50	50	40	40	

VS-N	C	12	12	15	5	15	11	11	15	10	10	5	6.5
	N	25	25	30	25	25	25	25	25	25	25	25	20
Fe-mica	C	8	12	15	11	20	12	15		18	11.0	15	13.5
	N	14	14	11	10	9	6	6		7	22	6	4
Mg-mica	C	12	11	14	10	13	15	12		9	11	12	
	N	6	12	11	11	6	6	6		15	15	6	
SY-1	C	12	11	8	10	10	18	10	8	12	14	12	16
	N	25	40	40	48	35	30	30	30	30	30	25	8
SY-2	C	20	10	13	10				15	12	12	15	
	N	6	9	7	9				7	6	6	4	
SY-3	C		14	17	10	16	20	12	10	10	10	10	
	N	5	16	10	10	15	5	5	5	6	5	5	
BCS-267	C		9.5	20	10	11	20	20					8
	N		6	6	6	6	6	6					6
BCS-269	C	18	15	16	15	16	15	13		15	8.5		10
	N	16	9	8	11	6	7	6		9	10		8

[a] C, coefficient of variation; N, number of determinations.

Table 4.29 Volatilization Rate of Al_2O_3 and TiO_2 Matrices with and without 10% Added Fluorides[a]

Matrix	Volatilization rate,[b] Q_j^{ch} (mg s^{-1})
Al_2O_3	0.04–0.08
Al_2O_3 + 10% AlF_3	0.029–0.10
TiO_2	0.035–0.06
TiO_2 + 10% Teflon[c]	0.095–0.1

[a] 6.0-mm arc gap at 13 A; arcing time, 40 s. Al_2O_3, charge 30 mg; TiO_2, charge 50 mg.
[b] Minimum and maximum values; 10 samples each matrix.

heated at the same temperature and time. Titanium metal or alloys, obtained in chip form, were transformed in TiO_2 by roasting air at 1100°C for 2 h.

The cathode-region method for common trace elements was developed after the investigation of the volatilization rate of each matrix and its behavior on the dc arc variables.

4.4.6.1 Volatilization Rate

The volatilization rates for Al_2O_3 and TiO_2 matrices were measured chemically (Q_j^{ch}). The values of Q_j^{ch} for Al_2O_3 were found to be erratic (i.e., for the same arcing time the spread of results was within a factor of 3). For a constant volatilization rate the charged electrodes (anode) were heated on a hot plate for several minutes before arcing and 10% of fluorides was added to each refractory matrix. Aluminum fluoride and Teflon spray "kingerflon" were added, respectively, to Al_2O_3 and TiO_2. Table 4.29 shows the influence of the added fluorides on a more constant volatilization rate. Furthermore, the 10% fluorides increases the values of Q_j^{ch} by a factor of 2.

Based on the results reported in Tables 4.1a and 4.1b, one assumes that the volatilization rate of the trace elements follows that of the refractory matrix. In other words, a factor-of-2 greater volatilization rate of the matrix will enhance the volatilization rate of the trace elements.

4.4.6.2 Plasma Variables

Figure 4.18 shows the axial distribution of temperature (Zn and Cu spectral lines) in a graphite-air plasma containing Al_2O_3 and TiO_2 refractory matrix. The cathode region, compared to the anode and central regions, shows the maximum values of temperature.

Figure 4.26 shows the axial distribution of electron density (Mg and Mn ion

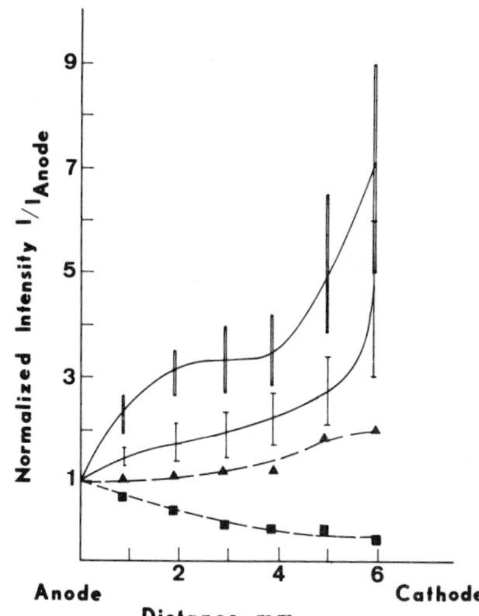

Fig. 4.43 Axial distribution of relative line intensity normalized to the anode region in Al_2O_3 and TiO_2 with 10% fluoride matrices at 6.0-mm gap, 13 A. ▯, Common trace elements in TiO_2 matrix; ■, Ti; I, Common trace elements in Al_2O_3 matrix; ▲, Al.

and atom spectral lines) of the plasma with the refractory matrices. Again, the maximum n_e values were obtained in the cathode region.

Based only on temperature and electron density (no other plasma variables were measured for Al_2O_3 and TiO_2) and their similarity of the cathode region to other refractory matrices such as U_3O_8, ThO_2, ZrO_2, and rare earth oxides, it is assumed that the cathode region is the favorable region for spectrochemical analysis of trace elements. This conclusion can be granted by the axial line intensity of trace elements in the cathode region (Fig. 4.43).

Figure 4.43 shows the axial distribution of the normalized line intensities (I/I_{anode}) of trace and matrix elements. The majority of trace elements have their maximum intensity value in the cathode region. For B, Ba, Ca, and Sr traces, the maximum line intensity was obtained in the anode region; therefore, their spectrochemical analysis was done by exposing the anode region (~0.6 mm above the anode) instead of the cathode.

4.4.6.3 Analytical Results—Cathode Region

4.4.6.3a Detection Limits and Working Curves. Table 4.30 shows the detection limits of each trace element in each matrix, together with the slope value of the working curves (see Section 4.4.2.3). For common trace elements, the cathode region (0.2 mm beneath the cathode) was exposed for 40 s for the Al_2O_3 matrix and 35 s for the TiO_2 matrix. It is evident from the coefficient of correlation values (Table 4.30) that the method yields good reliability without the use of internal standards. The standards were prepared from specpure oxides.

Table 4.30 Detection Limits, Working Curves, and Coefficient of Variation of Trace Elements in Al_2O_3 and TiO_2 Matrices in the Cathode Region[a]

Element[b]	Detection limits (ppm)		Working curves: Slope value		Working curves	Al_2O_3 and TiO_2	
	Al_2O_3	TiO_2	Al_2O_3	TiO_2	Coefficient of correlation, R^c	Number of determinations	Coefficient of variation
Ag	0.5	0.5	0.80	0.78	0.95	10	9
Al	–	1	–	0.75	0.96	10	15
B	0.5	0.5	0.83	0.82	0.96	10	10
Ba	25	25	0.70	0.65	0.92	10	20
Bi	0.5	0.5	0.72	0.80	0.92	10	10
Cd	0.5	0.5	0.80	0.85	0.98	10	12
Co	0.5	1.0	0.69	0.70	0.95	10	15
Cr	2	5	0.70	0.70	0.94	10	15
Cu	0.5	0.1	0.78	0.75	0.92	10	15
Fe	0.5	2	0.76	0.75	0.96	10	12
Ga	0.1	0.1	0.80	0.75	0.98	6	8
In	0.1	0.1	0.80	0.75	0.98	8	9
Mg	1	2	0.80	0.75	0.95	10	13
Mn	0.5	2	0.72	0.70	0.95	10	12
Mo	1	3	0.65	0.60	0.95	10	15
Nb	50	50	0.52	0.55	0.90	10	20
Ni	3	3	0.58	0.60	0.96	10	15
Pb	1	1	0.80	0.70	0.98	10	10
Sb	5	5	0.50	0.52	0.96	10	10
Si	5	0.5	0.68	0.65	0.98	10	12
Sn	0.5	1	0.72	0.70	0.98	10	8
V	5	5	0.69	0.70	0.96	10	15
Zn	10	10	0.65	0.65	0.95	10	20

[a] 6.0-mm arc gap at 13 A; 40-s exposure.
[b] Wavelength as indicated in Table 4.24.
[c] See Table 4.16 for formula.

4.4.6.3b Statistics of Results. Table 4.30 shows the coefficient of variation for each trace element analyzed, based on 10 determinations of each element in each sample. The C.O.V. varies from 10 to 20% for trace elements contents up to 500 ppm.

4.4.7 Molybdenum and Tungsten Oxides

Trace elements in molybdenum and tungsten matrices are analyzed mainly by the use of dc arc sources.[158–162] The spectrochemical techniques employed for MoO_3 and WO_3 matrices are chemical separation prior to arcing,[158,159] carrier distillation,[160,161] and matrix mixtures with carbon.[162] The samples of molybdenum and tungsten, obtained in chip form, were transformed into their respective oxides by roasting in air at 1100°C for 2 h.

The variables investigated were only volatilization rate and axial distribution of the normalized spectral line intensities of common trace elements in each matrix.

The cathode-region method was applied in the spectrochemical analysis of common trace elements in NoO_3 and WO_3 matrices.

4.4.7.1 Volatilization Rate

The volatilization rate for the oxides above, measured chemically, showed erratic values for the same arcing time. Owing to the sublimation behavior of molybdenum and tungsten oxides, the spread of Q_j^{ch} values varied as much as a factor of 5. The addition of fluoride up to 10% to the oxides of molybdenum and tungsten did not have a favorable influence on the Q_j^{ch} values (i.e., the value varied widely for the same arcing time). Addition of graphite was found to be favorable; constant volatilization rates were obtained for 35 s of arcing. The minimum concentration of graphite necessary for constant volatilization rates of MoO_3 and WO_3 was 66% and 50%, respectively (Table 4.31). This table shows

Table 4.31 Volatilization Rate of MoO_3 and WO_3 Matrices with and without Graphite[a]

	Ratio WO_3/graphite			Ratio MoO_3/graphite		
	1:0.5	1:1[b]	1:2[b]	1:1	1:2[b]	1:3[b]
Volatilization rate, Q_j^{ch} (mg s^{-1})	0.70–0.30[c]	0.22	0.15	0.80–0.35[c]	0.22	0.13

[a] 6.0-mm arc gap at 13 A; arcing time, 35 s. WO_3 ; MoO_3 charge 35 mg.
[b] Mean values from 10 samples each matrix.
[c] Minimum and maximum values obtained from 10 samples.

the favorable influence of various additions of graphite on the volatilization-rate values of each matrix. The graphite reduced the Q_j^{ch} values as compared to the values of the matrix without graphite (Table 4.31).

4.4.7.2 Normalized Line Intensity

The axial distribution of the line intensities of common trace elements was measured only for the following matrix mixtures with graphite: $MoO_3:C = 1:2$ and $WO_3:C = 1:1$. Figure 4.44 shows the axial values of the normalized line intensity in the three plasma regions. The maximum intensity values for almost all the common trace elements were obtained in the cathode region. Furthermore, in such mixtures the spectral lines of the matrix elements, either Mo or W, were not detected. In the 6.0-mm arc gap, photographic plates exposed for 35 s show no trace of the matrix-element spectrum, thus leaving the trace-element spectrum undisturbed.

The undetected matrix-element spectrum indicates the following:

1. The molecules of the matrix were not decomposed in Mo and W atoms or ions; or

2. The oxides of molybdenum and tungsten react with graphite to form high-melting carbides (MoC, mp 2692°C; WC, mp 2870°C), as described by Nickel.[12] The carbide volatilization rate, as well as its decomposition to the respective atoms, could be very low.

For common trace elements, Fig. 4.44 shows that the cathode region is the favorable region for their spectrochemical analysis in the refractory matrices of molybdenum and tungsten.

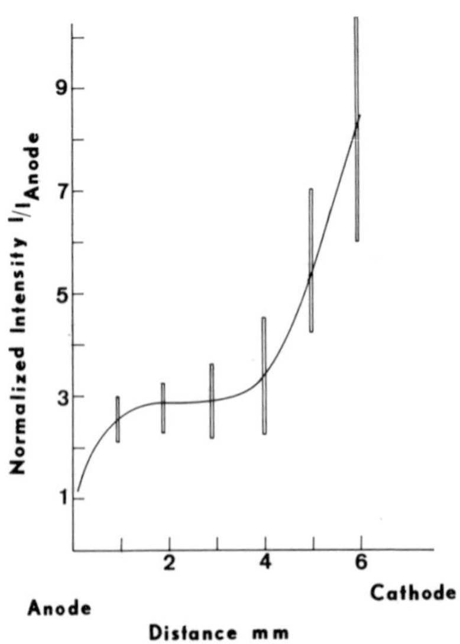

Fig. 4.44 Axial distribution of relative line intensities of common trace elements, normalized to the anode region in MoO_3 and WO_3 matrices with graphite.

Table 4.32 Detection Limits, Working Curves, and Coefficient of Variation of Trace Elements in Molybdenum and Tungsten Matrices in the Cathode Region[a]

Element[b]	Detection limits (ppm)		Working curves			MoO_3 and WO_3	
			Slope value				
	MoO_3	WO_3	MoO_3	WO_3	Coefficient of correlation, R^c	Number of determinations	Coefficient of variation
Ag	1	1	0.88	0.85	0.99	5	10
Al	2	2	0.75	0.70	0.95	10	15
B	3	1	0.83	0.82	0.98	10	10
Bi	1	1	0.72	0.80	0.98	10	10
Ca	10	10	0.60	0.65	0.90	10	25
Cd	0.5	1	0.80	0.82	0.97	10	15
Co	10	10	0.69	0.70	0.95	10	15
Cr	3	3	0.65	0.68	0.95	10	15
Cu	2	2	0.70	0.70	0.96	10	15
Fe	2	3	0.68	0.70	0.95	10	17
Ga	1	2	0.72	0.75	0.98	5	12
In	1	1	0.80	0.82	0.97	5	10
Mg	1	1	0.85	0.85	0.97	7	15
Mn	1	1	0.72	0.75	0.95	7	12
Ni	3	3	0.65	0.70	0.92	12	17
Pb	1	2	0.80	0.75	0.99	10	10
Sb	5	5	0.53	0.55	0.89	10	15
Si	5	2	0.68	0.65	0.90	10	20
Sn	2	2	0.75	0.80	0.96	10	10
Sr	25	20	0.60	0.62	0.90	10	25
Ti	15	15	0.60	0.55	0.95	10	18
Zn	10	10	0.65	0.65	0.95	10	25

[a] 6.0-mm arc gap at 13 A; 35-s exposure.
[b] Wavelength as indicated in Table 4.24.
[c] See Table 4.16 for formula.

4.4.7.3 Analytical Results—Cathode Region

4.4.7.3a Detection Limits and Working Curves. Table 4.32 shows the detection limits of the trace elements in each matrix, their concentration range, and the slope values of working curves. For common trace elements, the cathode region (0.2 mm beneath the cathode) was exposed for 35 s. The coefficients of correlation values show good reliability of the working curves. The standards were prepared from specpure oxides.

4.4.7.3b Statistics of Results. Table 4.32 shows the C.O.V. for each trace-element analysis, based on 10 determinations of each element in each sample. Up to a concentration of 1000 ppm for the trace element, the C.O.V. varies from 10 to 20%.

ACKNOWLEDGMENTS

I wish to extend my gratitude to Z. Goldbart, A. Harel, U. Carmi, J. Taieb, and I. B. Brenner for their assistance and contributions on various topics of this work. I also wish to acknowledge my indebtedness to Mr. A. Seroussi and M. J. Tulipman, Director of the N.R.C. Negev, for their interest and continuous support.

4.5 REFERENCES

1. J. A. Smit, Thesis, State University, Utrecht (in Dutch) (1951).
2. R. L. Roes, Thesis, State University, Utrecht (in Dutch) (1962).
3. L. Huldt, Thesis, University of Uppsala (1948).
4. W. Finkelenburg and H. Maecker, eds., *Handbuch der Physik*, Vol. 22, p. 254, Springer-Verlag, Berlin (1956).
5. P. W. J. M. Boumans, *Theory of Spectrochemical Excitation*, Hilger & Watts Ltd., London (1966).
6. W. Elenbaas, *Physica 1*, 673 (1934).
7. G. Heller, *Physics 6*, 389 (1935).
8. A. G. Gaydon and H. G. Wolfhard, *Flames: Their Structure, Radiation, and Temperature*, 3rd ed., Chapman & Hall Ltd., London (1970).
9. C. T. J. Alkemade, Thesis, State University, Utrecht (1954).
10. T. Hollander, Thesis, State University, Utrecht (1964).
11. H. Nickel, *Spectrochim. Acta 21*, 363 (1965).
12. H. Nickel, *Spectrochim. Acta 23B*, 323 (1968).
13. J. Brill, *Spectrochim. Acta 23B*, 55 (1968).
14. R. Rautschke, *Spectrochim. Acta 23B*, 55 (1968).
15. R. Rautschke, and M. Holdefleiss, *Spectrochim. Acta 24B*, 125 (1969).
16. R. Rautschke, G. Amelung, N. Nada, P. W. J. M. Boumans, and F. J. M. J. Maessen, *Spectrochim. Acta 30B*, 397 (1975).
17. R. Avni and Z. Goldbart, *Spectrochim. Acta 28B*, 189 (1973).
18. P. W. J. M. Boumans and F. J. M. J. Maessen, *Spectrochim. Acta 24B*, 585 (1969).
19. L. de Galan, *J. Quant. Spectrosc. Radiation Transfer 5*, 735 (1965).
20. H. Maecker, *Ergeb. Exakt. Naturw. 25*, 293 (1951).
21. J. Meixner, *Z. Naturforsch. 7a*, 553 (1952).
22. V. Vukanović, N. Ikonomov, and B. Pavlović, *Spectrochim. Acta 26B*, 95 (1971).
23. B. Pavlović, V. Vukanović, and N. Ikonomov, *Spectrochim. Acta 26B*, 109 (1971).
24. N. Ikonomov, B. Pavlović, V. Vukanović, and N. Rakieević, *Spectrochim. Acta 26B*, 117 (1971).
25. N. Ikonomov, B. Pavlović, and V. Vukanović, *Spectrochim. Acta 26B*, 127 (1971).
26. L. T. Steadman, *J. Opt. Soc. Amer. 38*, 1100 (1948).
27. R. C. Hirt and N. H. Nachtrieb, *Anal. Chem. 20*, 1077 (1948)
28. H. G. Short and W. L. Dutton, *Anal. Chem. 20*, 1073 (1948).
29. R. Cypres, *Bull. Centre Phys. Nucl. Univ Libre Bruxelles 40* (1953).
30. R. S. Vogel and J. R. Nelms, *U.S. At. Energy Comm. Rept. MCW 1948* (1966).
31. R. A. Edge, J. D. Dunn, and L. H. Ahrens, *Anal. Chim. Acta 27*, 551 (1962).
32. M. Z. Kharkover, M. A. Desyatkova, V. F. Barkovskii, N. A. Mitropol'skaya, and T. A. Ganopol'skaya, *J. Anal. Chem. U.S.S.R. (English Transl.) 21*, 77 (1966).
33. H. N. Barton, *Anal. Chem. 38*, 1077 (1966).
34. J. R. Butler, *Spectrochim. Acta 9*, 332 (1957).
35. B. D. Joshi and B. M. Patel, *Indian At. Energy Comm. Ref. BARC 441* (1969).
36. R. R. Brooks, *Anal. Chim. Acta 24*, 456 (1961).
37. H. J. Hettel and V. A. Fassel, *Anal. Chem. 27*, 1311 (1955).
38. S. Held, *Bull. Res. Council Israel 9A*, 173 (1960).

39. J. P. Faris, *Appl. Spectrosc. 12*, 157 (1958).
40. D. L. G. Smith and K. E. Smith, *At. Energy Res. Estab. (Gt. Brit.) Anal. Method. AM 62* (1960).
41. J. Walkden and K. E. Heatfield, *At. Energy Res. Estab. (Gt. Brit.) Anal. Method. AM 34* (1959).
42. Z. Radwan, B. Strzyzewska, and J. Minczewski, *Appl. Spectrosc. 17*, 60 (1963).
43. J. A. Carter and J. A. Dean, *Appl. Spectrosc. 14*, 50 (1960).
44. R. Ko, *Appl. Spectrosc. 16*, 157 (1962).
45. J. K. Brody, J. P. Faris, and R. F. Buchanan, *Anal. Chem. 30*, 1909 (1958).
46. R. A. Edge and L. H. Ahrens, *Anal. Chim. Acta 26*, 355 (1962).
47. R. A. Edge, *Anal. Chim. Acta 29* 321 (1963).
48. K. Govindaraju, *Centre Natl. Rech. Sci. Ref. 923* (1970).
49. A. N. Zaidel, N. J. Kaliteevski, L. W. Lipis, and M. P. Chaika, *U.S. At. Energy Comm. Tr 5745 Chem.* (1963).
50. F. J. Flanagan, *Geochim. Cosmochem. Acta 28,* 447 (1969).
51. K. Govindaraju, *Bull. Soc. Franc. Ceram. 85*, 31 (1969).
52. M. M. Sine, W. O. Taylor, G. R. Webber, and C. L. Lewis, *Geochim. Cosmochim. Acta 33*, 121 (1969).
53. L. H. Ahrens and S. R. Taylor, *Spectrochemical Analysis*, Pergamon Press Ltd., London (1961).
54. R. Avni, A. Harel, and I. B. Brenner, *Appl. Spectrosc. 26*, 641 (1972).
55. I. B. Brenner, H. Elded, L. Argov, A. Harel, and M. Assous, *Appl. Spectrosc. 29*, 82 82 (1975).
56. I. B. Brenner, L. Gleit, and A. Harel, *Appl. Spectrosc. 30*, 335 (1976).
57. I. B. Brenner, H. Elded, and S. Erlich, *Geol. Surv. Geochem. Div. (Jerusalem) Ref. GD. 1/76* (1976).
58. R. Breckport, *Congr. Avan. Method. Anal. Spectrograph. Produits Met. (Paris) 8*, 33 (1947).
59. G. Rossi, *Spectrochim. Acta 16*, 25 (1960).
60. J. Vilant and I. Voinovitch, *Proceedings of the Ninth Colloquium Spectroscopicum Internationale*, Groupement pour l'Avancement des Méthodes Spectrographiques, Paris (1961) p. 1552.
61. J. C. Cotteril, *At. Energy Res. Estab. (Gt. Brit.) C/R 2456* (1958).
62. F. Hegemann, *Ber. Deut. Keram. Ges. 29*, 68 (1952).
63. H. Okuda and M. Inagaki, *Chem. Abstr. 47*, 8576d (1953).
64. S. Landergren and W. Muld, *Mikrochim. Acta 2-3*, 245 (1955).
65. O. I. Joensuu and N. H. Suhr, *Appl. Spectrosc. 16*, 101 (1962).
66. G. R. Webber and J. U. Jellema, *Appl. Spectrosc. 16*, 133 (1962).
67. K. Govindaraju, *Publ. Group Avan. Methodes Spectrog. 22* (1960).
68. K. Govindaraju, *Centre Natl. Rech. Sci. Rep. 923*, 133 (1970).
69. A. Martin and M. Quintin, *Centre Natl. Rech. Sci. Rep. 923*, 187 (1970).
70. P. W. J. M. Boumans and F. J. M. J. Maessen, *Appl. Spectrosc. 24*, 241 (1970).
71. F. J. M. J. Maessen, Doctoral thesis, University of Amsterdam (in Dutch) (1974).
72. M. Roubault and J. Sinsou, *Rept. XIX Congr. Group. Avan. Methodes Spectrog.* 35 (1956).
73. J. P. Willis, M. Kaye, and L. H. Ahrens, *Appl. Spectrosc. 18*, 84 (1956).
74. P. A. Koka and G. A. Salomantina, *Izv. Akad. Nauk. SSSR Ser. Fiz. 18*, 294 (1954).
75. G. Holdt, *Appl. Spectrosc. 16*, 96 (1962).
76. G. Holdt, F. Moritz, and C. J. de Mark, *Proceedings of the Ninth Colloquium Spectroscopicum International*, Groupement pour l'Avancement des Méthodes Spectrographiques, Paris (1961), Vol. 2, p. 572.
77. F. R. Maritz and A. Strassheim, *Appl. Spectrosoc. 18*, 97, 185 (1964).
78. B. F. Scribner and H. R. Mullin, *J. Res. Natl. Bur. Std. 37*, 379 (1946).

79. C. E. Pepper, *U.S. At. Energy Comm. NLCO 999* (1967).
80. C. Feldman, *U.S. At. Energy Comm. ORNL-TM-1950* (1966).
81. E. J. Spitzer and D. D. Smith, *Appl. Spectrosc. 6*, 9 (1952).
82. J. R. Nelms and R. S. Vogel, *U.S. At. Energy Comm. MCW 1495* (1966).
83. N. Belegisanin, *Rec. Travaux Inst. Rech. Struct. Matiere (Belgrade) 2*, 27 (1953).
84. C. Cecarelli and G. Rossi, *Euratoru Report* 4541 f. (1970).
85. Anonymous, *U.K. At. Energy Authority Prod. Group PG Rept. 464(s)* (1964).
86. B. Strzyzewska, Z. Radwan, and R. Minczewski, *Polish Rept. PAN/IBJ 522/VIII* (1963).
87. A. B. Whitehead and H. H. Heady, *U. S. Bur. Mines Rep. Invest.* 6091 (1962).
88. J. Janda, J. Shausberger, and E. Scroll, *Mikrochim. Acta 122* (1963).
89. B. Strzyzewska, Z. Rodman, and J. Minczewski, *Appl. Spectrosc. 20*, 236 (1966).
90. A. Mykyink, D. S. Russel, and S. S. Berman, *Talanta 13*, 175 (1966).
91. H. G. King and C. M. Neff, *Appl. Spectrosc. 17*, 51 (1962).
92. G. M. Russel, *S. Africa N.I.M. Ref. NIM 450* (1968).
93. G. T. Day, P. A. Serin, and K. Heykoop, *Anal. Chem. 40*, 805 (1968).
94. R. Avni, *Spectrochim. Acta 23B*, 619 (1968).
95. Y. D. Raikhbaum and V. D. Malykh, *Opt. Spectrosc. 10*, 524 (1961).
96. R. Avni and M. Chaput, *C.E.A. Ref. 1908* (1961) (French); translated by F. T. Birks, *At. Energy Res. Estab. (Gr. Brit.) Transl. Trans 875* (1961).
97. Z. N. Siemenova, *Opt. Spectrosc. 12*, 466 (1962).
98. E. E. Vainshtein and Y. J. Belayev, *Intern. J. Appl. Radiation Isotopes 4*, 179 (1959).
99. E. E. Vainshtein, *Proceedings of the Ninth Colloquium Spectroscopicum Internationale*, Groupement pour l'Avancement des Méthodes Spectrographiques, Paris (1961), Vol. 1, p. 105.
100. G. Atwell, C. E. Pepper, and G. L. Stukenbroecker, *U.S. At. Energy Comm. TID-7568* (Part 1), 287 (1958).
101. B. H. Goldfarb and E. V. Ilina, *Opt. Spectrosc. 11*, 243 (1961).
102. Z. N. Samsonova, *Opt. Spectrosc. 12*, 466 (1962).
103. V. M. Vukanović, *Proceedings of the Eighth Colloquium Spectroscopicum Internationale*, Verlag H. R. Sauerländer Co., Aarau (1960), p. 62.
104. V. Vukanović, *Emissionspektrosckopie*, Tagung der Physik Gesselschaft DDR, Akademie Verlag, Berlin (1964), p. 9.
105. J. L. Daniel, *U.S. At. Energy Comm. HW-64299* (1960).
106. R. Avni, *Spectrochim. Acta 24B*, 133 (1969).
107. R. Avni, and A. Boukobza, *Appl. Spectrosc. 23*, 483 (1969).
108. S. L. Mandelshtam, *Bull. Acad. Sci. U.S.S.R. 26*, 850 (1962).
109. L. de Galan, Thesis, University of Amsterdam (1965).
110. R. Avni and A. Boukobza, *Spectrochim. Acta 24B* 515 (1969).
111. W. Lochte-Holtrgeven, ed., *Plasma Diagnostics*, North-Holland Publishing Co., Amsterdam (1968), p. 135.
112. R. Avni and F. S. Klein, *Spectrochim. Acta. 28B*, 319, 331 (1973).
113. R. J. Decker and P. A. McFadden, *Spectrochim. Acta 30B* 1 (1975).
114. G. Ecker, *Ergeb. Exakt. Naturw. 33*, 1 (1961).
115. R. Avni, *Spectrochim. Acta 23B*, 597 (1968).
116. J. W. Mellichamp, *Anal. Chem. 37*, 1211 (1965).
117. L. S. Orenstein and H. Brinkman, *Physica 1*, 797 (1934).
118. J. W. Pearce, in: *Optical Spectrometric Measurements of High Temperature* (P. J. Dickerman, ed.), p. 125, University of Chicago Press, Chicago (1961).
119. W. Lochte-Holtgreven, *Temperature, Its Measurement and Control in Science and Industry*, Reinhold Publishing Corp., New York (1955), p. 413.
120. R. J. Clark, *J. Opt. Soc. Amer. 53*, 1314 (1963).

121. P. W. J. M. Boumans, *Proceedings of the Fourteenth Colloquium Spectroscopicum International*, Adam Hilger Ltd., London (1967), Vol. 1, p. 23.
122. P. W. J. M. Boumans and L. de Galan, *Anal. Chem 38*, 674 (1966).
123. L. A. Ginsel, Thesis, State University, Utrecht (1933).
124. H. Bavinck, *Math. Centre Amsterdam Ref. TW 98* (1965).
125. R. Avni and Z. Goldbart, *Spectrochim. Acta 28B*, 241 (1973).
126. V. D. Malyhh and M. A. Serd, *Opt. Spectrosc. 16*, 368 (1964).
127. P. W. J. M. Boumans, in: *Analytical Spectroscopy Series* (E. L. Grove, ed.), Vol. 1, Part II, Marcel Dekker, Inc., New York (1971).
128. H. Kaiser, *Spectrochim. Acta 3*, 40 (1947).
129. H. Kaiser, *Spectrochim. Acta 3*, 159 (1948).
130. P. W. J. M. Boumans, *Proceedings of the Seventh Colloquium Spectroscopicum Internationale, Rev. Univ. Mines 15*, 396 (1959).
131. J. C. de Vos, *Physica 20*, 690 (1954).
132. R. Avni, A. Boukobza, and B. Daniel, *Appl. Spectrosc. 24*, 406 (1970).
133. R. Avni, A. Harel, and A. S. Lourenco, *Preprint of the Sixteenth Colloquium Spectroscopicum Internationale*, Adam Hilger Ltd., London (1971), Vol. 1, p. 1.
134. W. van Tongeren, *Contribution to the Knowledge of the Chemical Composition of the Earth's Crust in the East Indian Archipelago I, II*, D. B. Centens, Amsterdam (1938).
135. R. L. Mitchell, *Commonwealth Bur. Soil. Sci. (Gt. Brit.) Tech. Commun. 44* (1948).
136. C. B. Dutra and K. J. Murata, *Spectrochim. Acta 6*, 373 (1954).
137. C. L. Waring and H. Mela, *Anal. Chem. 25*, 432 (1953).
138. R. T. O. Connor, *Ind. Eng. Chem. Anal. Ed. 13*, 597 (1941).
139. J. R. Butler, *Spectrochim. Acta 9*, 332 (1957).
140. H. Bastron, P. R. Barnett, and K. J. Murata, *U.S. Geol. Surv. Bull. 1084-G*, 165 (1960).
141. K. J. Murata, H. J. Rose, and M. K. Carron, *Geochim. Cosmochim. Acta 4*, 292 (1953).
142. F. J. Flanagan, *Geochim. Cosmochim. Acta 33*, 81 (1969).
143. H. de la Roche and K. Govindaraju, *Bull. Soc. Franc. Ceram. 85*, 31 (1969).
144. H. de la Roche and K. Govindaraju, *Assoc. Natl. Rech. Tech.* 904 (1970).
145. Spectroscopy Society of Canada Certificate (1968).
146. S. R. Taylor and P. Kolbe, *Geochim. Cosmochim. Acta 28*, 447 (1964).
147. National Bureau of Standards Certificate.
148. M. Fleisher, *Geochim. Cosmochim. Acta 33*, 65 (1969).
149. *Methods for Emission Spectrochemical Analysis*, American Society for Testing and Materials, Philadelphia (1957; 1963).
150. G. Bonisoni and M. Paganelly, *Met. Ital. 58*, 268 (1966).
151. R. C. Calkins, *Appl. Spectrosc. 20*, 146 (1966).
152. E. Cerrai, Z. Hainski, G. Rossi, and R. Trucco, *Energia Nucl. (Milan) 11*, 9 (1964); *12*, 406 (1965).
153. O. P. Killeen, *U.S. At. Energy Comm. Rept. Y-1532* (1966).
154. B. E. Balfour, D. Jukes, and K. Thornton, *Appl. Spectrosc. 20*, 168 (1966).
155. Z. Hainski and G. Rossi, *Met. Ital. 58*, 295 (1966).
156. W. J. Robert and Carpenter Lloyd, *U.S. Bur. Mines Rept. Invest. 6105* (1962).
157. O. F. Degtyareva, L. G. Sinitsyna, V. R. Negina, and L. S. Chikisheva, *Spekt. Khim. Met. Anal. Mat. 1964*, 7.
158. E. F. Spano and T. E. Green, *Anal. Chem. 38*, 1341 (1966).
159. G. L. Hubbard and T. E. Green, *Anal. Chem. 38*, 428 (1966).
160. B. Kucharzewski, *Chem. Anal (Poland) 7*, 349 (1962).
161. M. J. Peterson and C. L. Chaney, *U.S. Bur. Mines. Rept. Invest. 5903* (1961).
162. J. Dvorak and W. Wanek, *Chem. Anal. (Poland) 7*, 201 (1962).

Preparation and Evaluation of Spectrochemical Standards 5

A. H. Gillieson

5.1 INTRODUCTION

The determination of an element by emission spectroscopy is never, in practice, an absolute measurement. Although the energy required to excite a particular line of an element may be known with reasonable accuracy, the phenomena in a flame or electrical discharge are so complex that it is impossible to ensure that the energy supplied excites that line and only that line of the element. This is true when only one element is present, but the complexity of these phenomena of excitation is greatly increased when two or more elements are present. Since the emission of a spectral line results from an electron dropping back from an upper to a lower energy level in the atom, the atom must first be raised to the upper level before the possibility of emission exists, and then it may lose energy by collision or other nonradiative processes without the emission of the desired line.

Except in the case of the rare gases, the element must first be dissociated into atoms, and if the element is normally a solid or liquid, it must be volatilized into the flame or discharge.

If the material under examination is a solid, the more volatile elements tend to enter the discharge first and the refractory later. Thus, unless the discharge process is carried out for a sufficiently long time, the record of the emission presents a distorted picture of the elemental composition of the material.

Even when the elements present are of approximately equal volatility, the

---

A. H. Gillieson • Retired from the Department of Energy, Mines and Resources, Mineral Sciences Division, Ottawa, Ontario, Canada. Present address: 10 Thistledown Court, Ottawa, Ontario, Canada energy requirements for excitation may vary widely, and the emission lines of the more easily excited elements will predominate. A familiar example is the excitation of sodium chloride, where with conventional means of excitation only the sodium lines appear, because although chlorine is more volatile than sodium, its atom requires a very much higher energy for its excitation.

The volatility and excitation behavior of an element may also be affected by the physical state or nature of its chemical combination. A powdered material may display different properties in the discharge from those shown by the same material in solid form. An element in an alloy will differ in volatility, depending on whether it is in solid solution or occurs as part of an interstitial material. In minerals an element may show marked differences in its emission properties when it is combined as a silicate or a sulfide as compared with the emission properties of its oxide.

When three or more elements are present, interelement effects may occur. The addition of a third element to a binary system may affect the emission rate of the first two elements out of all proportion to the concentration of the added third element. Such an effect may result either in suppression or enhancement of the emission of one of the elements present in the original binary mixture, compound, or alloy. These factors make emission spectroscopy impractical for absolute measurements. As in the case of other instrumental methods which measure radiation emission or absorption, materials of unknown composition must be compared with reference materials of known composition and the comparison must be made under the same conditions. The reference materials used must be as similar to the unknown sample as possible in both physical constitution and chemical composition. The criteria necessary in their preparation and their physical and chemical evaluation will be discussed in detail in this chapter.

5.2 GENERAL REQUIREMENTS

Some general requirements for a good spectrochemical reference standard have already been indicated. However, a more complete list of the desirable properties of a standard are as follows:

1. The standard should be physically similar to the specimen to be analyzed.
2. Chemical similarity between the standard and the analysis specimen is required.
3. The constituents of the standard should be prepared from materials of the highest available purity.
4. The methods of preparation must be planned with care to avoid contamination during the manufacture of the standard.
5. The standard should be physically and chemically homogeneous to as high a degree as possible.
6. The standard should not change its physical properties on storage.
7. Chemical stability of the standard is essential.

8. The number of standards used to prepare the working curve should preferably be not less than five.

9. For primary reference standards, accurate analyses should be carried out by methods other than spectrochemical, and by four or five different analysts.

5.2.1 Physical Similarity

If the analytical sample is in powder form, such as in mineral or slag analysis, the standard should have the same range of particle size.

If the material to be analyzed is in bulk or solid form, as in the case of a metallic alloy, the size and shape of the unknown samples should be very similar to that of the standard. In addition, the surface finish of the standard and of the unknown in the areas to be examined should be alike and should preferably be produced by the same means and equipment.

5.2.2 Chemical Similarity

It is essential not only that the standard and the unknown have similar percentage composition, but also, as far as possible, that the elements be present in the same form or chemical combination. The boiling points or decomposition temperatures of the compounds of an element can vary widely. For example: B, bp, 2550°C; B_2O_3, bp > 1500°C; B_4C, bp > 3500°C; and BCl_3, bp, 13°C; Fe, bp, 3000°C; Fe_2O_3, decomposes at 1560°C; Fe_3C, mp, 1837°C, decomposes at higher temperatures; and $FeCl_3$, bp 315°C. The more volatile material will enter an arc stream first and to a greater extent, unless altered by a chemical reaction during the burning period. In addition, B from B_4C will show little sensitivity in a matrix of relatively low volatility or appear last in an arc-complete burn technique unless the B_4C is changed to a more volatile boron compound. Therefore, the ease of detection and reproducibility of a determination can be markedly affected by the other elements with which the required element is combined.

5.2.3 High-Purity Constituents

Small quantities of impurities can profoundly affect the properties of a material such as an alloy. This, in turn, can influence the spectra by producing possible line interference, changes in sensitivity and limits of detection, and possibly incorrect sampling by nonhomogeneous crystal structure or segregation. For this reason standards (powder, bulk solid, solution, or even gaseous) should be prepared from high-purity materials.

There are a reasonable number of suppliers of high-purity materials, and a fairly exhaustive list of materials, together with their stated purity and their source. A good reference for these high-purity materials is the American Society for Testing and Materials, *ASTM Spec. Tech. Publ. DS-2*, Table 5.1.[1]

5.2.4 Avoidance of Contamination during Preparation

Stringent precautions must be taken to avoid contamination of the high-purity starting materials during preparation of most types of standards. For example, in the preparation of standards for the spectrochemical control of high-purity production materials (e.g., in semiconductor or electroceramic technology), it is a virtual necessity to carry out these preparations in dust-free laboratories to prevent contamination by the normal constituents of airborne dust.

When contamination by the alkali metal, alkaline-earth metals, or silicon must be avoided, aqueous solutions (both the standards themselves and the solutions used in the preparation of the standards) should be handled in plastic-coated or plastic vessels. Plastics commonly used are Teflon (polytetrafluoroethylene) or Polythene (polyethylene). The former, although more expensive, is recommended because of its great chemical inertness and high resistance to heat and acids.

5.2.5 Homogeneity

A standard, whether its physical form is powder, solid, liquid, or gaseous, must be of the same composition throughout. The degree of homogeneity necessary will be dictated by the method in which it is to be used. The amount of material sampled in many widely used spectrochemical analytical methods is in the order of a few tenths of a milligram. Therefore, the reference standards employed with such methods must be homogeneous down to a weight of $\frac{1}{10}$ to $\frac{1}{100}$ of this amount.

In the evaluation of materials for standards, adequate testing of the homogeneity is essential. This testing should be done before the lengthy and exacting calibration or certification analyses are begun. The homogeneity analysis may be performed chemically or spectrochemically. Customarily, one or two of the minor constituents are selected as the analyte for this test. The samples should be selected as completely randomly as possible.

5.2.6 Physical Stability

A standard should not only have adequate homogeneity initially, but should retain its homogeneity in storage (i.e., it should not segregate). This is a common problem with powder standards, especially when light and dense constituents are present. Partial sintering of the mix is recommended to avoid this phenomenon.(2)

5.2.7 Chemical Stability

A satisfactory standard must be chemically stable in storage. The components should not react with the components of air; that is, none of the components should oxidize easily or absorb carbon dioxide or water. In addition, its initial constituents should not react with one another to form new chemical compounds.

5.2.8 Concentration Range

Since the prime function of a standard is its use as the basis for a working calibration, a single standard has much less value than a set of standards covering a range of element concentrations above and below each of the expected concentrations in the specimen or group of specimens to be analyzed. Such sets of four to six standards have a much greater value if they have all been prepared, chemically analyzed, and metallographically or mineralogically examined under nearly identical conditions by the same group of metallurgists or mineralogists and by the same teams of analysts.

The overall range of concentrations needed in a set of standards depends upon the degree of uncertainty as to the expected concentration in the test specimens. When the uncertainty is great, the concentration may be as high as 10. Normally, however, other information or semiquantitative analysis has provided an estimated value of the unknown concentration which permits the use of a smaller concentration ratio between the adjacent standards values. For mineral analyses, Ahrens and Taylor(2) recommend a factor of $\sqrt{10} = 3.16\%$, resulting in a set of ratios such as 1.0%, 0.31%, 0.1%, 0.0316%, 0.01%, and 0.00316%. Another common concentration ratio, which gives an even smaller range, is $\sqrt{2}$ (e.g., 1.0%, 0.70%, 0.5%, 0.34%, 0.25%, 0.177%, and 0.125%).

Such constant ratios of element concentration are easily achieved in the preparation of sets of standards by the dilution of powder standards, of standard solutions, and of standard gas mixtures. However, with bulk or solid metal alloy standards, this constant ratio may, in many alloys, be difficult to achieve. It is, therefore, customary to accept as satisfactory a set of standards where element concentration ratios of adjacent standards are only approximately equal. An example is the CAAS phosphor-bronze standards, in which the nominal tin content is 5, 6, 7, 8, and 10%.

5.2.9 Analyses of Reference Standards

The values for standards that have been prepared by methods other than the thorough mixing of accurately weighed or measured amounts of pure materials must be established by chemical analyses; examples include ores, slags, and alloys. These analyses must be carried out with the highest degree of precision and accuracy and, if possible, by more than one suitable method. In addition, these analyses should be carried out by more than one recognized competent analyst in a number of different analytical laboratories with an established reputation for quality of work. These standards, no matter how carefully prepared, are actually of no greater value than is represented by the precision and accuracy of their calibration analyses.

Secondary standards may be calibrated spectrochemically by comparison with good primary standards whose chemical calibration analyses have met the requirements above. In general, however, it is accepted that primary spectrochemical standards should not be calibrated spectrochemically.

Physical analytical methods of analyses, such as nuclear techniques, may be used in place of chemical where these methods are applicable and advantageous.

5.3 POWDER STANDARDS

Powder standards are necessary for materials which have poor electrical conductivity, since these materials must be intimately mixed with an electrical conductor such as pure graphite, or a copper or silver powder. Samples of other materials, even good conductors such as metallic corrosion materials, oxides, and slags, are usually reduced to powder form, and so will require powder standards.

5.3.1 Classes of Powder Standards

Powder standards may be classified as synthetic, part synthetic–part natural, and natural.

5.3.1.1 Synthetic Standards

Synthetic standards, those prepared from pure materials, are the most widely used class and are the easiest to prepare. Powders of the metal compound, usually the oxide, must be available in high purity to prepare such standards. The powder should be of uniform particle size and of sufficient fineness to pass through a 325-mesh sieve. Nylon sieves should be used. Accurate weighing of the desired quantities followed by the thorough mixing of the components should produce adequate standards.

5.3.1.2 Part Synthetic–Part Natural Standards

Occasionally, the sample to be analyzed is nearly as pure as the high-purity material in the synthetic standard, or the matrix of the sample may differ widely from the synthetic standard. The analytical results are not considered reliable if such samples are compared with synthetic standards.

In these cases a standard addition procedure can be used which results in a series of standards that consist of part synthetic and part natural sample.

The first step in the preparation of these standards is semiquantitative analysis of the test samples to reveal the approximate element concentrations. Small

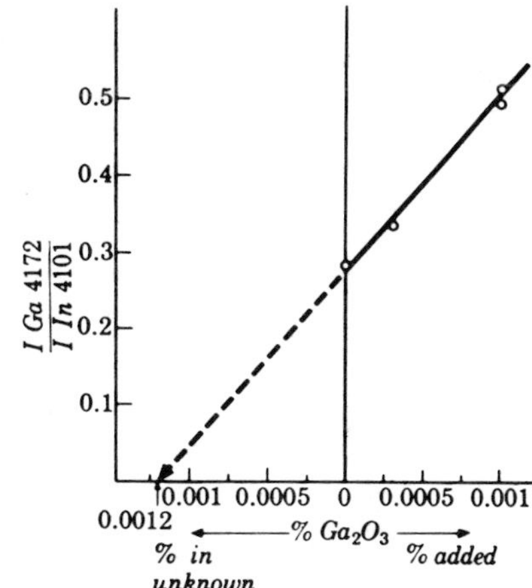

Fig. 5.1 Addition method of determining the concentration of an element in an unknown. (From Ahrens and Taylor.[2])

multiples of the estimated amount present of the required pure compound(s), usually as an oxide, are added to a series of the purest of the test samples. The resulting analytical curve, which should be linear, is then extrapolated to its intercept on the concentration axis. The numerical value of the negative concentration value at this intercept indicates the amount present in the original test sample. This technique is illustrated in Fig. 5.1. If the concentration sought in the unknown is $y\%$ and if small multiples of this element are added (i.e., $x\%$ and $2x\%$, etc.), then the standards will have $(y + x)\%$ and $(y + 2x)\%$ of the element sought. In the plot percent concentration added versus intensity, the amount present in the sample is equivalent to 0% material added.

If the physical and chemical nature of the test sample is not the same as that of the synthetic standard, the two may not behave in the same manner in the discharge. The properties of the synthetic standard should be very similar to those of the test sample. Thus, it is often wise to start with pure samples of the base materials for the preparation of standards. Ahrens and Taylor's work[2] on the determination of tin and indium in mica and the work by Oftedahl[3] on the determination of minor constituents in sphalerite are examples in which it was preferable to use pure natural specimens as the base material for the preparation of standards. Ahrens and Taylor point out that chemically analyzed minerals, rocks, slags, and so on, may be used as standards for the elements in relatively high concentration. However, in trace analyses it is necessary that the standards be at least partly synthetic.

5.3.1.3 Natural Standards

Natural standards are those naturally occurring or production samples in which the composition is known as the result of a number of accurate chemical analyses. If the composition of these samples covers the necessary range, the required series of standards can be selected.

5.3.2 Grinding and Mixing

The grinding and mixing of powder standards are two operations that present a contamination hazard. Agate or boron carbide mortars and pestles should be used for both purposes. Mullite, a synthetic ceramic, is satisfactory for softer materials but introduces much undesirable contamination when hard, abrasive materials are ground in it. Even when agate mortars are used, the retention of elements such as lithium, thallium, and indium has been noted.[2] When an estimation of these and other elements, such as aluminum, as trace impurities is involved, a fresh agate mortar should be used toward the end of the powder dilution procedure for making the low concentration standards. It is a useful rule of thumb for grinding by hand in a mortar that the operation should be continued for twice as long as the time it takes for the sample to appear uniform.

Fig. 5.2 Spex-Crescent Wig-L-Bug.

Grinding and mixing may also be carried out using devices such as the Wig-L-Bug (Fig. 5.2), shatterbox (Fig. 5.3), mixer mill (Fig. 5.4), or the Bleuler mill (Fig. 5.5). With the first instrument the container should be made of plastic, and hardened steel, tungsten carbide, or plastic-coated steel grinding balls are recommended. The last two machines are more useful for larger amounts of material, but since their grinding parts are of metal, there is a greater possibility of contamination.

Naturally occurring materials such as rocks and ores that are ground and mixed in quantities of 500 lb or more for the production of uniform powdered standards can be prepared by "autogenous" grinding, in which the lumps of material act as their own grinding balls. Chances of contamination is thus reduced, but the process is to a certain degree wasteful, since only about 80% of the original material can be converted into "fines." Separation of the fines by airflow or the use of nylon sieves is a necessary sequel to the grinding process.

Because grinding machines will not accept lumps of hard material, such as rocks larger than a person's fist, such lumps must first be reduced in size by the use of sledgehammers or rock crushers. The former method is preferred, since contamination is not as great as when crushers are used. When crushing, pulverizing, or the grinding of hard materials has to be done in iron or steel equipment,

Fig. 5.3 Spec shatterbox.

the resulting material should be passed at least twice through a magnetic separator to remove iron particles.

In general, all equipment and all operations involved in the preparation of standards should be considered as sources of adventitious impurities, and care is necessary to minimize possible contamination. This includes the grinding, crushing, polishing, turning, sieving, and mixing operations and such equipment as furnaces and molds.

5.3.3 Particle Size

Powders for use in the preparation of standards should be the size that passes through at least a 325-mesh sieve. The sizing should be done by air-flow methods

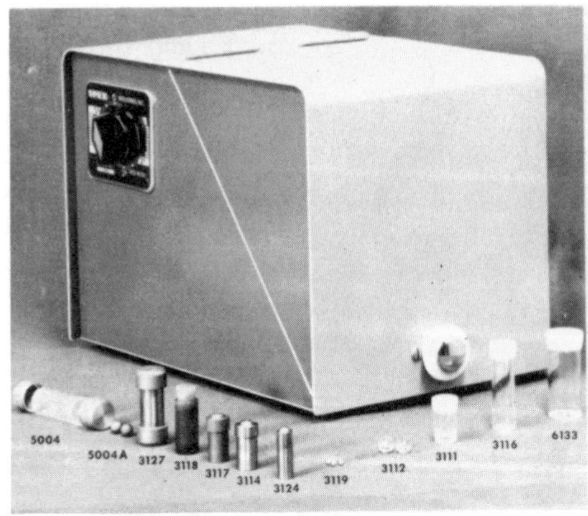

Fig. 5.4 Spec 5000 mixer mill (top) and Spec 8000 mixer mill (bottom).

Fig. 5.5 A. R. L. Bleuler rotary mill.

if possible. If sieves are to be used, they should be made of nylon, and only one sieve should be used on each class of material.

5.3.4 Segregation

When the components of a powder standard are of very different density, segregation of the powder may occur in storage. To avoid this, the freshly prepared standard may be lightly sintered and then reground.

5.3.5 Typical Preparations of Powder Standards

Numerous methods have been recorded in the literature for the preparation of powder standards for the analyses of specific materials. These methods can be found by referring to the material to be analyzed or elements to be determined in the spectrochemical abstracts[4,5] or in the chemical abstracts.[6]

The U.S. National Bureau of Standards has made and provides a number of powder standards for spectrochemical use. These are listed in *ASTM Spec. Tech. Publ. DS-2*[1] and in *Natl. Bur. Std. (U.S.) Misc. Publ. 260.*[7]

In the field of rock analysis, the U.S. Geological Survey prepared the now famous Standards G-I (a granite) and W-I (a diabase), which have for a number of years been the worldwide mainstay of rock analysis calibration, and their preparation is described by Fairbairn.[8] The supply of this standard has recently been exhausted.

As an example of the production of typical reference rock standards, the following paragraphs give a brief summary of the procedures involved in the preparation of a syenite rock standard and a sulfide ore standard by the Non-metallic Standards Committee of the Canadian Association for Applied Spectroscopy during the years 1957–1960.[9,10]

The syenite rock (180 lb) was collected from four different locations in the Bancroft Area of the Province of Ontario. These samples were crushed and then pulverized in a Braun pulverizer with the plates set to yield at least 100 mesh. The product was not screened or passed through the magnetic separator. It was placed in a small drum and was turned for 24 h on rolls at a speed calculated to give cascade action. The resulting powder was then placed on a new sheet of plywood and was split in a riffle to obtain four hundred $\frac{1}{4}$-lb portions. These portions were bottled in wide-mouth, 4-oz glass jars with Bakelite screw tops.

The sulfide ore (500 lb) was an accumulation of sample rejects provided by the Falconbridge Nickel Mines Ltd. After reduction, a screen test of a grab sample of the product gave the following results: less than 100 mesh, 93.4%; greater than 100 but less than 65 mesh, 5.3%; and larger than 65 mesh, 1.3%. To ensure homogeneity, the reduced product was poured in two equal piles on a new sheet of plywood, and each pile was riffled through a sample splitter, first to yield four 125-lb portions and then these portions to yield eight 62.5-lb portions. One portion from each of the original 250-lb piles was combined to make 125-lb sample representative of the original 500 lb.

The 125-lb sample was placed into a small drum and turned in the same manner and the same time as described for the syenite rock. The same procedure was followed for the splitting and bottling as was described for the syenite rock.

5.4 BULK (SOLID) STANDARDS

5.4.1 General Considerations

The bulk or solid samples are generally metallic in nature, since they must conduct an electrical current. Therefore, these standards are prepared by specially designed applications of normal metallurgical procedures for melting and casting, usually followed by working and annealing. The materials used should be of the highest purity available, because unknown amounts of trace elements affect planned concentrations and possible metallurgical processes.

It is customary to make up "master alloys," each containing the major element together with small amounts of the required minor elements. After careful analysis of the composition of these master alloys, the required quantities are added to the molten major element to give the desired range. Direct introduction of the minor elements can cause sizable errors in the desired composition due to (1) poor precision in the weighting of the small quantities required, and (2) the difficulty of dispersing homogeneously a small amount of material in a large volume of melt. Certain of the more volatile constituents should be added only a short time before pouring the melt to avoid undue losses by volatilization.

The melting of these alloys is usually done in an induction furnace. In this type of furnace, the heat is concentrated in the melt; the furnace walls are heated only by conduction from the melt, so they remain at a relatively low temperature. The possibility of contamination of the melt by elements present in the furnace wall is thus greatly reduced.

In addition, the melt is thoroughly stirred by the induced electrical current producing a homogeneous melt. No stirring mechanism is required that may induce contamination. The use of an inert or hydrogen gas furnace atmosphere above the melt presents no serious difficulties or hazards.

When a gas-fired furnace is used, the combustion gases, the mechanical stirrer, and the externally heated hot crucible are probable sources of contamination.

Mold design and materials have to be chosen with care. Blow-holes and scum inclusions in the cast metal can be avoided by well-designed entry ports and exits for the displaced air and any gases produced from the mold. The molding sand or other mold materials must not contaminate the castings.

The standards should be produced in a manner similar to that of the production alloys for which they are to serve as reference standards, because the behavior of many alloys in the spectrographic discharge is greatly influenced by the metallurgical history of the sample. The much smaller scale of operation makes it difficult at times to achieve close similarity to the production samples, but the procedure should ensure that the degree of homogeneity is at least as good as that of the production material, and that the crystalline structure and phase compositions are as nearly identical as is possible.

The very small castings, as represented by the cast standards, will not have adequate uniformity of composition or crystalline structure in all portions of the casting, so it is important to determine and define those portions suitable for use as the spectrographic standard. In two of the examples discussed below, only a portion of the block or disc (i.e., for a certain distance below the face of the casting in contact with the chilled face of the mold) was considered of adequate physical and chemical homogeneity for spectrographic use. In order to remove possible or actual contamination from the mold, the surface of the standard is milled or ground to a depth sufficient to remove all surface casting irregularities and mold inclusions.

In the case of cast standards, the desired chemical homogeneity and crystalline structure has to be achieved during the melting and casting processes; but in the case of wrought standards, the portions of the original casting that do not show adequate homogeneity can be removed before forging, and the homogeneity of the remainder can be preserved and improved by subsequent heat treatment during working. On the other hand, not all alloys can be forged or extruded, so the types of wrought standard alloys are limited for metallurgical reasons.

5.4.2 Typical Preparation of Bulk (Alloy) Standards

To illustrate the preparative factors discussed above, three examples of alloy standard preparation will be described in some detail.

5.4.2.1 White Cast-Iron Standards

The first example is the preparation of a set of white cast-iron standards. This is one of the more recent developments by the U.S. National Bureau of Standards (NBS). It is fully described in one of a series of NBS monographs on standard reference materials.[11] The following summary is based on this monograph.

The usual form of cast iron is known as gray iron because of the characteristic appearance of the fracture. The material can be regarded as an iron–carbon alloy, 1.7–4.5% C, which contains more carbon than can be retained in the austenite at the eutectic temperature. On slow cooling the graphite flakes separate and appear dispersed in a pearlite matrix.

Unfortunately, investigation at NBS and elsewhere showed that the amount, nature, and distribution of the graphite flakes had a marked effect on the volatility of certain elements during emission spectrochemical analysis.

There is essentially no free carbon in "white cast iron," and, although industrial application of the material is limited, this form of cast iron was considered to be better for spectrochemical standards, for the following reasons:

1. It would provide a uniform metallurgical structure for a series of standards that would cover wide concentration ranges of the elements of interest.

2. It would be a casting material that could be prepared with the required chemical homogeneity.

3. Investigation had shown that test samples of most types of iron could be prepared in the foundry with an induced white structure, and that these test samples could be satisfactorily compared with the white cast-iron standards in

spectrochemical analysis. The white structure can be produced in at least a portion of each test sample either by using a rapid chill-casting procedure or by adding elements such as tellurium or bismuth, which are known to produce whiteness.

4. The important determination of total carbon in white cast iron can be made by vacuum optical emission analysis, a procedure that cannot be used if much uncombined carbon is present.

Five commercial cast-iron compositions were chosen as the basis for the set of eight standards: piston ring, wear plate, die, brake drum, and mold. Three special compositions were added to cover the desired range of element concentrations. The balancing of eight compositions to promote a white structure had to be carefully studied. Silicon is the main factor in controlling depth of chill. Chromium and vanadium markedly promote chilling properties, because they readily form carbides. Phosphorus and nickel reduce the depth of chill. Manganese and sulfur combine, and if in stoichiometric quantities have little effect; but if sufficient manganese is present to react with all the sulfur, the excess has a pronounced stabilizing action on the cementite, Fe_3C, and thus promotes chill. Because the "carbon equivalents" of several of the compositions favored graphitizing, tellurium and bismuth were added to ensure a heavy chill.

As a result of work at Watertown Arsenal on the casting of unidirectionally solidified slabs on a water-cooled copper plate, a massive water-cooled copper mold assembly was designed and constructed at NBS for the unidirectional and simultaneous casting of a large number of white cast-iron standards.

The mold assembly shown in Fig. 5.6 consisted of a 5-ft^2 steel plate to which were edge-welded 5-in.-high steel strips to form water passages which spiraled out from the center to the periphery of the plate. A plate of oxygen-free, high-conductivity (OFHC) copper, 5 ft square and 1 in. thick, was bolted to the steel structure with 1-in.-diameter bolts of the same material. At the periphery a water seal was made with a lead gasket. The cooling water from a high-pressure supply was fed to the center of the steel plate by a 2-in. hose line, and left the periphery tangentially by a 3-in. outlet equipped with a valve to aid in purging air from the cooling chamber.

The casting surface of the copper plate was protected by a thin layer of colloidal graphite in alcohol.

Nine grid mold assemblies, 18 in. square, were mounted on the copper plate. Each grid mold could produce 64 white cast-iron blocks separated by $\frac{1}{4}$ in. at the chill face. On the grid assembly was mounted a $\frac{3}{4}$-in.-thick plate with a tapered in-gate located over the center of each of the nine grid molds. The size of the in-gate proved critical. Too small an opening caused lack of fill, "cold shuts," and spatter, while too large an opening produced a riser difficult to subsequently break off without damage to the cast block. Good castings were obtained with in-gates $\frac{1}{2}$ to $\frac{5}{8}$ in. in diameter at the lower end, with a taper of $\frac{1}{8}$ in.

Fig. 5.6 Water-cooled copper mold assembly designed and constructed at the National Bureau of Standards (U.S.) for the preparation of cast spectrochemical standards. (From Michaelis and Wyman.[11])

The top section of mold was provided with nine interconnected basins $1\frac{7}{8}$ in. deep. Above this array of basins was a pouring basin with a 2-in. diameter and stoppered holes, one over each of the nine grid bases. In the casting procedure more than enough metal to fill the entire mold was poured into this basin, and then all nine stoppers were pulled out simultaneously.

The melting procedures in the high-frequency induction furnace were carefully controlled with respect to time, temperature, and the amount and order of additions to the furnace. A typical melt log is given in Table 5.1 for the white cast-iron brake drum standard (NBS No. 1179). The total melt time averaged about $1\frac{1}{2}$ h.

Before pouring the molten metal into the mold, two ladle samples for preliminary analysis were cast into containers mounted on the copper plate. These samples were similar to the cast standard in size and shape. One of these samples was fractured and quickly examined before casting the main heat to ascertain whether the white iron structure had been obtained throughout the thickness of the casting. Had the test sample from any heat shown evidence of free carbon, a further addition of tellurium to the molten metal in the ladle was planned. The recoveries of the elements added, summarized for 10 elements in Table 5.2, were, in general, quite satisfactory.

The resulting cast-iron standard blocks were of adequate homogeneity for a distance of $\frac{5}{16}$ in. from the chill cast surface. The blocks as cast were very susceptible to cracking and, as a result of exploratory tests, were heat-treated for 1 h at 595°C.

Table 5.1 Log of Melt for Brake Drum Standard[a]

Charge:	Master alloy melting stock (305.23 lb), Ni (4.50 lb), Cu (1.00 lb).
Furnace on:	Time 9:30 A.M.
Half melt down:	Time 10:15 A.M., added C (1.88 lb), Si (none), Mn (2.68 lb).
Melt down:	Time 10:43 A.M., temperature 2500°F.
Skim:	Time 10:45 A.M., temperature at end 2475°F.
Additions:	(Nonvolatile) Temperature to 2500°F, time 10:46 A.M., P (2.78 lb), S (1.92 lb), Cr (0.70 lb), Mo (1.05 lb), V (0.22 lb), Co (0.07 lb), Ti (0.44 lb), Zr (0.18 lb); time 10:48 A.M., temperature 2520°F.
Stir:	Time 10:52 A.M., temperature 2550°F.
Additions:	(Volatile) Temperature to 2500°F, time 10:54 A.M., As (0.17 lb), Sb (0.39 lb), Su (0.39 lb), B (1.79 lb), Pb (0.88 lb), Te (0.26 lb), Bi (0.019 lb).
Stir:	Temperature to 2600°F rapidly, time 10:58 A.M.
Tap:	Stir in ladle, time 10:59 A.M.
Mold water on:	
Sample:	Two specimens.
Pour:	Time 11:00 A.M.
Remarks:	No apparent spattering of additions; 535 of 536 casting block filled.

[a]Natl. Bur. Std. (U.S.) Standard 1179, Jan. 8, 1960.

Table 5.2 Recovery Values for 10 Elements in the Eight NBS White Cast-Iron Standards

		Average recovery values	
Element	Concentration range	Expected %	Found %
C	2.0–4.0	90	88
Mn	0.3–1.5	95	95
P	0.05–0.8	100	99
S	0.03–0.2	80	68
Si	0.5–3.2	95	90
Cu	0.2–1.5	100	101
Ni	0.1–3.0	100	100
Cr	0.08–2.0	100	100
V	0.02–0.025	100	103
Mo	0.15–1.5	100	100

5.4.2.2 Copper Alloy Standards

The National Bureau of Standards also produced a series of seven copper standards making use of the same massive, water-cooled, copper-mold assembly employed in the casting of the white cast-iron standards. This work is described in the *Natl. Bur. Std. (U.S.) Misc. Publ. 260-2.*[12]

An alternative casting procedure that was chosen by the metallurgists of the Mines Branch, Canadian Department of Mines and Technical Surveys, for the production of five compositions of phosphor-bronze disc standards will be described.[13]

The five compositions of phosphor-bronze chosen were those with 5, 6, 7, 8, and 10% nominal tin content. To ensure homogeneity in the molten state, 250-lb melts were prepared in a lift coil induction furnace, with a charcoal cover to minimize oxidation. Before pouring, melts were flushed with dry nitrogen for degassing and the subsequent elimination of hydrogen porosity on cooling.

Flat disc spectrographic standards are customarily cast in iron book molds, so that the metal is rapidly chilled and relatively homogeneous. The chilling is desirable but casting 250-lb melts by this procedure would have required too much time. To reduce pouring time, sand molds were used. The advantages were (1) that all the mold cavities could be prepared beforehand, (2) that the cast samples need not be stripped from the mold until pouring was completed, and (3) that the cavities could be ganged together for simultaneous casting of many samples.

Vertical stacking of the cavities in the mold presented some serious disadvantages because the cavities are not filled simultaneously. There is no way to ensure that only clean metal enters the cavity, and there are also difficulties from sand penetration and "flash" due to the high hydrostatic pressure.

The mold design chosen for the molding of these standards employed horizontal stacking and consisted of an assembly of 10 cores on each side of a central sprue, each core containing two cavities, to give 40 discs from each mold.

Figure 5.7 shows the core box on the right and an individual core, 8 by 6 by 1 in., with the iron chills embedded in the sample cavity. A casting made from 20 of the individual cores cemented to a sprue block is illustrated in Fig. 5.8.

Each cast disc is $2\frac{1}{4}$ in. in diameter, $\frac{1}{4}$ in. thick, and weighs $5\frac{1}{4}$ ounces (6 cm by 0.7 cm, 160 g). One face of the sample cavity was formed by the back of the adjacent core, while the other face consisted of an iron chill of the same dimensions as the sample disc embedded in the core sand. On the face of the cast disc which had not been in contact with the chill, a small raised arrow was cast, showing both the position of the "gate" and identifying the nonchilled face.

To preserve symmetry and thus to yield identical pouring conditions, only two sample cavities were made in each core. Each was vented to the atmosphere

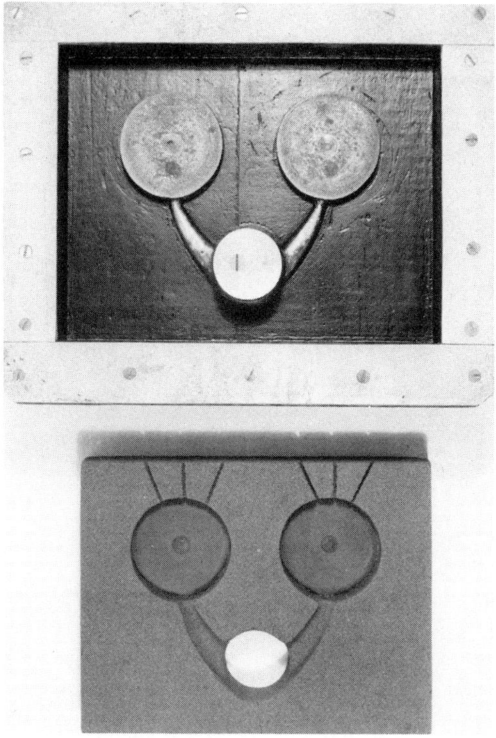

Fig. 5.7 Individual core and core box for CAAS phosphor-bronze standards. Note cast-iron chills in sample cavity and vents cut into top of each cavity.

and was jointed to the central runner by a single secondary runner in the vertical plane. These secondary runners were tapered to give streamlined flow, and each was set at an angle of 30° from the vertical in the core plane in an attempt to trap along the upper edge any dross which might enter from the main runner. The secondary runners were half oval in cross section, and at the gate their cross-sectional area was 0.06 in.^{2}.

A massive horizontal main runner ($1\frac{1}{2}$-in. diameter, 1.75 in.^{2} cross-sectional area) was used so that there would be sufficient time and adequate metal volume to even out any temperature gradients or segregation in the metal first poured, before the metal rose into the secondary runners and sample cavities. The exits to the secondary runners were cut off $\frac{1}{4}$ in. before the top of the main runner, so that the resulting space in the top of the main runner acted as a dross trap.

The sprue was of standard pattern, 9 in. high and tapering in rectangular cross section from $1\frac{3}{4}$ by 1 in. at the top to $1\frac{7}{16}$ by $\frac{11}{16}$ in. at the foot. This design was known to inhibit vortex formation and the aspiration of air and mold gases. Both the sprue mold and the pouring basin fixed to its top were molded from core sand.

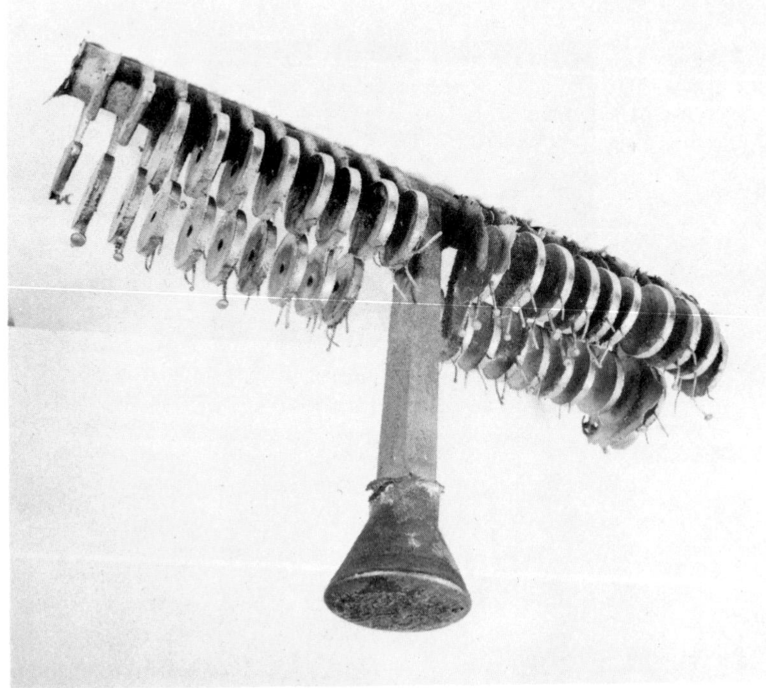

Fig. 5.8 Block of spectrographic standard castings made by cementing 40 of the cores shown in Fig. 5.7 onto a central sprue block.

The mold assemblies were preheated for 3 h at 204°C (400°F) before pouring took place. Seven molds yielding 280 discs (90 lb of sample) were poured in less than 2 min, thus minimizing changes in analysis due to oxidation, volatilization, or segregation. Pouring temperatures were kept as low as possible.

Metallographic examination of the cast discs showed, in general, a fine-grained columnar structure extending from the chill face to the sample midline. Beyond this level the crystals were equiaxed, slight shrinkage porosity became apparent, and the distribution of constituents such as lead particles was more irregular, but in no case were these effects severe. The portion of the disc from the chill face to the half-thickness level was considered to have adequate homogeneity.

5.4.2.3 Pure Copper Standards

In order to be in the same form as the production material for which they are intended to serve as reference, the pure copper standards have to be prepared as rod or wire. Thus, they represent a third type of procedure, in which casting is followed by extrusion. The particular procedure to be described is that employed by the Non-Ferrous Standards Committee of the Spectroscopy Society of Canada.

The choice of base material for these pure standards presents certain peculiar difficulties, in that a relatively large amount of high-purity material is required, and such material is of necessity of the same order of purity as the production metal whose purity is to be assessed by reference to the standards. In these circumstances the best procedure is to select material from commercially pure copper, which, by analysis, represents the highest level of purity in large-scale production.

A batch of high-purity copper was obtained, one fifth of which was retained unchanged for drawing into the purest wire standard. The other four fifths were cast into four wire-bar ingots, each from a separate fifth of the original metal. To each of the four ingots the chosen amounts of impurity elements were added in the melting process. Before drawing, $\frac{1}{2}$ in. of metal was milled from the top of the ingot as a precaution against the introduction of any dross into the final wire. Commercial experience had shown that any vertical segregation in the ingot can persist throughout the drawing operations. A small portion was also milled off one end of the bar and chemically analyzed to ensure that the composition was reasonably close to the intended composition. The other end of the sire bar ingot was machined into a "nose" which could be gripped by the jaws of the drawing machine.

The four ingots were drawn down to $\frac{5}{16}$-in.-diameter wire and each cut into about 600 1-ft lengths. Every twenty-fifth length was selected, marked, and reserved for homogeity testing and chemical certification analysis. The homogeneity of the wire thus produced was satisfactory for use as standards. The standards are marketed in the unusual form of 1-ft lengths of $\frac{5}{16}$-in. diameter, so that the user has sufficient length of wire to make possible drawing down to lesser diameter if his spectrographic procedure requires a rod diameter less than $\frac{5}{16}$ in.

5.5 LIQUID STANDARDS

Spectrochemical liquid standards find their main application in the following types of analytical procedures:

1. Solution residue–dc arc method
2. Solution residue–spark method
3. Porous cup–spark method
4. Vacuum cup–spark method
5. Rotating electrode–spark method
6. Plasma arc
7. Flame emission and atomic absorption

Viscosity of the liquid is a factor that must be considered in all except the first two methods. In some form they all use capillaries or capillary action, and in their operation the solution viscosity tends to play a major role in the rate of supply of the liquid to the discharge. It is considered good practice to work with solutions of as low viscosity as possible, and for this reason it is recommended that suitable small additions of surface-active agents—preferably pure organic reagents—added to the solution. Octyl alcohol or one of the more complex proprietary wetting agents is commonly used.

In the preparation of liquid standards, high purity of the materials used is as important as in the manufacture of solid standards. The liquids employed as solvents (e.g., water, acids, etc.) should be doubly or triply distilled. As for normal chemical standard solutions, spectrochemical liquid standards should also be stable if they are to be stored for any prolonged period, or made up freshly for each analytical run. If there is any reason to believe that the solutions have only short-term stability, they must be freshly prepared for each analytical run.

The containers used for storage of such solution standards can be a source of contamination (e.g., alkalies and silica from glass vessels). Such standards should preferably be stored in clear polythene containers fitted with airtight closures. It is to be noted that the screw caps for most polyethylene bottles do not adequately seal.

5.6 GASEOUS STANDARDS

The chief precautions in making gaseous standards are concerned with the high purity of starting materials, adequacy of mixing, and minimizing contamination during preparation and storage.

A wide range of high-purity gases are available from the commercial suppliers of liquified gases, and at least one manufacturer of spectrographic equipment offers a number of spectroscopically pure gases.

A good vacuum line is a virtual necessity in handling these gases. In the most stringent requirements, not only should a very high vacuum [e.g., 10^{-8}–10^{-10} Torr (mm Hg)] be employed, but provision should be made for baking certain parts of the equipment to ensure the removal of adsorbed gases or liquids from the surfaces exposed to the high-purity gases. Thus, rubber and Tygon-type tubing should be avoided. Vacuum taps and vacuum greases must be excluded from many parts of the apparatus, for even the inert gases have an appreciable solubility in these greases, and the more reactive gases may react with them. In glass or quartz apparatus ground-glass or quartz valves actuated and sealed by mercury, making use of the Torricellian vacuum, have proved adequate.

It is not always realized how slow a process the diffusion of one gas into another can be. To ensure homogeneity of the mixed gases, it is preferable to carry out repeated transfer of the mixed bases from one vessel to another by means of a Toepler pump. The action of such a pump can be made automatic by means of contacts, relays, and valves coupled with a compressed air line to avoid the laborious and tedious manual operation.

5.7 ANALYSIS OF STANDARDS

Since emission spectrochemical analyses are performed by comparison with reference standards, the composition and homogeneity of these reference standards must be known to greater accuracy and precision than that desired for the unknown material(s). Thus, the care exercised in the selection of high-purity starting materials and in the preparation of the standards to ensure homogeneity and freedom from contamination will be wasted unless the standards are properly evaluated. This consists of homogeneity testing and composition analyses carried out by the most reliable and refined methods available.

The certification evaluation is customarily a lengthy process as compared with the preparation, and normally includes the following operations:

1. Preliminary physical inspection
2. Preliminary analysis
3. Microscopic or metallographic examination
4. Homogeneity testing
5. Multiple compositional analysis
6. Statistical analysis of the elemental determinations
7. Evaluation and weighting of the analytical results
8. Determination of final certified composition

It should be clear from this list that the complete certification of a spectrochemical standard requires the services of a number of experts in various fields (e.g., mineralogy, metallurgy, analytical chemistry, physics, and mathematics). The whole operation necessitates close cooperation with well-organized teamwork from a number of groups and a number of laboratories, all of high standing in their respective fields. In the past, most of this effort has been by free contribution, and in consequence the price charged for most of these reference standards has represented only a small fraction of the total real costs of production.

5.7.1 Preliminary Physical Inspection

This essential operation, such as examining for gross inhomogeneities, is applied almost exclusively to powder and bulk (solid) standards. For example, cast standards are examined for poor mixing of components, gross imperfections and segregations, visual inclusions, and blow-holes. The gross nature and distribution of crystal structure is determined by fracture or rapid sectioning of a few samples.

5.7.2 Preliminary Analysis

Although a preliminary sampling is not expected to be truly representative of solid standards, a "snap" sample of the material is analyzed chemically or, where applicable, by optical emission or x-ray fluorescence analysis.[11,12] This provides a check to determine that the desired composition exists within at least approximate required limits. For cast standards, special preliminary samples have been cast [11,12] and rapidly analyzed before pouring the standards melt to make sure that no required element has been omitted or lost from the melt.

5.7.3 Microscopic or Metallographic Examination

A microscopic examination of powder standards quickly reveals any lack of uniformity of particle size or inequalities in the distribution of the various components of the mixture. For this examination a randomly selected number of test samples must be taken from the gross lot of powder, and a sufficiently large number of microscopic fields must be examined to ensure that the results of the test truly mirror the nature of the powder standard. To assist in the correlation of results, these samples should be taken at the same time or be a part of those taken for homogeneity analysis. Not less than five samples should be taken, but 10 is considered more adequate for a 300- to 500-lb standard.

Fig. 5.9 Etched longitudinal section of cast brass. (From Michaelis *et al.*[12])

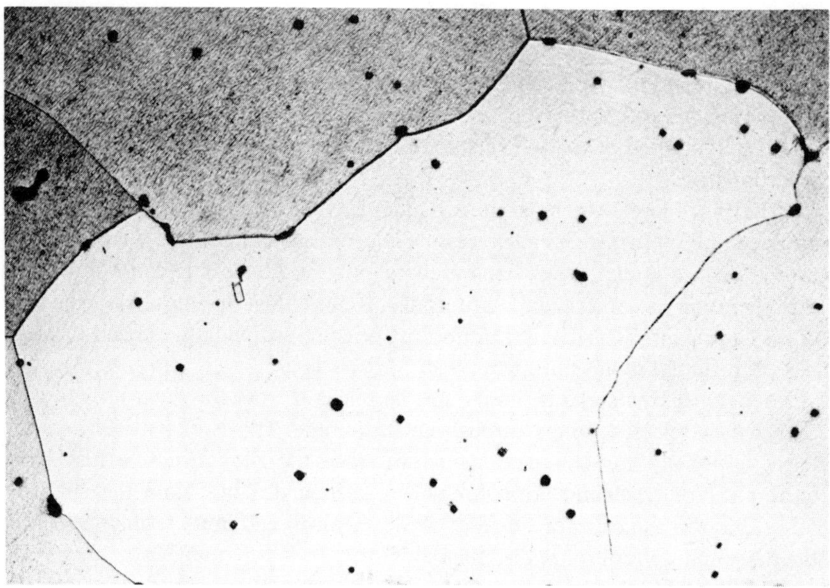

Fig. 5.10 Blow-holes in cast copper-base alloy at ×175 (NBS C1102 cast brass). (From Yakowitz *et al.*[20])

The microscopic or metallographic examination of bulk (solid) standards contributes detailed information with respect to the nature and distribution of the crystal structures, the number and variety of inclusions, and the number of small-scale imperfections, and blow-holes. Typical micrographs of a cast copper-base alloy in etched longitudinal section and in polished cross section are shown in Figs. 5-9 and 5.10. From such micrographs it is possible to obtain an indication of what portions of a cast standard possess sufficient uniformity of crystal structure. In the brass standard whose etched longitudinal section is shown in Fig. 5.9, the columnar structure indicative of rapid solidification extended from the chill-cast surface of the standard to about half its thickness, a distance of about $1\frac{1}{4}$ in. In Fig. 5.10, typical blow-holes, which occur particularly at grain boundaries in cast copper alloys, can be seen.

5.7.4 Homogeneity Testing

Before undertaking the major task of the certification analysis, it is necessary to make sure that the standard, whether powder or bulk (metal solid), is of adequate homogeneity to justify the long and detailed certification program.

The three operations described previously will have already given a quite reliable indication of the homogeneity of the powder standard. In general, it will be necessary to take only an adequate number of random samples from the gross lot for the homogeneity analysis. The uniformity of distribution of the elements present in major and minor percentages should be checked.

Because a large number of samples are usually involved, it is customary to use a rapid method of analysis such as optical spectroscopy or x-ray fluorescence spectrometry.

The procedure with bulk solid standards, such as metal alloys, is a much more complex process. Samples must be selected at random from a lot produced in one casting operation and analyzed to prove that the standards formed in this operation are of adequate uniformity. If more than one mold has been used in the casting process, adequate uniformity of composition between and throughout the different molds must be proved.

A number of standards should also be selected from the gross lot and each analyzed at selected points throughout its volume. This is of particular importance, especially for chill-cast standards, since the alloy structure may show uniformity of crystalline form for only a certain distance from the chill-cast face. This was the case for the NBS white cast-iron and brass standards and for the CAAS phosphor-bronze disc standards.

The homogeneity testing of these brass and phosphor-bronze samples[14,15] will be described to illustrate the typical procedure and the statistical treatment of the results. The metal was cast in five molds, each containing 40 discs. A disc was selected at random and analyzed spectrochemically for nickel, tin, zinc, and lead. Fifty observations were made using positions near the center of the disc but at successive levels through the thickness of the disc. Successive layers were machined off to produce fresh surfaces for analysis.

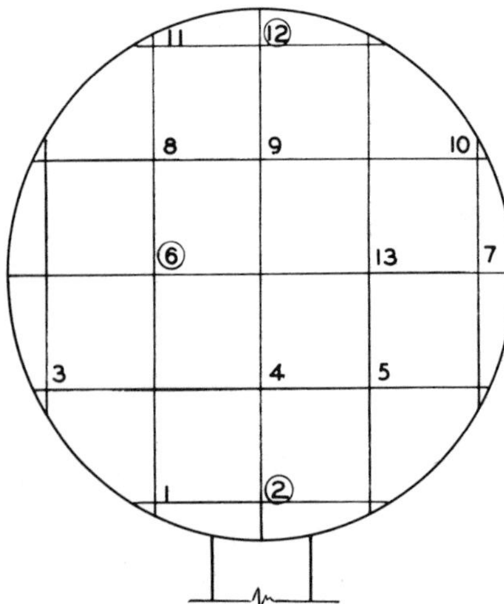

Fig. 5.11 Sparking positions for homogeneity testing of phosphor-bronze disc.

Tests were also carried out on six discs, three taken from positions adjacent to the sprue in the first, third, and fifth molds, and three taken from positions at the ends of the runners in the same molds (Fig. 5.8). Spectrochemical analyses were performed on these discs at positions 6 and 13 in Fig. 5.11. On one disc chosen at random, homogeneity tests for nickel, tin, and zinc were made at positions 1-12 on the chill-cast surface and at positions 2, 6, and 12 at four levels approximately $\frac{1}{16}$ in., $\frac{1}{8}$ in., $\frac{3}{16}$ in., and $\frac{1}{4}$ in. from the chill-cast surface. All surfaces were freshly machined on a lathe.

Statistical methods were employed to:

1. Determine the distribution of the spectrographic data
2. Test for significant differences between molds and between individual discs with a mold
3. Determine what portions of a disc could not be considered homogeneous

5.7.4.1 Distribution Studies

The data from the 50 tests on the one disc were in the form of log intensity ratios. Histograms of these ratios were found to be quite symmetrical, and when these ratios were plotted cumulatively on normal probability paper, reasonably straight lines resulted.

5.7.4.2 Factorial Experiment to Determine the Significance of Differences between Molds and between Positions within Molds

Four identical factorial experiments, one for each of the elements nickel, tin, zinc, and lead were designed to test for the significance of any mold or position effects. The experiments were in the form of 3 X 2 X 2 factorials; the three referred to the three molds, the first, third, and fifth poured; the first two related to the two positions in each mold, one adjacent to the sprue, the other at the end of the runner; and the second two associated with the two analyses on each disc. Twelve observations were thus carried out for each element.

The mathematical model used was:

$$x_{ijk} = n + a_i + b_j + c_{ij} + e_{ijk}$$

where x_{ijk} referred to any individual observation, and n was a constant corresponding to the true mean of the population if the mold effect, a_i, the position effect, b_j, and their interaction, c_{ij}, were all equal to zero. The term e_{ijk} corresponds to the random variation due to sampling and analysis fluctuations.

The assumptions implied in this model were that:

1. The a_i, b_j, and c_{ij} terms, if these existed, were additive.
2. The position effect was a crossed rather than a nested classification.

3. The mold and position factors were random.
4. The variance was homogeneous.

The first assumption, that the factors were additive, was necessary to avoid a very complex and possibly indeterminate mathematical situation. Any deviation from this assumption usually results in an expanded estimate of the interaction effect.

The second assumption is essentially a concept rather than an assumption. That is, by considering the position as a cross classification, a position effect is conceived inherent in that *type* of mold rather than in any particular mold.

The third assumption implied that the factor levels (i.e., molds or positions used in this experiment) were random samples from an infinite population. The size of the molds and the amount of metal melted were the only factors governing the number of levels; therefore, this assumption was valid, since any number of molds or positions could have been obtained by varying the size of molds and/or the castings.

The data were presented in the form of log intensity ratios. Standard methods of analysis of variance were used; thus, assuming that the terms a_i, b_j, c_{ij}, and e_{ijk} were normally distributed with equal variances, the e_{ijk} has a mean value of zero. This assumption was at least partially warranted by the results given by the distribution study.

The 95% confidence level was used for all tests of significance in the analysis of variance; that is, an effect was judged to be significant only if it were so great that its probability of occurring *by chance alone* was less than 5%.

The results of the analysis of the variance effect did not prove significant for any of the four elements determined, for the mold, or for the position in a mold. The estimated standard deviation for each element was compared with that obtained from the 50 observations on one disc in the distribution study. By F tests at the 95% level, no significant differences were found between the corresponding estimates, and therefore the estimates of standard deviation obtained from the distribution study and from the four factorial experiments were equivalent. This equivalence was further confirmation of the lack of significance deduced from the analysis of variance of the factorial experiments.

If any of the assumptions underlying the analysis of variance had been invalid, the estimates of the standard deviations obtained from the two sources would have been expected to differ.

5.7.4.3 Homogeneity within a Disc

For the homogeneity study within a disc, analyses for nickel, tin, and zinc were carried out on a disc chosen at random. Twelve observations were made for each element at selected points on the surface of the disc, and three observations were made for each element at four different levels for three of the 12 points selected (Fig. 5.11).

The 12 results obtained from the surface of the disc were used to compute a mean value. The means obtained from the distribution studies and the analysis of variance were not used because their plate calibration results differed from that in this last test. However, the standard deviations computed from the distribution studies were employed to define the 95%–95% tolerance intervals about the means from the 12 points. (A tolerance interval defines a range of values within which 95% of all subsequent items from the same population can be expected to fall. The probability level within which this interval was computed was also 95%.)

If a value does not fall within the tolerance interval, it can be concluded that it probably does not belong to the same distribution as the other values, thus indicating heterogeneity. The chances of error in making this decision are fixed by the probability levels of the tolerance interval.

The tolerance intervals were used as criteria of the homogeneity for the 12 surface points as well as the sets of three points at the four levels within the disc.

The results of this study indicated that for all three elements, the gate areas of the disc were probably not homogeneous. In the thickness of the disc there was no definite heterogeneity, but there was some suspicion of segregation at the fourth level (i.e., three fourths of the thickness below the surface initially examined).

In summary, the results of the three studies were:

1. The log intensity ratios were normally distributed.
2. There were no significant effects due to molds or positions within a mold.
3. The discs possessed acceptable homogeneity except for the gate regions.

It must be emphasized that all tests or conclusions derived in this way are relative to the spread of the data. This spread is an inherent property of the data and depends on variations due both to the material tested and the method of analysis.

5.7.4.4 Other Methods of Homogeneity Testing and Evaluation

The method of homogeneity testing and evaluation described above is the type which workers with the customary equipment is a spectrochemical laboratory will be able to carry out. However, with recently developed and more complex apparatus, a more detailed and comprehensive examination is possible.

Such a refined method is described by Michaelis *et al.*[16,17] and by Moore *et al.*[18,19] where special equipment has been constructed to obtain a quantitative estimate of homogeneity by automated micrometry and a computer evaluation of the results. With this equipment it was possible to determine the mean and the distribution of the grain size of a low-alloy steel standard. The steel was shown to be structurally homogeneous at a 5-μm level, since the mean free path in the ferrite was 1.63 μm and in the pearlite was 3.83 μm.

In addition to the metallographic technique for determining structural homogeneity, the electronic microprobe x-ray analyzer is a powerful tool for determining both the homogeneity of composition at the micro level and the nature and occurrence of inclusions.[16,17,20] Thus for the NBS-1102 brass standard, line scans for copper and zinc were run at random on a carefully polished sample to reveal gross inhomogeneities. No variation greater than three times the standard deviation of the count rate was observed. The electron beam scan was then stopped and counts made twice at 42 points selected at random on the surface of the cast standard. The maximum zinc concentration was 1.3% higher than the mean and the minimum was 2.2% below. The standard deviation of the zinc content was 0.79%, and the standard deviation of the counting error was 0.37%. The analytical range, two standard-deviation limits of content at the 3-μm level of resolution, was, therefore, 26.7-27.5% zinc.

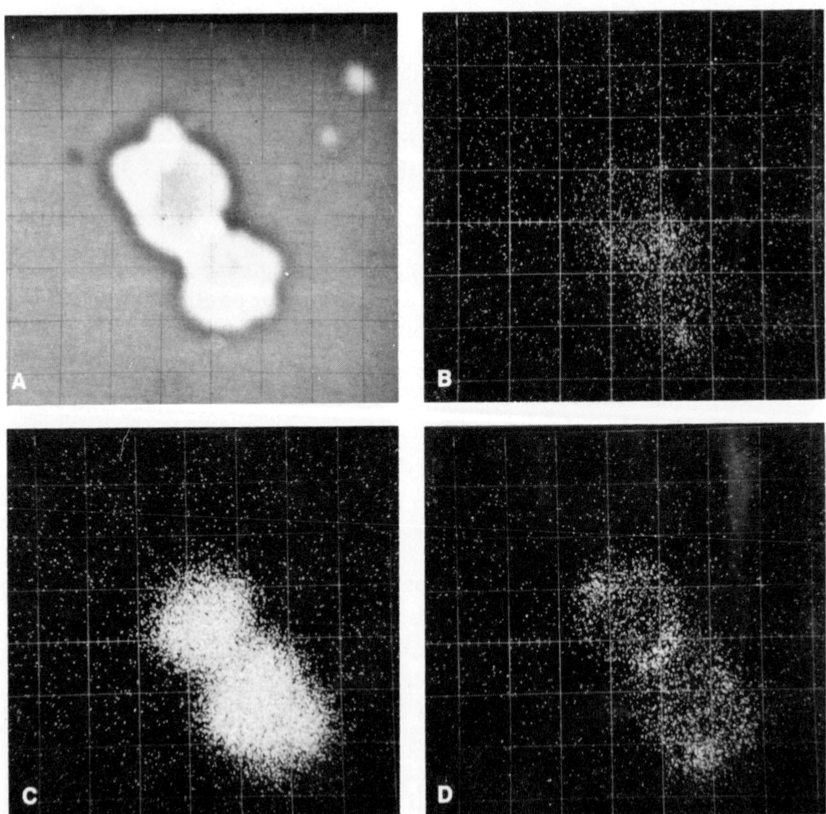

Fig. 5.12 Electron beam scanning of a typical brass inclusion: A, specimen current; B, Pb–Mβ x-ray; C, S–Kα x-ray; D, Si–Kα x-ray; E, Al–Kα x-ray; F, Zn–Kα x-ray; G, Cu–Kα x-ray. All ×430. (From Yakowitz *et al.*[20])

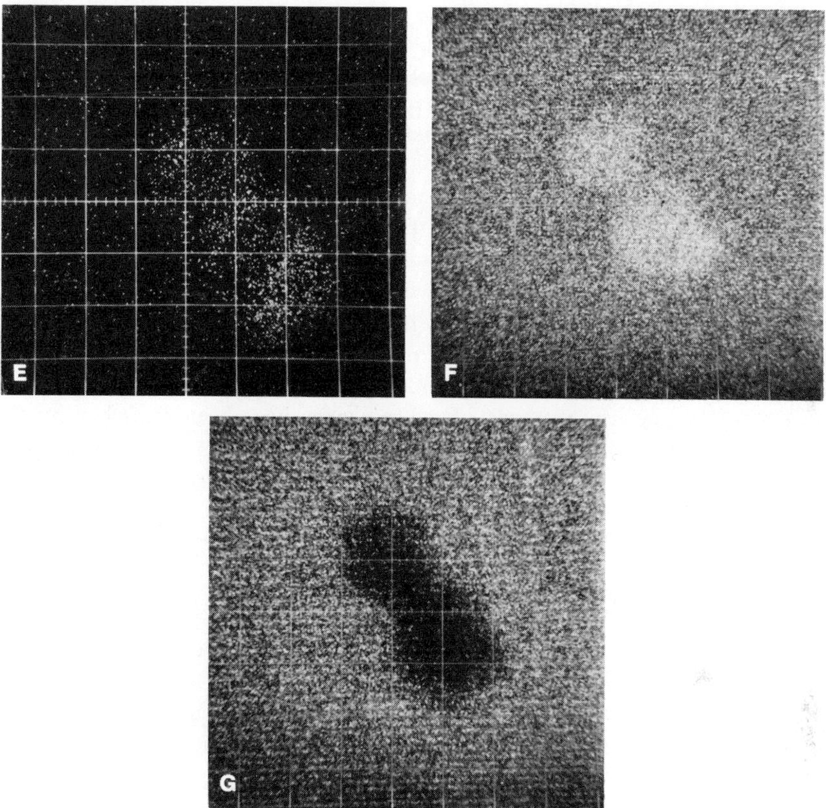

Fig. 5.12 (*Continued*)

Examination of the inclusions in this material revealed large amounts of lead and sulfur, variable amounts of zinc, silicon, and aluminum, but no copper. Microprobe scanning pictures of a typical inclusion are shown in Fig. 5.12a-g. Clearly, the elements appearing in the inclusion are not distributed homogeneously through the material, but the lack of uniformity is of practical significance only for lead, sulfur, and silicon.

The great value of electron beam microprobe analysis for homogeneity testing of materials lies in the rapidity with which the uniformity of composition can be checked at the micro level and the location of inhomogeneously distributed elements can be determined. As this instrument becomes more generally available, it is certain to be widely used in examination of materials for compositional homogeneity.

5.7.5 Multiple Compositional Analysis

When material prepared for use as a standard has been shown to possess adequate homogeneity, at least for the elements of major interest, it has to undergo a very precise analysis, preferably by methods other than spectrochemical.

These analyses should be carried out by a number of different laboratories, customarily not less than three. When possible, more than one analytical method should be used in each laboratory. Sufficient repetitions of each method should be carried out in each laboratory so that a reliable estimate of the standard deviation of the results for each method used by the laboratory can be calculated.

The analytical methods and the details of the results should be included in the report to the coordinating laboratory. The major requirement for the selection of the analytical method is their inherent capability for maximum precision and accuracy. Typical methods of proved value have been published by the National Bureau of Standards for their white cast-iron and copper-base spectrochemical standards.[21–23] A good source of tested methods for chemical analysis is the ASTM Methods for Chemical Analysis of Metals.[24]

5.7.6 Statistical Evaluation of Results

The homogeneity testing preceding the multiple compositional analysis by the various analytical methods in the three or more certifying laboratories must be statistically evaluated to determine the precision and accuracy of the analytical averages, both overall averages and those related to the different methods and to the same methods in different laboratories.

Standard tests on statistical analysis may be referred to for this evaluation; however, they are frequently too general in their treatment for application to this specific problem. One of the most useful treatments for conciseness and lucidity is that appearing as Section E-2 SM 2-4 in the ASTM volume *Methods for Emission Spectrochemical Analysis.*[25] A second useful reference for evaluating interlaboratory testing is the *ASTM Spec. Tech. Publ. 335.*[26] Although the statistical treatment in E-2 SM 2-4[25] is designed to evaluate a particular spectrochemical method, it can be employed equally well to determine the precision and accuracy of results obtained by other methods. The ASTM-E-2 Committee book on methods is probably the most widely used handbook in the field, so any group that plans to prepare spectrochemical standards has access to a basic reference. It is recommended, therefore, that any such group should consult this treatment in the basic statistical evaluation of results.

It gives information on the planning of cooperative testing and the consequent statistical computations necessary for estimation of the precision and accuracy of the analysis. The minimum number of samples, laboratories, and determinations are laid down for adequate statistical evaluation. Criteria are given for the rejection of results, and methods for the determination of systematic error are detailed.

Most methods of statistical analysis assume that the deviations from the mean fall into a normal or Gaussian distribution; this is also true of the ASTM method.[25] The assumption that a normal distribution of errors is obtained from spectrochemical analyses is justified from experience, and the literature quotes very few examples of distributions proved other than normal. An example is one by Ahrens and Taylor,[2] who have found that in the trace analysis of rocks, the logarithms of the concentrations found, rather than the concentrations themselves, fit a normal distribution.

When the number of results is small [i.e., less than 16 (some indicate less than 32)], the application of normal distribution statistics is not mathematically justified. When this is the case, nonparametric tests may be applied. The nonparametric methods involve no assumptions regarding the distribution of the results; in addition, they are arithmetically simpler and the calculations more rapid than for the normal distribution treatments.

One nonparametric test which has been found to be valuable in spectrochemical statistics is the Mann-Whitney or U test, as described by Siegel.[27] This text gives a thorough treatment, presenting both the advantages and disadvantages, stating when nonparametric tests may or may not be applied, and showing the tables necessary for the use of the U test for numbers of results less than eight.

5.7.7 Weighing and Evaluation of Analytical Results

After all the statistical evaluation has been completed, the group responsible for preparation of the standards has to exercise its judgment on the statistical data presented and decide whether the precision and accuracy of the results are adequate to certify the material as a standard. Such a judgment will take into consideration the nature of the standard and the uses for which it is designed. It may place greater weight on the results obtained by a particular method or by a particular laboratory. It will also determine what element percentages present in the standard should be certified and which elements, because of inhomogeneous distribution, should only be quoted as present. Although an element percentage is not certified, it is, nevertheless, of importance to the user to know that the element is present and its approximate concentration.

5.7.8 Final Certification

The certification finally provided with the standard should give the user of the standard as many details as is reasonably possible regarding the preparation and the certification. The cooperating laboratories may or may not be named, but the analytical procedures for each element should be listed for each laboratory. The laboratories should be indicated by a number. The final certification value should be quoted for each element certified, and a clearly indicated value should be given for those elements present which were not certified.

When it is applicable, the certification should indicate the area and depth of the sample that meets the requirements. With powder standards the particle-size distribution should be quoted.

Canadian Association for Applied Spectroscopy Non-Ferrous Standards Committee

555 Booth Street
Ottawa, Ontario.

Analysis Certificate
Spectrographic Standard Sample No. 304

Phosphor Bronze

Element	Analyst 1	2	3	4	5	Suggested value
Copper	–	a 86.63	a 86.34	a 86.47	a 86.54	86.48
Tin	b 9.56	c 9.66	d 9.64	c 9.72	d 9.70	9.67
Iron	e 0.033	f 0.0348	g 0.034	u 0.037	v 0.026	0.035
Lead	a 0.52	a 0.46	a 0.54	a 0.48 l 0.49	m 0.46	0.46
Zinc	n 2.87	n 3.03	o 3.02	p 2.99 l 3.03 n 3.12	p 2.90	2.99
Phosphorus	r 0.011	–	r 0.006	r 0.0045	r 0.009	0.007
Aluminum	–	w 0.0472	x 0.062	q 0.049	–	0.05 (tentative)

Methods of Analysis

(a) A.S.T.M. electrolytic.
(b) Iodometric titration, reduction with iron antimony alloy.
(c) Hypophosphite reduction, iodate titration.
(d) Aluminum-hydrochloric acid reduction, iodine titration.
(e) Ammonium hydroxide separation, $KMnO_4$ titration.
(f) Thiocyanate photometric.
(g) Thiocyanate photometric after extraction with 4-methyl-2-pentanone.
(l) Polarographic.
(m) Sulphate method A.S.T.M. E54.
(n) Oxide method.
(o) EDTA disodium salt titration extraction as zinc thiocyanate.
(p) A.S.T.M. volumetric ferrocyanide.
(q) Photometric Eriochrome Cyanine RA after mercury cathode heavy metal removal.
(r) Molybdovanadophosphoric acid method.
(u) A.S.T.M. NH_4OH separation, dichromate titration.
(v) Spectrographic copper oxide powder technique.
(w) Aluminon photometric method.
(x) Aluminon photometric method after mercury cathode heavy metal removal.

Fig. 5.13 Example of certificate of analysis.

As a typical example of a certificate of analysis, the certificate for a phosphor-bronze disc standard prepared by the Canadian Association for Applied Spectroscopy is shown in Fig. 5.13.

A recent publication by G. H. Faye and R. Sutarno,[(28)] restricted to the certification of powdered reference ores, describes up-to-date procedures for cooperative analytical testing and statistical evaluation of such materials.

5.8 CONCLUSION

In conclusion, the purpose of this account of the preparation and evaluation of spectrochemical standards was to show the great amount and high quality of effort and care needed to produce standards of adequate precision and reliability. Because most spectrochemical analyses are based on available standards, the results can only be as good and as precise as the certification analyses of the reference standards used.

ACKNOWLEDGMENTS

The author thanks Dr. Wayne Meinke of the National Bureau of Standards, Institute for Materials Research, for Figures 5.6, 5.9, 5.10, and 5.12, and for permission to quote from the NBS Miscellaneous Publications listed in the references; the Non-Ferrous Standards Committee of the Canadian Association for Applied Spectroscopy for Figs. 5.7, 5.8, 5.11, and 5.13, and for permission to quote from Reports A, B, and C of the Second Minutes, 1956: Mr. A. J. Mittledorf, SPEX Industries, Inc., for Figs. 5.2-5.4, and Mr. Victor N. Ryland, Bausch & Lomb Canada, for Fig. 5.5.

5.9 REFERENCES

1. R. Michaelis, Report on Available Standard Samples, Reference Samples, and High-Purity Materials for Spectrochemical Analysis, *ASTM Spec. Tech. Publ. DS-2* (1964).
2. L. H. Ahrens and S. R. Taylor, *Spectrochemical Analysis*, 2nd ed., Addison-Wesley Publishing Company, Inc., Reading, Mass. (1961), pp. 111, 112, 114, 115, 116, 159.
3. I. Oftedahl, The accessory constituents of ore minerals of the principal Norwegian sphalerite deposits, *Skrifter Norske Videnskaps-Akad. (Oslo) I: Mat. Naturv. Kl. 8*, (1940).
4. B. F. Scribner and W. F. Meggers, Index to the Literature on Spectrochemical Analysis, Parts I–IV, *ASTM Spec. Tech. Publ. 41, A–D* (1939–1959).
5. *Spectrochemical Abstracts*, Vols. 1-9, Hilger & Watts Ltd., London.
6. *Chemical Abstracts*, American Chemical Society, Washington, D.C.
7. Standard Reference Materials, *Natl. Bur. Std. (U.S.) Misc. Publ. 260* (1965).
8. H. W. Fairbairn, A Cooperative Investigation of Precision and Accuracy in Chemical Spectrochemical Analysis of Silicate Rocks, *U.S. Geol. Surv. Bull. 980* (1951).

9. Report of Nonmetallic Standards Committee, Canadian Association for Applied Spectroscopy, *Appl. Spectrosc. 15*, 159 (1961).
10. G. R. Webber, *Geochim. Cosmochim. Acta. 29*, 229 (1965).
11. R. E. Michaelis and L. L. Wyman, Preparation of NBS White Cast-Iron Spectrochemical Standards, *Natl. Bur. Std. (U.S.) Misc. Publ. 260-1* (1964).
12. R. E. Michaelis, L. L. Wyman, and R. Flitsch, Preparation of NBS Copper-Base Spectrochemical Standards, *Natl. Bur. Std. (U.S.) Misc. Publ. 260-2* (1964).
13. J. O. Edwards, Casting of Spectrographic Standards, Report A, Minutes of Second Meeting of Canadian Association for Applied Spectroscopy, Non-Ferrous Standards Committee (1956).
14. J. K. Hurwitz, Spectrographic Tests, Report B, Ref. 13.
15. R. C. Shnay, Analysis of Spectrographic Data, Report C, Ref. 13.
16. R. E. Michaelis, H. Yakowitz, and G. A. Moore, *J. Res. Natl. Bur. Std. 68A*, 343 (1964).
17. R. E. Michaelis, H. Yakowitz, and G. A. Moore, Characterization of a NBS Spectrometric Low-Alloy Steel Standard, *Natl. Bur. Std. (U.S.) Misc. Publ. 260-3* (1964).
18. G. A. Moore and L. L. Wyman, *J. Res. Natl. Bur. Std. 67A*, 127 (1963).
19. G. A. Moore, L. L. Wyman, and H. M. Joseph, in: *Quantitative Metallography* (F. N. Rhines, ed.), Chap. 15, McGraw-Hill Book Company, New York (1964).
20. H. Yakowitz, D. L. Vieth, K. F. J. Heinrich, and R. E. Michaelis, Homogeneity Characterization of NBS Spectrometric Standards, II, Cartridge Brass and Low-Alloy Steel, *Natl. Bur. Std. (U.S.) Misc. Publ. 260-10* (1965).
21. R. Alvarez and R. Flitsch, Accuracy of Solution X-Ray Spectrometric Analysis of Copper-Base Alloys, *Natl. Bur. Std. (U.S.) Misc. Publ. 260-5* (1965).
22. J. I. Schultz, Methods for the Chemical Analysis of White Cast-Iron Standards, *Natl. Bur. Std. (U.S.) Misc. Publ. 260-6* (1965).
23. Rosemond K. Bell, Methods for the Chemical Analysis of NBS Copper-Base Spectrochemical Standards, *Natl. Bur. Std (U.S.) Misc. Publ. 260-7* (1965).
24. *Chemical Analysis of Metals, Sampling and Analysis of Metal Bearing Ores*, Part 32 of *ASTM Standards*, American Society for Testing and Materials, Philadelphia (1967).
25. ASTM Committee E-2, *Methods for Emission Spectrochemical Analysis*, 4th ed. American Society for Testing and Materials, Philadelphia (1964), pp. 218–229.
26. Manual on Interlaboratory Testing, *ASTM Spec. Tech. Bull. 335* (1962).
27. S. Siegel, *Non-Parametric Statistics for the Behavioral Sciences*, McGraw-Hill Book Company, New York (1956).
28. G. H. Faye and R. Sutarno, The Canadian Certified Reference Materials Project: How Reference Ores Are Certified, *CANMET Mineral Sci. Lab. Rep. MRP/MSL 76-241* (August 1976).

Applications of Emission and X-Ray Spectroscopy to Oceanography

6

Geoffrey Thompson

6.1 INTRODUCTION

The aim of marine chemistry is to describe the oceans in terms of concentration, chemical state, pathways, and time constants for the elements and compounds therein. The ocean, its contents, and its boundaries form a complex, multidimensional system, and its study is approached through various scientific disciplines that span the range of observable phenomena. Thus the marine chemist's work must mesh with that of many scientists from other disciplines. This is not to say that the marine chemist should be pictured in the role of concubine or mistress–ever present and only offering required services. Rather, he/she should be seen on a more moral plane, as an active partner in a successful marriage of natural sciences.

The development of chemical oceanography closely parallels that of suitable experimental techniques and adequate sampling equipment. Although speculations regarding the saltiness of the sea were made by the ancient Greeks, systematic chemical measurements did not begin until the eighteenth and nineteenth centuries. Robert Boyle (in 1670) and Lavoisier (in 1776) were among the early pioneers in identifying the various salts present in the sea. Dittmar (in 1884) reported on the quantitative aspects of seawater analysis using gravimetric and volumetric techniques. He noted that, although the total salt content may be variable, the ratios of the major constituents remain very nearly constant. Only in the last 50 years have we begun to probe the mysteries of the oceans in more detail as scientists with modern instruments have applied the analytical techniques

Geoffrey Thompson • Woods Hole Oceanographic Institution, Woods Hole, Massachusetts of their various disciplines. The most recent chemical analyses[1] have actually confirmed those of Dittmar.

In 1882 Manet predicted that all elements would be found in seawater. K, Na, Ca, Mg, S, and Cl had been identified in seawater by 1819, Br in 1826, B in 1853, and Sr and F in 1865. The development of the optical emission spectrograph by Bunsen and Kirchoff in 1860 paved the way to the identification of other elements. However, it was not by direct analysis of seawater but by inference: the analysis of the ashes of marine organisms, the elemental composition of which is determined by uptake from seawater. Iodine was first recognized in marine algae, later Ba, Co, Cu, Pb, Ni, Ag, and Zn were noted in marine organisms analyzed by emission spectroscopy. Recently, from many different analytical techniques, mostly involving some form of spectroscopy, chemists have confirmed the presence of 73 elements in seawater. Manet's prediction may well be true.

Measurements made in marine chemistry are not uniquely oceanographic, only the origin and nature of the samples are unusual; the high salt content, low concentrations, and complex multielement assemblages make the choice of analytical technique more difficult than in the freshwater system. Marine chemistry, of course, covers all aspects—organic, inorganic, and physical chemistry—with all the concomitant analytical techniques. In this chapter I shall consider three oceanographic analytical problems where emission spectroscopy and x-ray fluorescence spectroscopy play fundamental roles: seawater—analysis of the dissolved components; marine organisms—analysis of the nondissolved components of seawater; and marine sediments—analysis of the sea floor. First, however, I should point out that much of marine chemistry, although by no means all, is done in shore-based laboratories and not at sea. Analysis at sea is beset by many difficulties, particularly the rolling and pitching of the ship and vibrations from the engines and other machinery. The old adage "one hand for the ship and one for yourself" still holds. High-frequency motions and vertical accelerations make accurate weighing nearly impossible. Wide temperature variations, high humidity, and often poor voltage and frequency controls are common vicissitudes. Nevertheless, some analyses are feasible: spectrophotometry, titrimetry, and potentiometry are routinely done on-board ship. For the last two years the deep-ocean drilling ship *Glomar Challenger* has had a full-fledged x-ray fluorescence spectrometer—wavelength-dispersive type—on board and made measurements of the elemental composition of the igneous rocks drilled from the sea floor. In past years we have had a small 1-m spectrograph on-board ship and made routine analyses of sediments to help locate metal-rich deposits in the Red Sea. Even if the main workload is done back on shore, problems in sample collection, possible contamination, treatment of special samples before storage are such that most research chemists have to spend 1-2 months or more at sea each year collecting and preparing their samples for subsequent analysis back in the laboratory. Some of these aspects are treated in the following discussion.

6.2 SEAWATER

Seawater contains about 3.5% dissolved solids, but of these only 12 elements are present in amounts greater than 1 ppm. These are the major elements of seawater. Table 6.1 shows the average composition of seawater, including some of the trace elements, ranging in concentration from 0.5 ppm (nitrogen) to 4 parts per trillion (gold). The salts in seawater are mainly derived from the continents, where the rocks are continually being broken down, small quantities dissolved, and carried to the oceans via the rivers. Understanding the balance between input (rivers) and output (sediments) (i.e., the buffering of seawater composition) is a major goal of marine chemistry.[2]

Changes in temperature, pressure, and concentration of dissolved salts affect the physical properties of seawater and control its mixing. Thus these are the parameters of interest. Determination of the total salt content (salinity), rather than each individual major element, is usual. The ratios of the major elements (with the exception of Si) do not vary greatly–less than 1%–geographically or temporally, although, of course, the total salt content may vary appreciably. These major ions are commonly referred to as "conservative." Salinity determinations are generally done by measurement of the electrical conductivity or analysis of the chloride content (for many years, salinity = 0.03 + 1.805 Cl was the expressed relationship; now it is S = 1.80655 Cl; see Wilson[3] for a discussion of this concept). Dissolved oxygen and silica are additional measurements commonly made to help characterize a given water mass. These elements are nonconservative;

Table 6.1 Average Elemental Composition of Seawater (ppm)

		Major elements: (>1 ppm)			
Cl	19,000	K	380	Si	3.0
Na	10,500	Br	65	F	1.3
Mg	1,350	C	28		
S	885	Sr	8		
Ca	400	B	4.6		
		Trace elements: (<1 ppm)			
N	0.5	Cu	0.003	Cr	0.00005
Li	0.17	As	0.003	Ag	0.00004
Rb	0.12	U	0.003	Pb	0.00003
		Mn	0.002	Hg	0.00003
P	0.07	V	0.002	Bi	0.00002
I	0.06	Ni	0.002	Au	0.000004
Ba	0.03	Ti	0.001		
Al	0.01	Sn	0.0008		
Fe	0.01	Se	0.0004		
Zn	0.01	Co	0.0001		
Mo	0.01				

Table 6.2 World Production from Ocean Water, 1968[a,b]

Material	Tons/year	% of total world production
NaCl	35,000,000	29
Br	102,000	70
Mg (metal)	106,000	61
Mg (compounds)	690,000	6
Fresh water	142,000,000	59

[a] From Shigley.(53)
[b] Annual value $400,000,000.

that is, they are independent of the total salt content. All these measurements can be and are made routinely at sea. A full discussion of the salinity and major elements of seawater are found in Wilson.(3)

The major elements present in seawater have some direct commercial value, as well as importance in controlling basic physical and biological processes. The world production figures for recovery of major constituents from seawater are shown in Table 6.2. Future demands for fresh water will have to be met by desalination of seawater.

Trace elements in seawater are the constituents that vary most and cause the major analytical problems. They play an important role in the life systems: for example Zn in enzymes; Fe in the respiratory pigment hemoglobin in higher animals and Cu in hemocyanin, the oxygen-carrying molecule of mollusks and arthropods; and Si, Ca, and Sr in skeleton formation in marine plants and animals. C, N, P, and Si are important nutrient elements and determine the fertility and productivity of the oceans. Dissolved trace elements in seawater are also part of the world's mineral reserves. The oceans can be likened to a vast ore body (reserves of 50 quadrillion tons), the physical properties of which are itemized in Table 6.3. It is a unique ore body in that it is continuously being replenished; 40 billion tons of solids are added per year by rivers, 6 billion tons in solution. Commercial possibilities have traditionally excited many speculators: a $\frac{1}{4}$ cubic mile per year of seawater could provide U.S. needs for Si, Mg, Cl, Na, and Br; $1\frac{1}{2}$ cubic miles could supply all our needs for K and S; 5 cubic miles B and I; 50 cubic miles Th and F; greater than 100 cubic miles would be required to supply suffi-

Table 6.3 Physical Properties of the Ocean Ore Body

Average depth:	2.36 miles
Surface area:	140×10^6 miles^2
Volume:	330×10^6 miles^3
Mineral content:	165×10^6 tons/mile^3
Total reserves:	50×10^{15} tons

cient quantities of other metals. Recovery is the problem. To "mine" 1 cubic mile per year would require 2,100,000 gallons of seawater to be processed every minute, and with 100% efficiency would provide $1,000,000 worth of 17 critical metals: Sb, Bi, Cd, Cr, Co, Au, Pb, Mn, Hg, Mo, Ni, Ag, Sn, Ti, U, and Zn—an ore value of $0.20/hundred tons. At present, recoveries are neither economically nor technically feasible, but future depletions of the world continental reserve may leave us with no choice. A discussion of the extraction of economic inorganic materials from seawater can be found in McIlhenny.[4]

Variations in concentrations of the trace elements reflect biological and boundary interactions of the seas. The direct analysis of seawater for these components demands techniques with sensitivities in element determination from 10^{-6} to 10^{-11} g. The low levels of trace elements in seawater also present problems in the collection and storage of samples. For many element determinations, large volumes of seawater are required to provide sufficient measurable quantity of an element or one of its isotopes (see Fig. 6.1). Collecting containers have to be made of inert materials to avoid contamination. Storage of seawater may lead to adsorption of elements on the walls of the containers, or leaching of elements from the containers themselves. Seawater samples stored in glass or polyethylene bottles at their natural pH rapidly lose elements such as Sc, Fe, Co, Ag, and U onto the container walls.[5] Acidified samples stored in polyethylene containers show the least losses. Similarly, the sampling devices should be made of inert materials to avoid contamination. Many anomalous trace-element concentrations have been ascribed to contamination from such things as galvanized or epoxy-lined sampling devices, or even metal pumps used to transfer the samples.[6] An excellent discussion of the problems of determining minor elements in seawater can be found in Brewer.[7]

Many spectroscopic techniques are not sufficiently sensitive for direct analysis of seawater, and analysts are forced to resort to preconcentration techniques. The choice of technique is obviously influenced by such factors as equipment availability, cost, precision and accuracy required, sensitivity, number of samples, and speed of analysis. Some of the more common techniques are mentioned here; a fuller discussion of techniques, preconcentration, and recovery problems can be found in a classic work on the analytical chemistry of seawater by Riley [8] and also in Spencer and Brewer [9] and their respective bibliographies.

1. Neutron activation—this is one of the few techniques with sufficient sensitivity and precision for direct analysis of seawater. It also has the added advantage that after irradiation it is independent of contamination from other materials, and separations can thus be done easily. Elements determined by this method and λ spectroscopy include As, Ba, Sr, Au, Rb, Cs, Cu, Mn, Zn, Sb, Cr, Co, Se, Ag, and U.

2. Isotope dilution and mass spectrometry—this method is also feasible for direct analysis of seawater, because it has sensitivity limits of 10^{-7}-10^{-9} g. Elements commonly determined include Li, Ba, U, and Sr.

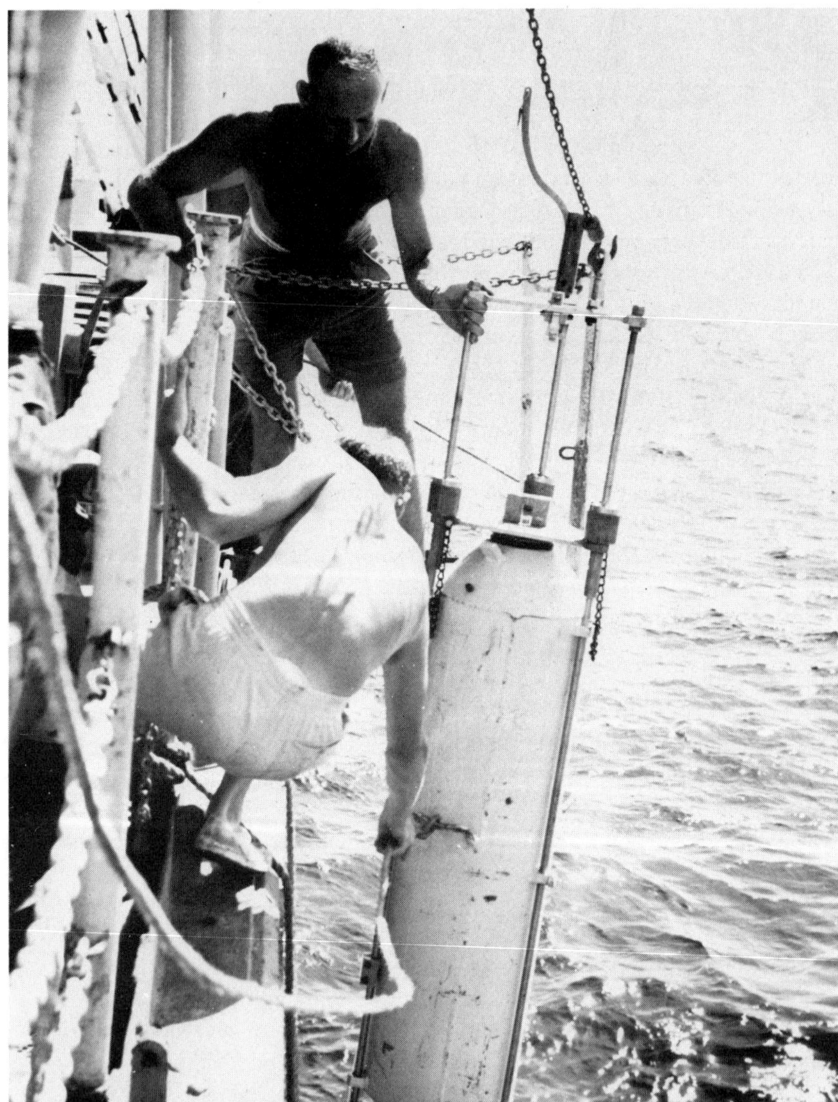

Fig. 6.1. Large-volume (35-gal, 130-liter) water-collecting bottle being lowered over the ship's side. The bottle is made of aluminum and is teflon-lined. Spring activated neoprene-lined stoppers close "on command" by a heavy weight sliding down the wire and releasing a retaining catch. This bottle was originally designed at the Woods Hole Oceanographic Institution for collecting large volumes of seawater to measure radioactive nuclides introduced into the oceans as "fallout" from atmosphere-exploded nuclear devices. These nuclides are used as tracers for studying ocean mixing processes as well as being monitored to ensure that their levels in the water and in organisms do not reach hazardous proportions.

3. Emission spectroscopy–this is one of the earliest used techniques but, except for some of the major elements, preconcentration is required. Procedures often used include coprecipitation with mercury or copper sulfide, ferric hydroxide, or 8-hydroxyquinoline. An ion-exchange concentration has also been used for such elements as Au, Bi, and Cd. Marine chemists are closely following the developments of plasma jet excitation sources for use with the emission spectrometer. In the past few months we have developed a technique for determination of Ba in seawater and in marine organisms, using the argon plasma excitation source and an emission spectrometer with echelle grating (Spectraspan model, Spectrametrics, Inc., Andover, Mass.).

4. Flame photometry–this method is commonly applied to the direct determination of Ca and Sr in seawater. It is also satisfactory for determination of Ba, Li, Rb, and Cs after ion-exchange concentration. Recent advances in instrument design and flame stabilization show promise for its further application to seawater analysis.

5. Atomic absorption spectroscopy–this technique is being increasingly adopted for seawater analysis in conjunction with concentration procedures. Good determinations of Co, Cu, Zn, Pb, Fe, Ni, and Mn have been reported with ammonium pyrrolidine dithiocarbamate (APDC) as the concentrating chelating agent, and methyl isobutyl ketone (MIBK) as the extracting solvent. Rb and Li have been determined by direct aspiration of untreated seawater. Nonflame devices (e.g., the graphite furnace) are also of great value in the analysis of seawater and are increasingly being utilized. The determination of Cu, Ni, and Cd by flameless AA spectrometry has recently been reported.[10]

6. Spectrophotometry–a commonly adopted technique for measurement of the nutrient elements, for example nitrite (strychnidine or diazonium color), P, and Si (molybdate color). It is also a common technique for certain trace elements, for example Cu (diquinolyl color), Cd (dithizonate color), and B (carmine red color). It is one of the few techniques commonly used at sea.

7. Fluorometry–this technique has better sensitivity than spectrophotometry but is less commonly used because of interference effects. Analyses done include Al (pantachrome blue-black R), Be (morin), Ga (8-hydroxyquinoline), and U (sodium fluoride).

8. X-ray fluorescence spectroscopy–this technique has not found widespread use even though it has good precision and sensitivity. Determinations of Cr, Cu, Co, Mn, Fe, Ni, V, and Zn, after preconcentration, have been reported by Morris.[11]

9. Polarography–this technique is becoming more popular whether as cathode ray, anodic stripping, or square-wave polarography. Elements nicely determined by these techniques include Cu, Pb, Cd, Zn, Mn, Al, and U. Some information on element speciation in seawater can also be obtained by this method.

Table 6.4 Trace Elements in Seawater: Intercalibration Study[a,b]

Element	Coefficient of variation	Abundance ($\mu g/kg$)
Sr	2.5	8100
F	2.8	1350
Rb	5.4	121
Cs	5.5	0.3
U	5.6	3.3
Sb	10.2	0.4
Zn	18.4	5
Cu	20.37	3
Co	22.0	0.1
Mn	24.5	1.5
Fe	25.8	14
Pb	29.2	5
Ni	33.5	2

[a] From Brewer and Spencer.(12)
[b] Elements ranked in order of increasing coefficient of variation of the grand means, calculated from the pooled standard deviations.

The analytical techniques used, however, do not always have the required precision or accuracy at the low concentrations of the elements in seawater. A recent laboratory intercalibration study(12) revealed very large disagreements among the 26 participating laboratories. Table 6.4 shows the coefficients of variation for the elements determined; only five elements (none of them transition metals) were determined with a coefficient of variation of $<10\%$. This study also lists the analytical techniques and details of the methods used. Our present understanding of the distribution and concentration of many trace metals in seawater is thus very limited and uncertain. A good review of our present knowledge and an assessment of the factors controlling metal speciation and distribution is found in Brewer.(7)

6.3 MARINE ORGANISMS

Plants and animals in the sea are responsible for some of the most dramatic changes in seawater composition. Many elements, particularly nutrients and trace metals, are taken up and concentrated by the organisms that inhabit the surface waters of the oceans. The "fixing" of these elements depletes the surface waters and inhibits further productivity. After the death of the organisms, their remains sink into the deep waters, where they are oxidized and the elements returned to circulation. The return of these nutrients to the surface governs the organic productivity and hence the fisheries in the various regions of the sea.

The study of the elemental composition of marine organisms is thus important in terms of productivity, in the movement and transfer of ions from surface to deep waters, and in the concentration of elements from seawater. This latter aspect provides a remarkable and interesting phenomenon. The actual concentration by the organism, compared to the abundance of the element in seawater, can be as high as a millionfold. Moreover, organisms can be very specific for which element they concentrate. For example, Carlisle [(13)] has shown that some tunicate species concentrate V, others Nb, some neither element, but that none concentrate both elements. Ti may be concentrated by some ascidians over 1 million times.[(14)] Nicholls *et al.*,[(15)] employing emission spectrographic techniques for analysis of plankton (small, often microscopic, plants and animals that live at or near the sea surface and are the primary link in the food chain), predicted from their early results that "for any given chemical element there will be at least one planktonic species capable of spectacularly concentrating it." Here at Woods Hole, we are continuing this work using direct-reading emission spectrometry.[(16)] We have not yet found it necessary to refute that original prediction as indicated by our data in Table 6.5. Phytoplankton (marine plants and photosynthesizing algae) are reported to accumulate such metals as Al, As, Ba, Be, Cd, Ce, Cr, Co, Cu, I, Fe, Pb, Mn, Ni, Nb, Pu, Sc, Ag, Zn, and Zr by factors of at least 10^3 from seawater.[(17)]

The possibility of "farming" certain marine species and "mining" them for their metal concentrates is not so futuristic as its sounds–in one instance it is fact–iodine from seaweed. From another point of view, concentration of certain toxic metals in species eaten by man is not at all welcome, for instance the concentration factors for certain elements in common shellfish (Table 6.6). Mercury poisoning, through concentration of that metal by shellfish, has already resulted in 52 deaths and 168 serious illnesses to the residents of Minimata, Japan.[(18)] Even now, mercury levels in some swordfish and tuna, and shellfish in certain regions, is close to our own levels of tolerance.[(19)] The surprising evidence of biological metal concentrations emphasizes how little we know of the natural cycles in the environment and how the difference between tolerable natural background levels and levels harmful to man and other animals may be very small.

Analysis of marine organisms, as with continental biological materials, involves drying and ashing. Low-temperature ashing with oxygen plasma, or refluxing with acid–peroxide mixtures, prevents loss of the volatile elements.[(20)] For many species, the body tissues and fluids are composed almost entirely of the combustible elements C, H, O, and N, and only a few milligrams or less of ash may remain from a large volume of original sample.

In selecting an analytical technique, the marine chemist is bound by constraints: (1) a large number of elements need to be quantitatively analyzed (samples are hard to come by and we have little knowledge concerning most element distributions); (2) there are often only small amounts of sample available

Table 6.5 Elemental Composition of

		Ag	Al	B	Ba	Bi	Cd	Co	Cr	Cu
Coelenterata										
Scyphomedusae										
Cyanea capillata	No. 7b	<1	75	175	2	<1	5	6	<1	70
Pelagia noctiluca	22b	<1	24	260	4	<1	2	<1	<1	210
Periphylla hyacinthina	88	<1	50	80	50	<1	2	<1	2	165
Siphonophora										
Velella sp.		<1	34	120	6	<1	5	3	4	195
Siphonophore agolina	65	2	85	75	15	<1	<1	<1	2	140
Leptomedusae										
Aequorea vitrina	11c	<1	70	325	2	<1	5	<1	<1	205
Ctenophore										
Mnemiopsis sp.	13	1	5	65	1	3	<1	<1	<1	125
Mnemiopsis sp.	15	<1	5	100	1	<1	<1	<1	<1	150
Beroe cucumis	20	1	5	70	1	<1	<1	<1	<1	270
Phytoplankton										
Mainly Diatomeae	No. 33A	<1	3380	120	44	<1	2	<1	4	210
	33b	2	1130	105	140	<1	<1	<1	<1	145
	34a	<1	3200	115	57	<1	<1	<1	<1	22
	85	<1	5465	140	60	<1	17	<1	6	95
Mixed collection	34b	1	1410	90	82	<1	3	<1	<1	195
Tunicata										
Salpida										
Salpa fusiformis	31a	<1	245	100	14	<1	4	<1	1	95
Salpa sp.	27	<1	1130	65	19	<1	<1	<1	1	60
Doliolida										
Doliolid sp.	CH 17 I Kt9	<1	27	70	8	<1	10	<1	1	220
Phaeophyceae										
Fucaceae										
Sargassum sp.	36b	<1	120	760	75	<1	14	1	3	980
Sargassum sp.	973	<1	18	660	170	<1	<1	8	3	31
Sargassum sp.	973-O	<1	33	1600	90	<1	8	1	9	19
Sargassum sp.	973-Y	<1	90	1710	140	<1	9	1	6	60
Crustacea										
Amphipoda										
Euthemisto sp.	No. 73	1	60	110	14	<1	3	3	2	42
Parathemisto gaudi	68	<1	495	360	31	<1	10	<1	<1	170
Phronemia sp.	CH 17 1 Kt9	<1	12	170	155	<1	28	<1	<1	220
Copepoda										
Copepods (mixed)	78	<1	120	125	6	<1	6	<1	3	220
Centropages typicus and *hamatus*	76	<1	1435	120	16	<1	210	<1	12	375
Centropages typicus and small phytoplankton	87	<1	7155	215	155	<1	34	<1	12	75

Marine Plankton Species (ppm in Ash)[a]

Fe	Ga	La	Li	Mn	Mo	Ni	Pb	Rb	Si	Sn	Sr	Ti	V	Y	Zn	Zr
55	<1		23		3	5	9	<100	150	<5	125		<5		55	
60	<1	<20	33	<1	<1	<1	<1	7	55	<5	<200	15	<5	<20	110	<1
150	<1	<20	55	9	<1	3	4	<20	2400	<5	335	<5	11	<20	35	5
350	<1		45		5	65	<1	<100	145	<5	145		1		40	
140	<1	<20	16	7	<1	14	<1	9	470	<5	<200	30	16	<20		<1
13	<1	<20	48	2	<1	<1	6	<20	85	<5	275	7	<5	<20	65	4
10	<1	<20	38	<1	<1	<1	<1	8	37	<5	250	<5	<5	<20	<1	<1
10	<1	<20	39	<1	<1	2	<1	8	18	<5	>200	<5	<5	<20	<1	<1
12	<1	<20	42	3	<1	4	<1	8	70	<5	295	<5	<5	<20	120	<1
2550	1	<20	3 3	67	2	5	3	<20	<5000	<5	245	190	<5	<20	125	22
725	<1	<20	5 5	19	<1	6	6	10	<5000	<5	>200	38	<5	<20	90	<1
1825	<1	<20	5 5	47	<1	8	<1	10	<5000	<5	>200	165	<5	<20	55	19
5665	2	<20	7 7	220	<1	13	<1	12	<5000	<5	>200	270	6	<20	120	9
730	<1	<20	7 7	19	<1	8	<1	10	<5000	<5	>200	55	<5	<20	170	1
215	<1	<20	7	23	<1	4	7	<20	1685	<5	435	20	<5	<20	55	6
550	<1	<20	7	70	6	6	445	14	>1000	<5	275	44	<5	<20	120	<1
30	<1	<20	55	1	<1	9	2	<20	255	<5	955	14	16	<20	30	7
1125	<1	<20	6	140	<1	19	95	37	8100	<5	5350	42	21	<20		1
65	<1	<50	13	95	5	18	105	90	345	<5	12850	29	<5	<20		19
290	<1	<50	3	130	<1	36	260	40	550	<5	7580	16	<5	<20		2
710	<1	<50	3	125	<1	60	380	50	455	<5	12330	19	<5	<20		2
1045	2	<50	2	46	<1	4	38	<20	1770	<5	1030	<5	<5	<20		14
1455	<1	<20	38	115	<1	22	30	<10	1580	<5	<200	31	<5	<20		<1
345	<1	<20	40	19	<1	17	9	<10	685	<5	1035	18	<5	<20		1
265	2	<20	68	24	<1	13	11	<20	1860	<5	405	325	<5	<20		13
3685	2	<20	19	90	9	17	85	<20	<5000	<5	340	365	<5	<20		18
1000	<1	<20	14	170	<1	25	53	<10	<5000	<5	13320	<500	<5	<20		16

(*Continued*)

Table 6.5 Elemental Composition of

		Ag	Al	B	Ba	Bi	Cd	Co	Cr	Cu
Decapoda										
Pandalid shrimp	86	2	10	75	10	<1	8	4	2	540
Acanthephyra sp.	CH 17 1 Kt*	9	23	135	26	<1	34	4	2	440
Acanthephyra multispina	69	4	10	175	22	<1	29	1	3	380
Paleomonetes sp.	71	4	2605	185	32	<1	15	<1	2	325
Euphausiacea										
Euphausia krohnii	70	1	315	110	20	<1	11	<1	1	255
Euphausids	25b	<1	100	110	22	<1	12	18	68	370
Meganyctiphanes norwegica	CH 17 St. 3	<1	<10	130	29	<1	17	<1	1	275
Meganyctiphanes norwegica	79	<1	170	115	16	<1	<1	5	6	190
Meganyctiphanes norwegica	74	1	125	150	18	<1	<1	2	2	260
Lopoda										
Idotea (fresh specimen)		7	65	20	17	<1	3	2	1	105
Mollusca										
Cephalopoda										
Loligo pealii	9	14	29	43	3	<1	4	<1	<1	580
Squid eggs	37a	<1	620	90	5	<1	2	<1	1	105
Squid eggs	37b	1	590	85	5	<1	<1	<1	<1	120
Pteropoda										
Limacina sp.	72	<1	125	13	15	<1	<1	5	<1	18
Limacina retroversa	75	<1	210	20	13	<1	3	7	13	28
Diacria trispinosa	10	1	40	20	16	<1	7	1	<1	12
Clione linacina	77	1	170	65	13	<1	17	<1	15	485

[a] From Thompson *et al.*[16]

(<1–100 mg); (3) the matrices in which the elements are present are complex and variable both with respect to major and trace element compositions; (4) the heterogeneous nature of the material requires careful sampling and preparation to ensure that the analyzed sample is representative and free from contamination; and (5) many of the elements are present in vanishingly low concentrations (<1 ppb–100 ppm).

These considerations favor spectroscopic instrumental analysis, and in the Chemistry Department at Woods Hole we combine several methods. For extremely small samples (<10 mg) we prefer neutron activation with λ spectrometry; a limited number of elements, those requiring high precision and accuracy, are better suited to atomic absorption and/or spectrophotometry. For the majority of samples and elements, we rely on direct-reading emission spectrometry.

The emission technique we have developed is for powdered ash samples and dc arc excitation; see Thompson and Bankston.[21] Methacrylate or agate grinding vials are used to avoid contamination.[22] Because of the large number of elements we determine, and the wide range of volatilities in the dc arc, we chose

Marine Plankton Species (ppm in Ash)[a] (*Continued*)

Fe	Ga	La	Li	Mn	Mo	Ni	Pb	Rb	Si	Sn	Sr	Ti	V	Y	Zn	Zr
325	<1	<50	9	36	4	4	7	<20	535	<5	1060	<5	<5	<20		13
485	<1	<50	12	42	4	8	45	<20	640	<5	1425	33	<5	<20		12
350	<1	<20	3	50	<1	4	33	<20	1485	<5	710	<5	<5	<20		<1
2305	<1	<20	4	55	<1	20	5	<20	>2000	<5	>200	285	<5	<20		42
770	<1	<20	120	60	<1	20	25	<10	2515	<5	345	45	<5	<20		1
585	<1	<20	16	40	<1	15	62	<10	1575	68	790	15	<5	<20		<1
380	<1	<20	175	25	<1	10	59	<10	1635	<5	1335	4	<5	<20		2
645	<1	<20	5	54	4	5	72	<20	>2000	<5	725	4	<5	<20		2
1225	<1	<20	6		3	6	70	<20	2950	<5	1000		<5	<20		13
425	<1	<50	47	31	5	15	26	16	>2000	<5	2755	40	<5	20		<1
110	<1	<20	6	29	<1	3	14	60	114	<5	135	4	<5	<20	375	10
480	<1	<20	2	25	<1	130	3	<20	>2000	<5	260	65	10	<20	105	29
280	<1	<20	2	12	<1	5	4	<20	>2000	<5	255	75	<5	<20		6
1385	<1	<50	21	35	3	9	12	<20	1705	<5	1025	53	<5	<20		15
1185	<1	<50	6	38	5	9	65	<20	>2000	<5	1025	130	<5	<20		15
175	3	<50	<5	<10	7	18	65	<20	1010	<5		<10	<5	<20		<1
1215	<1	<20	9	34	21	17	50	13	630	<5	320	16	9	<20		<1

Table 6.6 Concentration Factors for Some Metals in Commonly Eaten Shellfish[a,b]

Element	Scallop	Oyster	Mussel
Ag	2,300	18,700	330
Cd	2,260,000	318,000	100,000
Cr	200,000	60,000	320,000
Cu	3,000	13,700	3,000
Fe	291,000	68,200	196,000
Mn	55,500	4,000	13,500
Mo	90	30	60
Ni	12,000	4,000	14,000
Pb	5,300	3,300	4,000
V	4,500	1,500	2,500
Zn	28,000	110,300	9,100

[a] From Brooks and Rumsley.[54]
[b] Element concentration in the organism/element concentration in seawater.

a split-burn technique with buffered samples. The elements currently determined (on a 1.5-m Paschen–Runge grating-mount direct-reading spectrometer) are Ag, Al, As, B, Ba, Be, Bi, Ca, Cd, Co, Cr, Cu, Fe, Ga, Hg, In, K, La, Li, Mg, Mn, Mo, Na, P, Pb, Pd, Rb, Sb, Si, Sn, Sr, Te, Ti, V, Y, Zn, and Zr (see Fig. 6.2). For standardization we prepare artificial mixes as close to the composition of the natural samples as possible. This means that we need a wide range of standards, both in composition and elemental concentrations (see Table 6.7). The speed of analysis with the direct reader, the speed of data processing—we have developed our own computer program for this[23]—plus the improved accuracy from using standards similar to the samples offset the disadvantages of working with multiple standards. The data in Table 6.5 were obtained using our direct-reading emission spectrometric technique. Other published techniques used in medical agriculture or geologic research are applicable also, for example Ahrens and Taylor,[24] Boumans,[25] Helz,[26] Mitchell,[27] Goles *et al.*,[28] Bedrosian *et al.*,[29] Fassel and Kniseley,[30] and methods reported in the ASTM Committee E-2 reports published by the American Society for Testing and Materials.

A detailed discussion of the elemental composition of marine organisms and some of the analytical techniques used can be found in Vinogradov,[31] Goldberg,[32] and Martin and Knauer.[33] The elemental composition of

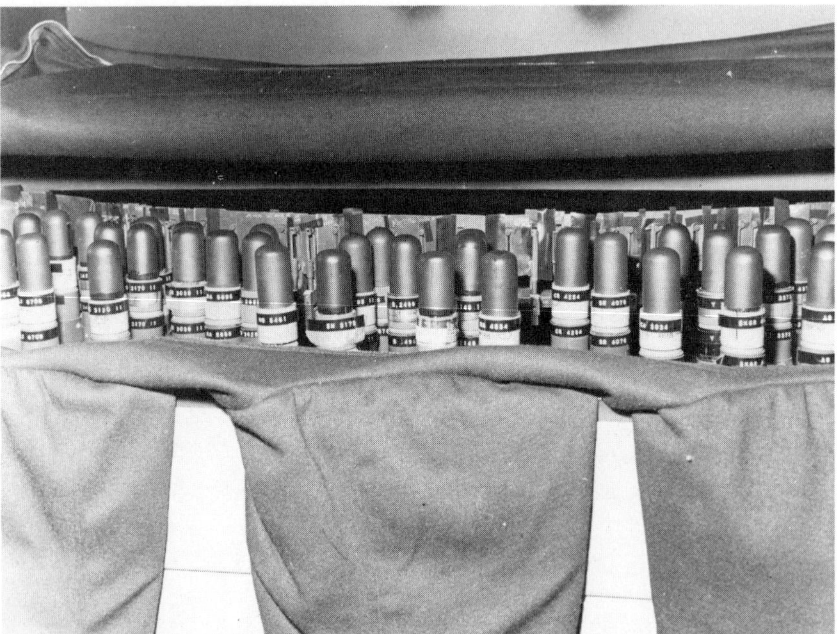

Fig. 6.2. Crowded interior of the direct-reading spectrometer. There are 43 exit slits and photomultiplier tubes to use in the analytical programs for seawater, marine organisms, and rocks and sediments from the sea floor.

Table 6.7 Some Matrices Used in Standardization with Emission Spectrometric Analysis of Marine Organisms

Organism	Matrix composition (wt%)
Calcareous (*Foraminifera*-type)	$CaCO_3$ 100%
Siliceous (*Radiolaria*-type)	SiO_2 100%
Ctenophore (*Mnemiopsis*-type)	NaCl 90.66%; MgO 6.90%; KCl 2.44%
Cephalopod (*Loligo*-type)	NaCl 39.70%; KCl 36.35%; $Ca_3(PO_4)_2$ 23.95%
Composite plankton	$CaCO_3$ 60.0%; Al_2O_3 10.0%; KCl 3.0%; Na_2CO_3 20.0%; SiO_2 5.0%; MgO 2.0%

marine organisms reflects a multitude of complex controlling factors involving the composition of the medium (seawater), taxonomy (kind of organism), food supply, method of feeding, physiology, ecology, and toxicity tolerances. Although there is a relationship between the metal concentration in the medium and in simple plant organisms–the phytoplankton,[(33)]–this will vary depending on the presence of organic chelators, which alter the activity of the free metal ion and thus its "availability." Zooplankton, organisms grazing on the plants or eating other organisms, show much greater interspecies differences. One of the factors is the response of organisms to varying concentrations of metals. The degree of toxicity varies with the metal species and the organism, but in general the observed order of toxicity is Hg > Ag > Cu > Cd > Zn > Pb > Cr > Ni > Co. The lethal-dose concentration can be measured experimentally; see the data in Table 6.8. However, sublethal effects of exposure to heavy metals can be most marked

Table 6.8 Approximate Static 48-h 15°C LD-50 Value for Toxicity of Heavy Metals in Seawater to Various Marine Organisms[a]

	Metal concentration (mg/liter)			
Organism	Hg	Cu	Zn	Ni
Pink shrimp (*Pandalus montagui*)	0.1	0.2	10.0	200.0
Brown shrimp (*Crangon crangon*)	6.0	30.0	100.0	150.0
Shore crab (*Carcinus maenus*)	1.0	100.0	12	300.0
Cockle (*Cardium edule*)	10.0	1.0	200	500.0

[a]From Portmann.[(55)]

and deleterious and may involve growth and reproductive inhibition or cessation. The uptake of metals can be by various routes (e.g., absorption or ingestion) and can be regulated by various metabolic processes (e.g., secretion, excretion, or even storage). These all vary from species to species, and no simple conclusions can be made. A review of this subject can be found in Bryan[34] and Eisler.[35]

6.4 MARINE SEDIMENTS

Marine sediments are mainly composed of inorganic aluminosilicate minerals brought into the sea by rivers, glaciers, winds, and volcanic eruptions, and by biogenous skeletons and shells (mainly calcium carbonate and silica) that reach the bottom when organisms die. In addition, new minerals may be formed on the sea floor from certain metal salts precipitating from seawater. The most abundant sediments and the range and complexity of their major element composition are indicated in Table 6.9. Any given sample may consist of a mixture of these three principal types, and the proportions may vary greatly with areal distribution and depth of the deposit.

Analytical studies of marine sediments contribute to the understanding of pathways and residence times of elements in the oceans, the formation and distribution of various sediments, and the possibilities of discovering and mapping economic mineral deposits. In shallow water on the continental shelves, for example, are exploitable deposits of sand and gravel, calcareous shell depos-

Table 6.9 Chemical Composition of the Three Principal Types of Deep-Sea Sediments (wt%)[a]

Composition	Red clay	Calcareous ooze	Siliceous ooze
SiO_2	53.93	24.23	67.36
TiO_2	0.96	0.25	0.59
Al_2O_3	17.46	6.60	11.33
Fe_2O_3	8.53	2.43	3.40
FeO	0.45	0.64	1.42
MnO	0.78	0.31	0.19
CaO	1.34	0.20	0.89
MgO	4.35	1.07	1.71
Na_2O	1.27	0.75	1.64
K_2O	3.65	1.40	2.15
P_2O_5	0.09	0.10	0.10
H_2O	6.30	3.31	6.33
$CaCO_3$	0.39	56.73	1.52
$MgCO_3$	0.44	1.78	1.21
C	0.13	0.30	0.26
N	0.016	0.017	–

[a]From El Wakeel and Riley.[56]

its of value in cement and chemical industries, and deposits of phosphates, $Ca_3F(PO_4)_3$, for fertilizer and chemical uses.

Other fascinating and, if recovery techniques are perfected, commercially viable deposits that occur on the sea floor are the "manganese nodules." These ores are actually composed of oxides and hydroxides of iron and manganese precipitated from seawater. They often contain variable but economically worthwhile amounts of such metals as Co, Ni, Cu, Mo, Zn, and Pb. Deposits occur as earthy, black, round-to-ovoid nodules from pea to cannonball size, as coatings up to several centimeters thick on other rocks, or as large slabs or pavements over 1 m in length. The Blake Plateau off the eastern United States has 3000 square miles of ferromanganese-covered floor; the Pacific Ocean has an estimated 1.5 trillion tons of nodules in surface deposits.

The composition of the nodules varies with geographical distribution and water depth; Co-rich nodules are found in the mid-Pacific; Co-poor, Mn-rich ones are found near shore (see Table 6.10). Atlantic Ocean deposits close to the continents are relatively impoverished in Ni, Cu, and Co. Commercial mining of the Co-, Cu-, and Ni-rich nodules of the central Pacific is now feasible, and companies are already testing various mining methods. A detailed discussion of these deposits and the factors controlling their distribution, composition, and growth can be found in the Reports of the Inter-University Program of Research on ferromanganese deposits.[36]

Collection and analyses of the different marine sediments, with their complex and variable compositions, is no easy or routine task. Although good major-element analyses can be done by classical gravimetric and volumetric techniques, these are neither rapid nor suited for large numbers of samples. The heterogeneity of the samples and the need to plot their distributions both spatially and in time often necessarily involve numerous samples and favor rapid instrumental analyses.

Table 6.10 Average Composition of Manganese Nodules from the Pacific Ocean (wt%)[a]

Element	Min	Max	Mean
Fe	2.4	26.6	14.0
Mn	8.2	52.2	24.2
Ba	0.08	0.64	0.18
Co	0.014	2.3	0.35
Cu	0.028	1.6	0.53
Pb	0.02	0.36	0.09
Mo	0.01	0.15	0.05
Ni	0.16	2.0	0.99
Ti	0.11	1.7	0.67
V	0.021	0.11	0.09
Zn	0.04	0.08	0.05

[a] From Skornyakova *et al.*[57]

X-ray fluorescence is nicely suited for the major elements. Being nondestructive, it allows trace-element determinations to be made on the same aliquot by the same or other techniques. As this method becomes more popular, we are beginning to cope with matrix effects, even in these multielement silicate samples, and are able to apply theoretical or empirical corrections. Good examples of the application to major element analyses, and for certain trace elements, are described by Norrish and Hutton,[(37)] Leake *et al.*,[(38)] and Brown *et al.*[(39)]

In our laboratory we use pressed powder discs for both major and trace-element analyses by x-ray fluorescence spectrometry and use only slight modifications of the technique described by Brown *et al.*[(39)] (see Tables 6.11 and 6.12 for instrumental parameters and results). In those cases where we have only small amounts of material, we have used the technique of Rose *et al.*[(40)] for major

Table 6.11 Instrumental Parameters for X-Ray Fluorescence Analyses of Powdered Geologic Materials

Element	Analyzing crystal	X-ray tube Target	X-ray tube kV	X-ray tube mA	Line	Peak angle
Major						
K	PE	Cr	60	24	Kα	50.67
Fe	LiF (200)	Cr	60	24	Kα	57.51
Ti	LiF (200)	Cr	60	24	Kα	86.12
Si	Pe	Cr	60	24	Kα	109.21
Ca	LiF (200)	Cr	40	8	Kα	113.07
Al	PE	Cr	60	24	Kα	145.07
Mg	ADP	Cr	50	40	Kα	136.69
P	Ge	Cr	50	40	Kα	89.51
Mn	LiF (200)	W	60	32	Kα	95.49
Na	ThAP	Cr	40	50	Kα	54.27
Trace[a]						
V	LiF (200)	W	55	40	Kα	123.18
Cr	LiF (200)	W	55	40	Kα	69.34
Ce	LiF (200)	W	55	40	Lβ	71.59
Nd	LiF (200)	W	55	40	Lα	72.08
Ba	LiF (200)	Mo	55	40	Lβ	79.18
Rb	LiF (200)	Mo	55	40	Kα	38.06
Y	LiF (200)	Mo	55	40	Kα	33.94
Zr	LiF (200)	W	55	40	Kα	32.13
Sr	LiF (200)	W	55	40	Kα	35.89
La	LiF (200)	W	55	40	Kα	82.87
Zn	LiF (200)	W	55	40	Kα	60.70
Ni	LiF (200)	W	55	40	Kα	71.44
Cu	LiF (200)	Mo	55	40	Kα	45.02
Sc	LiF (200)	Cr	55	40	Kα	97.69

[a]When the major-element composition of the samples is not known, we use the techniques of Reynolds[(58, 59)] involving the Compton scatter as a measure of the relative absorption.

elements, involving digestion of the sediments in acid and absorption on cellulose. To help understand the rates of sedimentation and relative inputs of various materials, we need to know the composition of the fine particulate matter presently in the water column and in the process of being "sedimented" to the ocean floor. Analyses of this particulate matter, milligram amounts on fine filter papers, can be done very nicely using thin-film x-ray fluorescence techniques, even for determining minor element contents. We use the technique of Price and Angell[(41)] using natural reference rock powders pressed on to adhesive strips for standardization. Instrumental conditions for analysis of a number of metals are shown in Table 6.13 with results for Sr analyses of selected reference materials. A similar technique for major and trace elements has also been reported by Cann and Winter.[(42)]

Emission spectroscopy is also widely favored as an analytical tool. For major elements we prefer spark excitation because we require precision and accuracy better than ±5%. For reasons of homogenization, reduction of matrix effects, and better standardization, we have developed a lithium borate fusion process followed by dissolution of the fusion mix in dilute nitric acid appropriately spiked with internal standards. These solutions are sparked in a vacuum-cup electrode and can also be aspirated, after further dilution, into the atomic absorption spectrometer for determinations of such elements as Na, K, Ca, Mg, and Sr.[(43)] This is similar to the techniques described by other workers.[(44, 45)] We are presently experimenting with the plasma jet excitation of these solutions in conjunction with an emission spectrometer for both major- and trace-element analyses.

Preparation of the samples for analysis whether by x-ray or emission spectrometry necessarily involves a crushing and grinding operation to ensure uniform grain size and homogeneity. The aliquots taken for analysis must be representative of the whole. Techniques using very small aliquots of powder can lead to vary large sampling errors; this has been well demonstrated for spark-source mass spectrometry.[(46)] We routinely grind our samples until all passes through a 47-μm (200-mesh) nylon sieve. We have exhaustively tested our grinding and sieving procedures for contamination and nonrepresentative sampling.[(32)] We have found that tungsten carbide devices introduce large amounts of Co and Ti (from the bonding agent); alumina mortars introduce Al, Cr, Fe, Ga, and Zr; ceramic grinding vials introduce Al, Cu, Fe, Ga, Li, Ti, B, Ba, Co, Mn, Zn, and Zr. Boron carbide mortars (except for B), agate grinding devices, and methacrylate (Lucite) vials introduce little or no contamination. Sifting with nylon sieves is recommended; brass or stainless steel introduces appreciable levels of Co, Cu, Fe, Mn, Ni, Pb, Sn, and Zn.

Marine sediments generally contain anywhere from 20 to 70 elements in the range <1–1000 ppm. The concentration of these trace elements varies greatly depending on the type of sediment and environment of deposition (Table 6.14). X-ray fluorescence, neutron activation, and atomic absorption spectroscopy are commonly used analytical techniques. We have chosen dc arc excitation and

Table 6.12 Major-Element Analyses of U.S. Geological Survey Standard Rocks by X-Ray Fluorescence Technique Compared with Recommended Values[a]

	G-1		W-1		GSP-1		AGV-1	
Oxide	Found	Recommended	Found	Recommended	Found	Recommended	Found	Recommended
SiO_2	73.37	72.64	51.97	52.64	67.72	67.38	60.54	59.00
Al_2O_3	13.92	14.04	14.88	15.–0	15.18	15.25	17.03	17.25
Fe_2O_3	1.91	1.94	11.07	11.09	4.40	4.33	7.11	6.76
MgO	0.28	0.38	6.93	6.62	1.04	0.96	1.60	1.53
CaO	1.36	1.39	11.04	10.96	2.07	2.02	5.02	4.90
Na_2O	3.48	3.32	1.95	2.15	3.08	2.80	4.08	4.26
K_2O	5.36	5.48	0.68	0.64	5.49	5.53	2.90	2.89
TiO_2	0.24	0.26	1.07	1.07	0.68	0.66	1.12	1.04
MnO	0.02	0.03	0.19	0.17	0.02	0.04	0.10	0.097
P_2O_5	0.06	0.09	0.22	0.14	0.32	0.28	0.50	0.49

[a]From Brown *et al.*(39) Recommended values from Flanagan.(60)

Table 6.13 Instrumental Conditions for Thin-Film X-Ray Analysis of Certain Metals[a]

Element	Line	Peak (Å)	Scatter background (Å)	Crystal	Collimator	kV	mA	Sr analyses: Sample	Rock type	Thin-film x-ray fluorescence			Mean	Recommended value[b]
Sr	Kα	0.875	±0.030	LiF (220)	Coarse 480	55	28	USGS G-1	Granite	260	250		255	250
Rb	Kα	0.926	±0.026	LiF (220)	Coarse 480	55	28	USGS W-1	Diabase	180	190		185	190
Ba	Kα	0.385	±0.420	Topaz	Fine 160	50	20	USGS AGV-1	Andesite	740	750	750	745	657
Ni	Kα	1.658	±1.618	LiF (220)	Coarse 480	55	28	USGS G-2	Granite	550	550	540	545	479
Zn	Kα	1.435	±1.405	LiF (200)	Fine 160	50	20	USGS BCR-1	Basalt	340	340		340	330
Th	Kα	0.956	±1.017	LiF (200)	Fine 160	60	20	NBS 1a	Limestone	2000	1970	1930	1970	1945
Pb	Kα	1.175	±1.192	LiF (200)	Fine 160	50	20	NBS 1014	Portland cement	2140	2140	2120	2130	2200
Zr	Kα	0.786	±0.810	LiF (200)	Fine 160	60	20							

[a] From Price and Angell.[41]
[b] From Flanagan[60] for U.S. Geological Survey (USGS) materials, and for National Bureau of Standards (NBS) materials, certified values for Sr.

Table 6.14 Average Trace-Element Content of Sediments from Various Marine Environments (ppm)[a]

Element	Near shore	Deep sea	Manganese nodule
Ba	750	2237	3100
Co	13	116	3400
Cr	100	77	10
Cu	48	570	3300
Ga	19	20	17
Ni	55	293	5700
Pb	20	162	1500
Sn	21	20	300
Sr	250	587	1000
V	130	330	590
Zr	160	145	340

[a]From Chester.(49)

direct-reading emission spectrometry as our main workhorse, using a method similar to that outlined for marine organisms. For the "involatile" elements (see Table 6.15) we use a buffered charge (1 part sample or standard to 3 parts graphite) with Pd as the added internal standard. The charge is pressed into a pellet (3.175 mm in diameter, 4.72 mm in length) and placed in "necked" crater electrodes (ASTM S-15). Arcing (anode excitation) is at 10 A for 60 s in a stallwood jet using 80% Ar, 20% O_2, flowing at 4 liters/min. For the "volatile" elements, 1 part sample is mixed with 2 parts of a sodium carbonate–graphite buffer (1 part Na_2CO_3 to 3 parts carbon); indium is the added internal standard. The charge is packed in center-post electrodes (Ringsdorf RW0089) and arced at 6 A for 60 s in air. Details of the technique have been documented.[21] We obtain precision and accuracy of the order ±10% for most elements and sensitivity to 1 ppm or less; see Thompson *et al.*(47, 48) Standardization is from artificial mixes or, preferably where they exist, natural silicate or carbonate reference samples. To offset the disadvantages of the direct reader with its necessarily limited number of elements and wavelengths (Table 6.15) we have set up a 3-m littrow-type spectrograph so that light from a single excitation can be read by both instruments simultaneously (Fig. 6.3). The spectrograph provides a permanent record, and we can identify or measure wavelengths and elements not available on the spectrometer. The use of emission spectrometric techniques also allows us to analyze small aliquots–single phases such as individual shells–as well as bulk analyses of the homogenized sediment (Table 6.16). A detailed discussion of the analyses and composition of marine deposits can be found in Chester(49) and Riley and Chester.(50)

Understanding the geochemistry of deep-sea sediments does not end with the

Table 6.15 Analytical Line Wavelengths, Exit Slits, and Registration Details with Corresponding Internal Standard and Background-Correcting Wavelengths[a]

Element	Wavelength (Å)	Order	Slit and refractor plate[b] (μm)	Photomultiplier tube type	Internal standard	Background correcting position
Involatile						
Al	3961.53	2nd	75G[c]	931A	Pd 3481.15	3039.36, 4027.9
B	2497.73	2nd	10C	R106	Pd 3481.15	3039.36, 4027.9, 3130.42
Ba	4554.04	1st	75G	931A	Pd 3481.15	4027.9
Co	3453.50	2nd	75C	R106	Pd 3481.15	3039.36, 4027.9
Cr	4254.34	1st	75G	931A	Pd 3481.15	4027.9
Fe	2599.39	1st	75Q	IP28	Pd 3481.15	3039.36, 4027.9
La	3988.52	1st	75Q	R106	Pd 3481.15	3039.36, 4027.9
Mn	2576.10	2nd	75C	R212	Pd 3481.15	3039.36
Mo	3170.35	2nd	75C	R106	Pd 3481.15	3039.36
Ni	3414.76	2nd	75C	R106	Pd 3481.15	3039.36, 4027.9, 3130.42
Si	2881.58	2nd	75C	R106	Pd 3481.15	3039.36
Sr	3464.46	1st	75Q	R106	Pd 3481.15	3039.36
Sr	4077.71	1st	75G	931A	Pd 3481.15	4027.9
Ti	3349.03	1st	75Q	R106	Pd 3481.15	3039.36, 4027.9
V	4379.24	1st	75Q	R106	Pd 3481.15	3039.36, 4027.9, 3130.42
Y	3242.28	1st	75Q	R106	Pd 3281.15	3039.36
Zr	3391.97	2nd	75C[c]	R106	Pd 3481.15	3039.36, 4027.9
Volatile						
Ag	3280.86	1st	75Q	IP28	In 3039.36	3481.15, 4027.9
Bi	3067.72	1st	75Q	IP28	In 3039.36	3481.15
Cd	2288.02	1st	75Q	R106	In 3039.36	2385.76, 2068.38, 2349.84
Cu	3273.96	2nd	75C	R106	In 3039.36	3481.15
Ga	2943.64	1st	75Q	R106	In 3039.36	3481.15
Li	3932.61	2nd	75C	931A	In 3039.36	3481.15, 4027.9
Li	6707.84	1st	75C[c]	931A	In 3039.36	3481.15
Pb	2833.07	1st	75Q	R106	In 3039.36	3481.15
Rb	7800.23	1st	75U	R136	In 3039.36	3481.15, 4027.9
Sn	3175.02	1st	75C[c]	R106	In 3039.36	3481.15, 4027.9
Zn	2138.55	1st	75C[c]	R106	In 3039.36	2068.38, 2385.76, 2349.84
Zn	4810.53	1st	75G	R106	In 3039.36	4027.9

[a] From Thompson and Bankston.(21)
[b] G, glass; U, uranium glass; C, corex; Q, quartz.
[c] Horizontal mirror used to deflect light to photomultiplier tube.

bulk analysis of an homogenized aliquot. The elemental composition is governed by, and thus we must have information on, the following:

1. The relative proportions of the minerals and biogenic skeletons, their distribution, and their composition.

2. The paths by which elements and phases are introduced into the marine environment (i.e., the mechanisms and fluxes of element supply to the oceans).

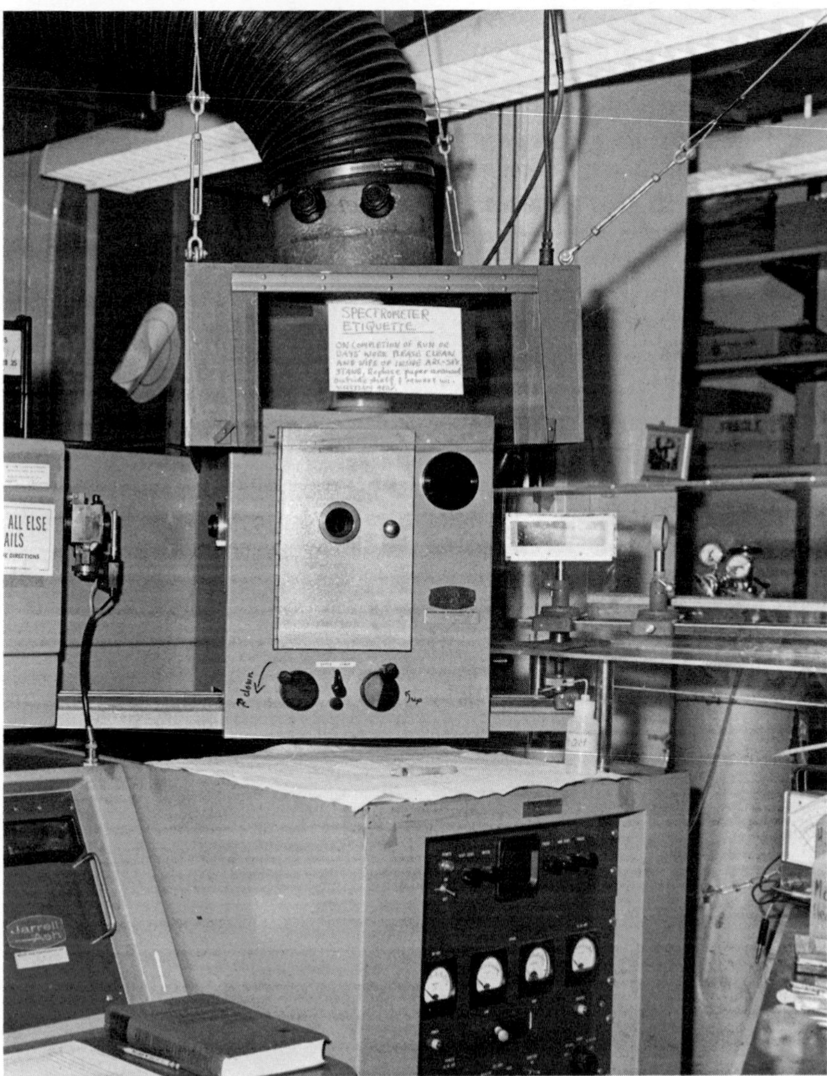

Fig. 6-3. Excitation stand for the emission spectrometer. The emitted light passes to the direct-reading spectrometer to the left; at the same time light is passed via the mirror and lens system to the right into a littrow-type spectrograph.

Table 6.16 Concentration of Some Elements in Carbonate Shell Material and Deep-Sea Sediments (ppm)[a]

Element	Average of foraminiferal shells from deep-sea sediments	Pteropod shell	Coccolith ooze	Average deep-sea carbonate sediments	Average deep-sea clay sediments
Sr	1,112	–	1,468	2,000	180
Ba	10–30	–	175	190	2,300
Mg	1,417	–	1,100	4,000	21,000
Fe	1,213	–	–	9,000	65,000
Mn	335	–	263	1,000	6,700
Cr	–	1	5	11	90
V	15	85	–	20	120
Ni	21	2	4	30	225
Co	–	20	4	7	74
Cu	23	30	13	30	250
Pb	138	200	–	9	80
Ti	525	5	113	770	4,600
Al	2,522	–	–	20,000	84,000
Si	7,536	–	–	32,000	250,000

[a]From Riley and Chester.(50)

3. The mechanisms by which the elements are incorporated into the sediments—the relative fluxes from organism dissolution, precipitation or adsorption from seawater, and diagenetic reactions in the sediments.

4. The rates of accumulation of various phases as a function of time and place, usually from radioisotopic measurements.

As yet, not all of these are fully understood or documented. However, our present thinking is that the deep-sea sediments are not an ultimate "sink." Elements are transferred into and out of the sea floor. Indeed, much of our current thinking and research is directed toward the hypothesis that the dissolved species in seawater are in equilibrium with the solid phases of the sediments. It is more likely, however, that the composition of seawater is in a steady state and controlled by a chemical mass balance between solid and dissolved phases.(51) At the present time we are actively studying the role of igneous volcanic rocks and their interaction with seawater in controlling seawater composition and the formation of ores in the marine environment.(52)

6.5 SUMMARY

The wide variety, heterogeneity, and complexity of samples the marine chemist may be asked to analyze present a continuing challenge to his analytical abilities. To meet the challenge, spectroscopy, and particularly emission and x-ray, have proved to be valuable and widely adopted techniques. The key to successful analysis in marine chemistry is to be flexible and adaptable in methods of sample collection, handling, and preparation, and in the choice of analytical technique. Every sample being unique, there is no place for fixed-mode, routine procedures.

More and more we recognize that it is on the ocean that we will depend for many of our future resources. As we do, we should realize that, although bountiful, it is not limitless and, moreover, it is no longer a virgin frontier. By our own neglect we are seriously polluting the oceans in many respects, thus not only sacrificing part of the potential of its resources, but increasing the cost of effectively harvesting these resources.

The only way to ensure wise exploitation of the ocean is through the acquisition of knowledge about all phases of marine life and ocean processes. The Marine Science Commission has summarized this very succinctly in a recent report to the President: "There is much to be learned about this planet Earth, and many keys to learning are in and under the sea. The total body of oceanic knowledge is advanced best by the pursuit of fundamental understanding of the biological, physical, geological and chemical characteristics of the oceans. Continuing and substantial support of basic marine science is a national investment which will provide an underpinning for all future activities in the sea."

6.6 REFERENCES

1. E. D. Goldberg, W. S. Broecker, G. M. Gross, and K. K. Turekian, in: *Radioactivity in the Marine Environment*, National Academy of Sciences, Washington D.C. (1971), p. 137.
2. L. G. Sillen, in: *Oceanography*, American Association for the Advancement of Science, Bailey, London (1961), p. 549.
3. T. R. S. Wilson, in: *Chemical Oceanography* (J. P. Riley and G. Skirrow, eds.), 2nd ed., Vol. 1, Academic Press, Inc., New York (1975), p. 365.
4. W. F. McIlhenny, in: *Chemical Oceanography* (J. P. Riley and G. Skirrow, eds.), 2nd ed., Vol. 4, Academic Press, Inc., New York (1976), p. 49.
5. D. E. Robertson, *Anal. Chim. Acta 42*, 533 (1968).
6. D. F. Schutz and K. K. Turekian, *Geochim. Cosmochim. Acta 29*, 259 (1965).
7. P. G. Brewer, in: *Chemical Oceanography*, (J. P. Riley and G. Skirrow, eds.), 2nd ed., Vol. 1, Academic Press, Inc., New York (1975), p. 415.
8. J. P. Riley, in: *Chemical Oceanography*, (J. P. Riley and G. Skirrow, eds.), 2nd ed., Vol. 3, Academic Press, Inc., New York (1975), p. 327.
9. D. W. Spencer and P. Brewer, in: *CRC Critical Reviews in Solid State Sciences* (R. W. Hoffman and D. E. Schuele, eds.), Vol. 3, CRC Press, Cleveland (1970), p. 409.
10. E. Boyle, Ph.D. thesis, Massachusetts Institute of Technology–Woods Hole Program (1976), 156 pp.
11. A. Q. Morris, *Anal. Chim. Acta 42*, 397 (1968).
12. P. G. Brewer and D. W. Spencer, Woods Hole Oceanographic Institution Technical Report 70–62 (1970).
13. D. B. Carlisle, *Nature 181*, 922 (1958).

14. E. P. Levine, *Science 133*, 1352 (1961).
15. G. D. Nicholls, H. Curl, and V. T. Bowen, *Limnol Oceanog. 5*, 472 (1960).
16. G. Thompson, V. T. Bowen, H. Curl, and G. D. Nicholls, *U.S. At. Energy Comm. Rept. NYO-2174-60* (1967).
17. V. T. Bowen, J. S. Olsen, C. L. Osterberg, and J. Ravera: in *Radioactivity in the Marine Environment*, National Academy of Sciences, Washington, D.C. (1971), p. 200.
18. L. T. Kurland, S. N. Faro, and H. S. Siedler, *World Neurol. 1*, 320 (1960).
19. M. Merlini, in: *Impingement of Man on the Oceans*, John Wiley & Sons, Inc. (Interscience Division), New York (1971), p. 461.
20. T. T. Gorsuch, *Analyst 87*, 112 (1962).
21. G. Thompson and D. C. Bankston, *Spectrochim. Acta 24B*, 335 (1969).
22. G. Thompson and D. C. Bankston, *Appl. Spectrosc. 24*, 210 (1970).
23. G. Thompson, F. T. Manheim, and K. Paine, *Appl. Spectrosc. 23*, 264 (1969).
24. L. H. Ahrens and S. R. Taylor, *Spectrochemical Analysis*, Addison-Wesley Publishing Company, Inc., Reading, Mass. (1961), 454 pp.
25. P. W. J. M. Boumans, *Theory of Spectrochemical Excitation*, Plenum Press, New York (1966), 383 pp.
26. A. W. Helz, *U.S. Geol. Surv. Prof. Paper 475-D* (1964), p. 176.
27. R. L. Mitchell, *Commonwealth Bur. Soil Sci. (Gr. Brit.) Tech. Commun. 44* (1948), 183 pp.
28. G. E. Goles, K. Randle, G. G. Goles, J. B. Corliss, M. H. Beeson, and S. S. Oxley, *Geochim. Cosmochim. Acta 32*, 369 (1968).
29. A. J. Bedrosian, R. K. Skogerboe, and G. H. Morrison, *Anal. Chem. 40*, 854 (1968).
30. V. A. Fassel and R. N. Kniseley, *Anal. Chem. 46*, 1110A (1974).
31. A. P. Vinogradov, *Mem. Sears Found. Mar. Res. 12*, 647 (1953).
32. E. D. Goldberg, in: *Review of Trace Elements in Marine Organisms*, Puerto Rico Nuclear Center, San Juan (1967), 535 pp.
33. J. H. Martin and G. A. Knauer, *Geochim. Cosmochim. Acta 37*, 1639 (1973).
34. G. W. Bryan, *Proc. Roy. Sec. (London) 177B*, 389 (1971).
35. R. Eisler, U. S. Environmental Protection Agency Report EPA-R3-007 (1973).
36. Inter-University Program of Research on Ferromanganese Deposits, Phase I Report, Seabed Assessment Program I.D.O.E. National Science Foundation, Washington, D.C. (1973), 358 pp.
37. K. Norrish and J. T. Hutton, *Geochim. Cosmochim. Acta 33*, 431 (1969).
38. B. E. Leake, G. L. Hendry, A. Kemp, A. G. Plant, P. K. Harvey, J. R. Wilson, J. S. Coates, J. W. Aucotte, T. Lunel, and R. J. Howarth, *Chem. Geol. 5*, 7 (1969).
39. G. C. Brown, D. J. Hughes, and J. Esson, *Chem. Geol. 11*, 223 (1973).
40. J. H. Rose, F. Cuttitta, and R. R. Larson, *U.S. Geol. Surv. Prof. Paper 5258* (1965), p. 155.
41. N. B. Price and G. R. Angell, *Anal. Chem. 40*, 660 (1968).
42. J. R. Cann and C. K. Winter, *Mar. Geol. 11*, M33 (1971).
43. G. Thompson and D. C. Bankston, *Appl. Spectrosc. 25*, 151 (1971).
44. N. H. Suhr and C. O. Ingamells, *Anal Chem. 38*, 730 (1966).
45. E. Jeanroy, *Chim. Anal. 54*, 159 (1972).
46. G. D. Nicholls, A. L. Graham, El Williams, and M. Wood, *Anal. Chem. 39*, 584 (1967).
47. G. Thompson, D. C. Bankston, and S. M. Pasley, *Chem. Geol. 5*, 215 (1969).
48. G. Thompson, D. C. Bankston, and S. M. Pasley, *Chem. Geol. 6*, 165 (1970).
49. R. Chester, in: *Chemical Oceanography* (J. P. Riley and G. Skirrow, eds.), Vol. 2, Academic Press, Inc., New York (1965), p. 23.
50. J. P. Riley and R. Chester, *Introduction to Marine Chemistry*, Academic Press, Inc., New York (1971), 465 pp.

51. E. D. Goldberg, ed., *The Sea*, Vol. 5, John Wiley & Sons, Inc., New York (1974), 861 pp.
52. S. E. Humphris and G. Thompson, *Oceanus 19*, 40 (1976).
53. C. M. Shigley, *University of Rhode Island Occasional Publication 4* (1968).
54. R. R. Brooks and M. G. Rumsley, *Limnol. Oceanog. 10*, 521 (1965).
55. J. E. Portmann, *Meeresuntersuchungen 17*, 247 (1968).
56. S. K. El Wakeel and J. P. Riley, *Geochim. Cosmochim. Acta 25*, 110 (1961).
57. N. S. Skornyakova, P. F. Andruschenko, and L. S. Fumina, *Deep-Sea Res. 11*, 93 (1964).
58. R. C. Reynolds, *Amer. Mineral. 48*, 1133 (1963).
59. R. C. Reynolds, *Amer. Mineral. 52*, 1493 (1967).
60. F. J. Flanagan, *Geochim. Cosmochim. Acta 37*, 1189 (1973).

Author Index

Subject Index